Planning and Installing
Solar Thermal Systems

A guide for installers, architects and engineers

Planning and Installing

Solar Thermal Systems

A guide for installers, architects and engineers

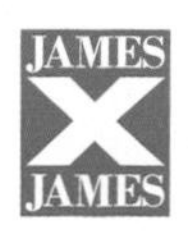

First published by James & James (Science Publishers) Ltd in the UK and USA in 2005
Reprinted 2006

ISBN-13: 978-1-84407-125-8
ISBN-10: 1-84407-125-1

Typeset by Saxon Graphics Ltd, Derby
Printed and bound in Croatia by Zrinski
Cover design by Paul Cooper Design

For a full list of publications please contact:

James & James / Earthscan
8–12 Camden High Street
London, NW1 0JH, UK
Tel: +44 (0)20 7387 8558
Fax: +44 (0)20 7387 8998
Email: earthinfo@earthscan.co.uk
Web: **www.earthscan.co.uk**

A catalogue record for this book is available from the British Library

Library of Congress Cataloging-in-Publication Data

Planning and installing solar thermal systems : a guide for installers, architects, and engineers / the German Solar Energy Society (DGS).
p. cm.
Includes bibliographical references and index.
ISBN 1-84407-125-1 (pbk.)
1. Solar heating--Installation. 2. Solar water heaters--Installation. 3. Solar thermal energy. I. Deutsche Gesellschaft für Sonnenenergie.
TH7413.P534 2005
697'.78--dc22

2004006812

Printed on elemental chlorine free paper

This guide has been prepared as part of the GREENPro project co-funded by the European Commission. Also available in the series:
Planning and Installing Bioenergy Systems: A Guide for Installers, Architects and Engineers
1-84407-132-4
Planning and Installing Photovoltaic Systems: A Guide for Installers, Architects and Engineers
1-84407-131-6

Contents

Foreword

The market for solar thermal systems is growing rapidly. Building owners, planning engineers, architects, fitters and roofers are increasingly being confronted with solar thermal technologies. Large amounts of additional new know-how are being asked for, in particular about the state of the art of the technology and the actual market situation, and this requires services in consultancy, planning and design, economics and marketing.

This guidebook will complement the services described above and support the decision-making process in offering up-to-date information on the latest technical developments based on best-practice experience. Finally, this guide should work as an aid for high-quality planning and careful system installation.

Highlights of the guide include the following:

- Detailed installation instructions for the large variety of on-site situations.
- Coverage of large solar thermal systems, including solar district heating networks.
- An overview of simulation software.
- Coverage of subjects such as solar swimming pool heating, solar air heating and solar cooling.
- A discussion of practical approaches for successful solar marketing.
- Results of market research giving an overview of the suppliers of solar thermal systems.

1 Solar radiation and arguments for its use

1.1 Solar radiation

1.1.1 Solar energy

The most important supplier of energy for the earth is the *sun*. The whole of life depends on the sun's energy. It is the starting point for the chemical and biological processes on our planet. At the same time it is the most environmentally friendly form of all energies, it can be used in many ways, and it is suitable for all social systems.

In the core of the sun a fusion process takes place in which pairs of hydrogen nuclei are fused into helium nuclei. The energy thus released is radiated into space in the form of electromagnetic radiation. As the sun is 143 million km from the earth, it radiates only a tiny fraction of its energy to the earth. In spite of this, *the sun offers more energy in a quarter of an hour than the human race uses in a whole year.*

The age of the sun is estimated by astrophysicists to be about 5 billion years. With a total life expectation of 10 billion years the sun will be available as an energy source for another 5 billion years. Hence from our human perspective the sun offers an unlimited life.

Figure 1.1. The sun: basis of all life on earth

1.1.2 Astronomical and meteorological bases

On the outer edge of the earth's atmosphere the irradiated power of the sun is virtually constant. This irradiated power or radiation intensity falling on an area of one square metre is described as the *solar constant*. This constant is subject to small variations influenced both by changes in the sun's activity (sunspots) and by differences in the distance between the earth and the sun. These irregularities are mostly found in the ultraviolet range: they are less than 5%, and hence not significant in application of the solar constant for solar technology. The average value of the solar constant is given as $I_0 = 1.367$ W/m^2 (watts per square metre).

Even based on the astronomical facts alone, the amount of solar energy available on the earth is very variable. It depends not only on the geographical latitude, but also on the time of day and year at a given location. Because of the inclination of the

IRRADIATED POWER, IRRADIANCE, HEAT QUANTITY

When we say that the sun has an irradiance, *G*, of for example 1000 W/m², what is meant here is the capability of radiating a given irradiated power, ϕ (1000 W), onto a receiving surface of 1 m². The watt is the unit in which power can be measured. If this power is referred, as in this case, to a unit area, then it is called the *irradiance*.

1 kW	(kilowatt)	=	10^3 W (1000 watts)
1 MW	(megawatt)	=	10^6 W (1 million watts) = 10^3 kW
1 GW	(gigawatt)	=	10^9 W (1 thousand million watts) = 10^6 kW
1 TW	(terawatt)	=	10^{12} W (1 million million watts) = 10^9 kW

When the sun shines with this power of 1000 W for 1 hour it has performed 1 kilowatt-hour of work (1 kWh) (Work = Power × Time).

If this energy were converted completely into heat, a heat quantity of 1 kWh would be produced.

Irradiated power,	ϕ (W)
Irradiance,	*G* (W/m²)
Heat quantity,	*Q* (Wh, kWh)

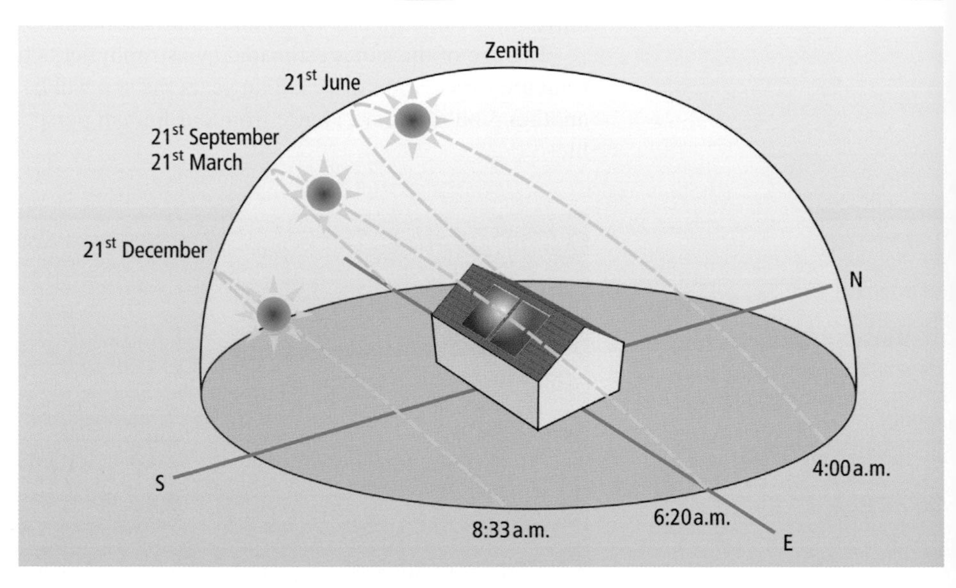

Figure 1.2.
The sun's path at different times of the year at central European latitude (London, Berlin)

earth's axis, the days in summer are longer than those in winter, and the sun reaches higher solar altitudes in the summer than in the winter period (Figure 1.2).

Figure 1.3 shows the sequence over a day of the irradiation in London on a horizontal receiving surface of 1 m² for four selected cloudless days over the year. It is clear from the graph that the supply of solar radiation, even without the influence of the weather or clouds, varies by a factor of about 10 between summer and winter in London. At lower latitudes this effect decreases in strength, but at higher latitudes it can be even more pronounced. In the southern hemisphere the winter has the highest irradiations, as shown in Figure 1.4, which shows the sequence over a day of the irradiation in Sydney on a horizontal receiving surface of 1 m² on three average days over the year.

Even when the sky is clear and cloudless part of the sun's radiation comes from other directions and not just directly from the sun. This proportion of the radiation, which reaches the eye of the observer through the scattering of air molecules and dust particles, is known as *diffuse radiation*, G_{dif}. Part of this is also due to radiation reflected at the earth's surface. The radiation from the sun that meets the earth without any change in direction is called *direct radiation*, G_{dir}. The sum of direct and diffuse radiation is known as *global solar irradiance*, G_G (Figure 1.5).

$$G_G = G_{dir} + G_{dif}$$

Unless nothing else is given, this always refers to the irradiation onto a horizontal receiving surface.[1]

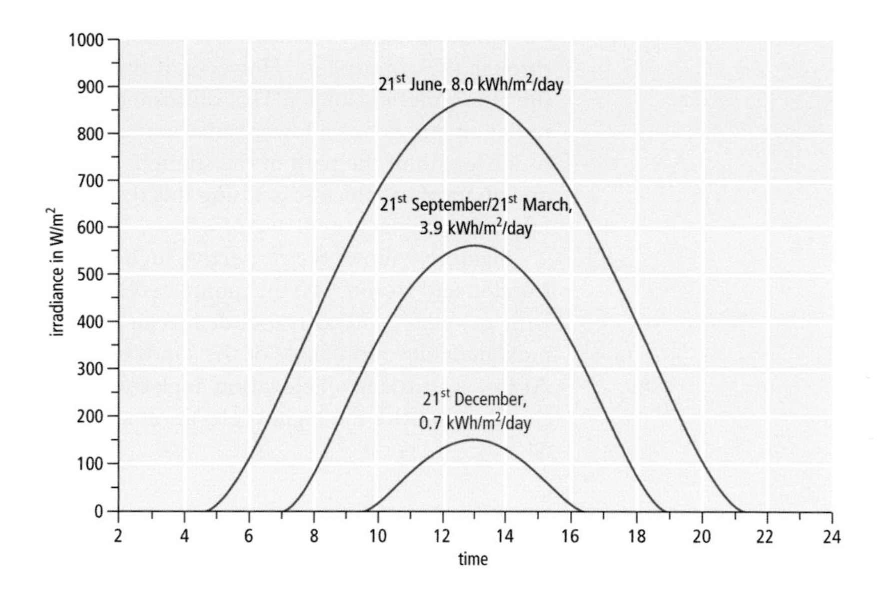

Figure 1.3. Daily courses and daily totals for irradiation in London

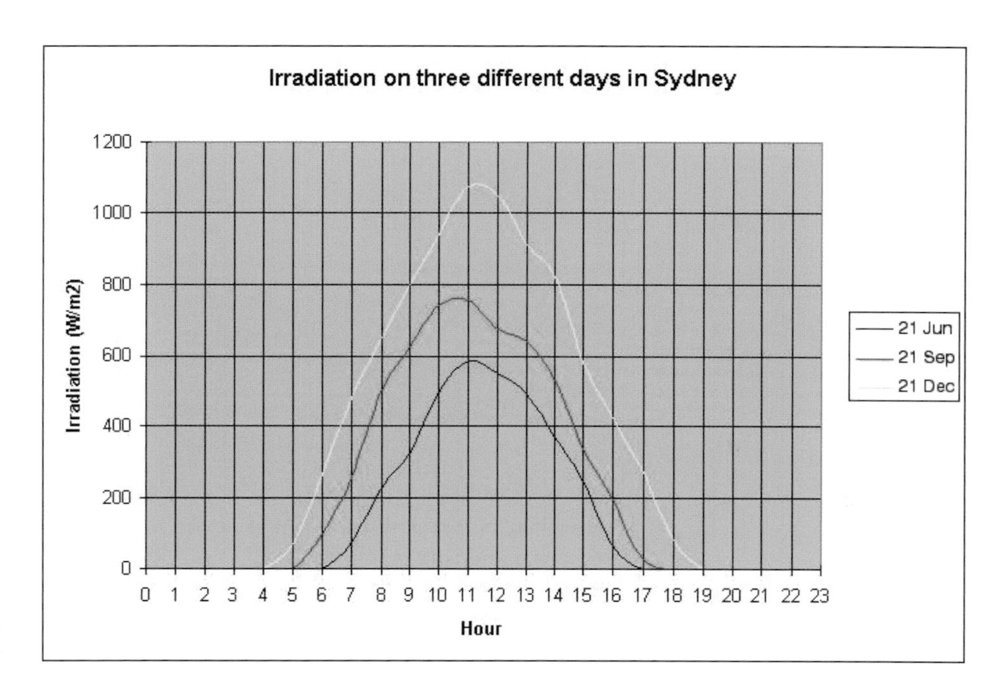

Figure 1.4. Irradiation on three different days in Sydney, Australia

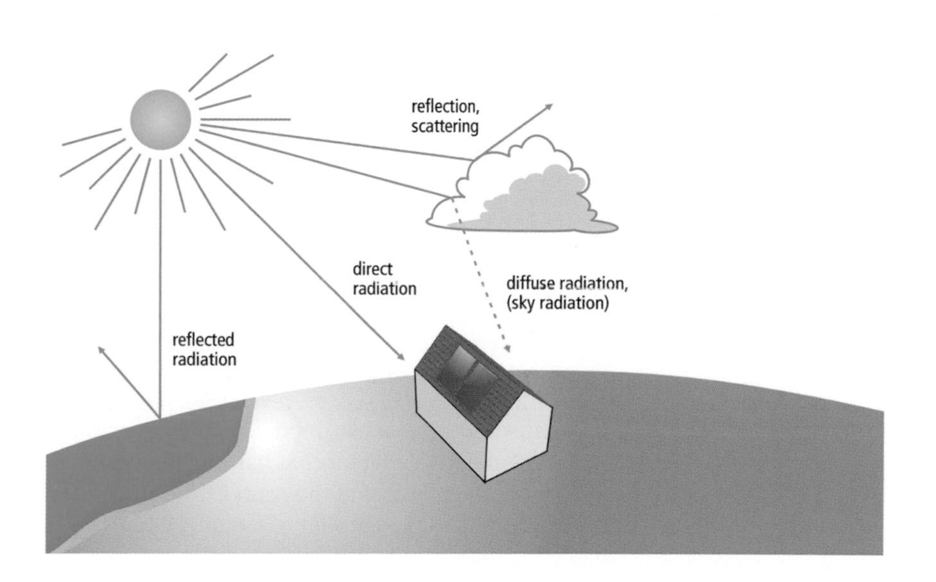

Figure 1.5. Global solar irradiance and its components

When the sun is vertically above a location the sunlight takes the shortest path through the atmosphere. However, if the sun is at a lower angle then the path through the atmosphere is longer. This causes increased absorption and scattering of the solar radiation and hence a lower radiation intensity. The *air mass factor* (*AM*) is a measure of the length of the path of the sunlight through the earth's atmosphere in terms of one atmosphere thickness. Using this definition, with the sun in the vertical position (elevation angle, $\gamma_S = 90°$), $AM = 1$.

Figure 1.6 shows the respective highest levels of the sun on certain selected days in London and Berlin. The maximum elevation angle of the sun was achieved on 21 June with $\gamma_S = 60.8°$, and corresponded to an air mass of 1.15. On 22 December the maximum elevation angle of the sun was $\gamma_S = 14.1°$, corresponding to an air mass of 4. At lower latitudes, all elevation angles will increase: for example, at a latitude of 32° (north or south), the highest elevation angle will be 80.8° and the lowest angle will be 34.1°.

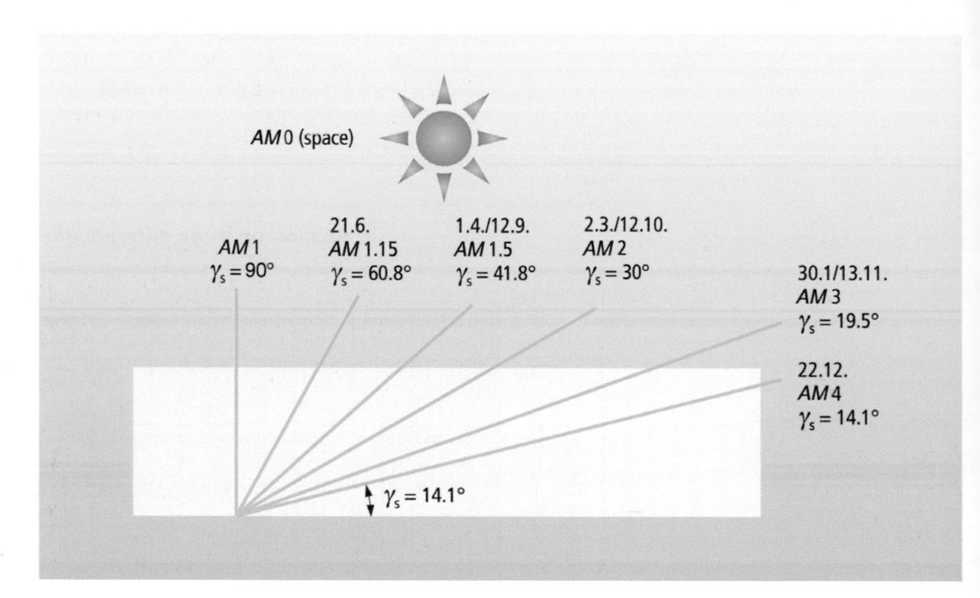

Figure 1.6.
Sun's level at midday within the course of a year in London and Berlin (latitude: 52°N)

The sun's radiation in space, without the influence of the earth's atmosphere, is described as *spectrum AM 0*. As it passes through the earth's atmosphere, the radiation intensity is reduced by:

- reflection caused by the atmosphere
- absorption by molecules in the atmosphere (O_3, H_2O, O_2, CO_2)
- Rayleigh scattering (scattering by the air molecules)
- Mie scattering (scattering by dust particles and contamination in the air).

See Figure 1.7.

Table 1.1 shows the dependence of the irradiation on the elevation angle, γ_S. Absorption and scattering increase when the sun's elevation is lower. Scattering by dust particles in the air (Mie scattering) is heavily dependent on the location. It is at its greatest in industrial areas.

Table 1.1.
Effect of elevation angle on attenuation of irradiation

γ_S	*AM*	Absorption (%)	Rayleigh scattering (%)	Mie scattering (%)	Total attenuation (%)
90°	1.00	8.7	9.4	0–25.6	17.3–38.5
60°	1.15	9.2	10.5	0.7–25.6	19.4–42.8
30°	2.00	11.2	16.3	4.1–4.9	28.8–59.1
10°	5.76	16.2	31.9	15.4–74.3	51.8–85.4
5°	11.5	19.5	42.5	24.6–86.5	65.1–93.8

After the general astronomical conditions, the cloud cover or state of the sky is the second decisive factor that has an effect on the supply of solar radiation: both the irradiated power and the proportions of direct and diffuse radiation vary greatly according to the amount of cloud (Figure 1.8).

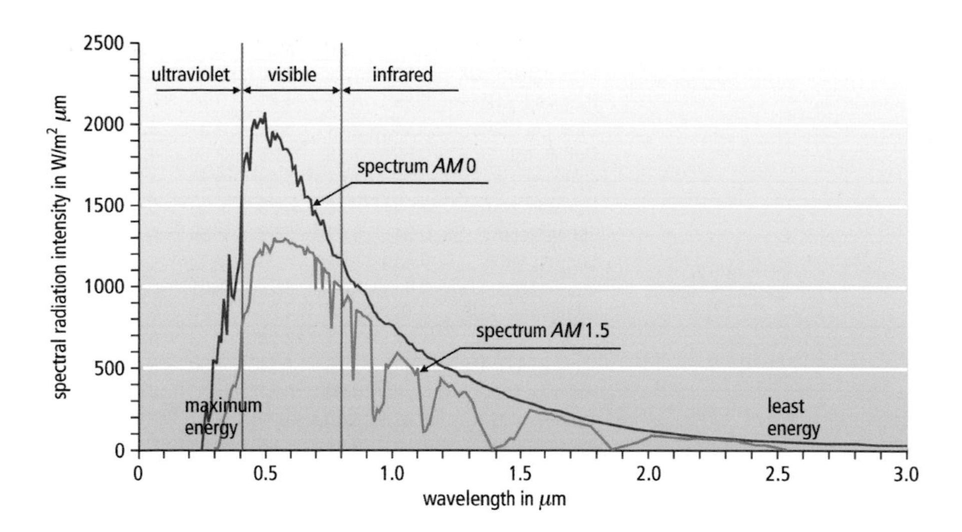

Figure 1.7. Sun spectrum AM 0 in space and AM 1.5 on the earth with a sun elevation of 41.8°

Over many years the average proportion of diffuse to global solar irradiance in central Europe has been found to be between 50% and 60%. In sunnier climates the fraction of diffuse radiation is lower; in the winter months this proportion is higher. See Figure 1.9.[2]

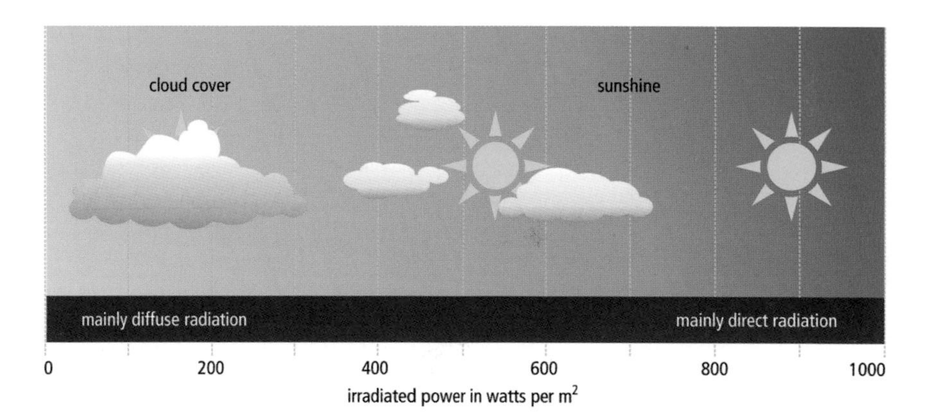

Figure 1.8. Global solar irradiance and its components with different sky conditions

The average annual global solar irradiance is an important value for designing a solar plant. It is significantly higher at lower than at higher latitudes, but for climatological reasons severe regional differences can arise. The maps in Figure 1.10 give an indication of the solar irradiation in different regions. Table 1.2 gives an overview of the monthly solar irradiation in a number of cities around the world (measured in kWh/m^2 per day on a horizontal surface). Over the course of a year global solar irradiance is subject to significant daily variations, especially in climates where cloudiness occurs regularly.

In addition to global solar irradiance, the *sunshine duration* is sometimes given: that is, the number of hours each year for which the sun shines. In the United Kingdom this value varies between 1300 and 1900 hours per year. However, the radiation is a far more reliable figure to use when designing or installing solar energy systems.

Table 1.2. Monthly solar irradiation (kWh/m² per day) around the world

City	Jan	Feb	Mar	Apr	May	Jun	Jul	Aug	Sep	Oct	Nov	Dec
Birmingham, UK	0.65	1.18	2.00	3.47	4.35	4.53	4.42	3.87	2.67	1.48	0.83	0.45
Brisbane, AUS	6.35	5.71	4.81	3.70	2.90	2.43	2.90	3.61	4.93	5.45	6.33	6.32
Chicago, USA	1.84	2.64	3.52	4.57	5.71	6.33	6.13	5.42	4.23	3.03	1.83	1.45
Dublin, IRL	0.65	1.18	2.26	3.60	4.65	4.77	4.77	3.68	2.77	1.58	0.77	0.45
Glasgow, UK	0.45	1.04	1.94	3.40	4.48	4.70	4.35	3.48	2.33	1.26	0.60	0.32
Houston, USA	2.65	3.43	4.23	5.03	5.61	6.03	5.94	5.61	4.87	4.19	3.07	2.48
Johannesburg, SA	6.94	6.61	5.90	4.80	4.35	3.97	4.26	5.10	6.13	6.45	6.57	7.03
London, UK	0.65	1.21	2.26	3.43	4.45	4.87	4.58	4.00	2.93	1.68	0.87	0.48
Los Angeles, USA	2.84	3.64	4.77	6.07	6.45	6.67	7.29	6.71	5.37	4.16	3.13	2.61
Melbourne, AUS	7.13	6.54	4.94	3.20	2.13	1.93	2.00	2.71	3.87	5.26	6.10	6.68
New York, USA	1.87	2.71	3.74	4.73	5.68	6.00	5.84	5.39	4.33	3.19	1.87	1.48
Philadelphia, USA	1.94	2.75	3.81	4.80	5.55	6.10	5.94	5.42	4.37	3.23	2.13	1.68
Phoenix, USA	3.29	4.36	5.61	7.23	8.00	8.17	7.39	6.87	5.97	4.84	3.57	2.97
Sydney, AUS	6.03	5.54	4.23	3.07	2.61	2.33	2.55	3.55	4.63	5.87	6.50	6.13
Toronto, CAN	1.58	2.54	3.55	4.63	5.77	6.30	6.29	5.45	4.03	2.68	1.37	1.16
Vancouver, CAN	0.84	1.75	3.00	4.27	6.03	6.50	6.52	5.42	3.80	2.06	1.03	0.65

Figure 1.9. Monthly sum of global solar irradiance (diffuse and direct)

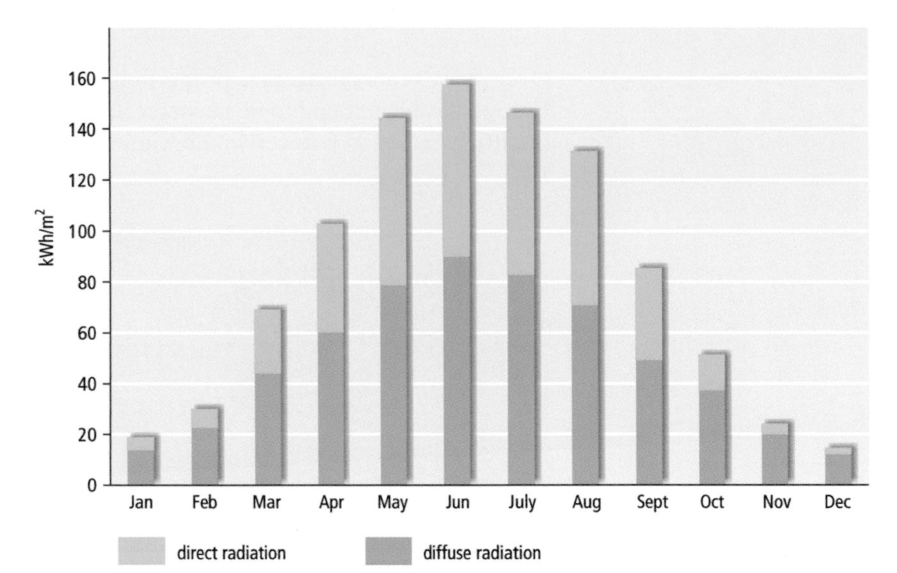

Figure 1.9a. Berlin, Germany

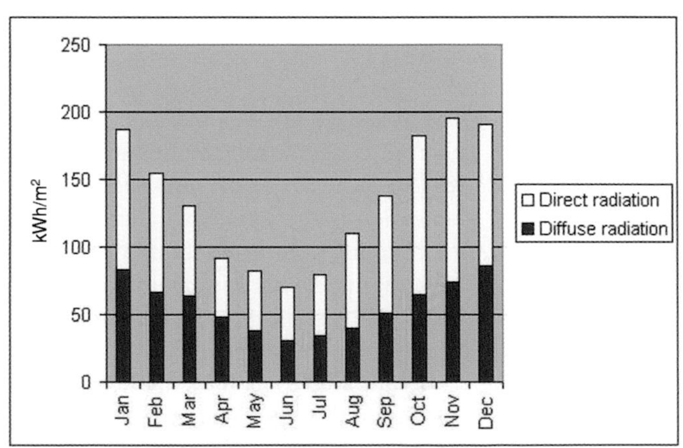

Figure 1.9b. Sydney, Australia

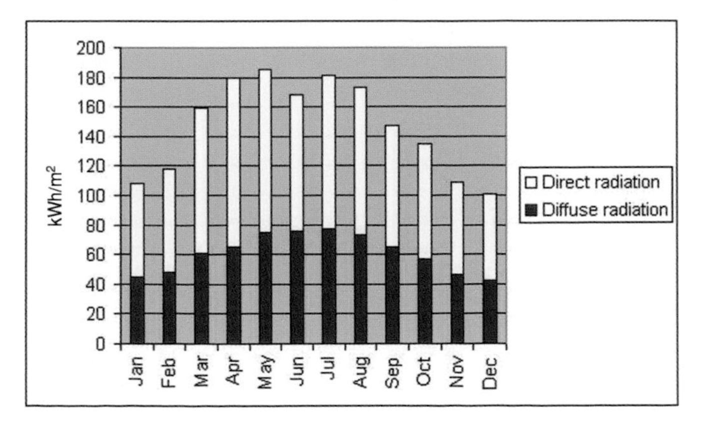

Figure 1.9c. Miami, USA

Figure 1.10. Total annual global radiation (on horizontal surface)

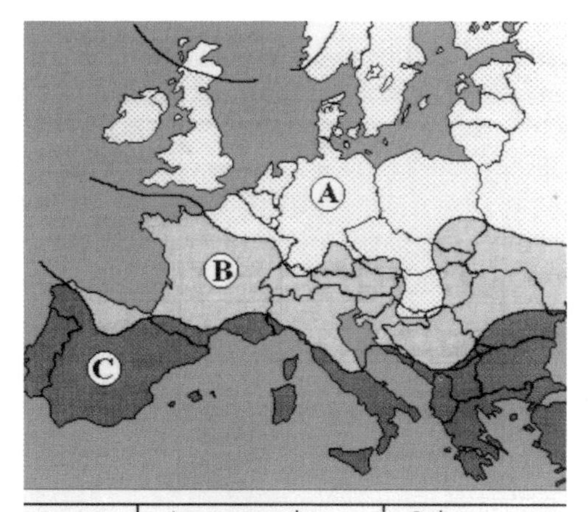

Zone (see map above)	Average solar radiation on collector surface [kWh/m² day]	Solar system heat yield [kWh/m²y]
A	2.4 to 3.4	300 to 450
B	3.4 to 4.4	400 to 550
C	4.4 to 5.4	500 to 650

Figure 1.10a. Europe

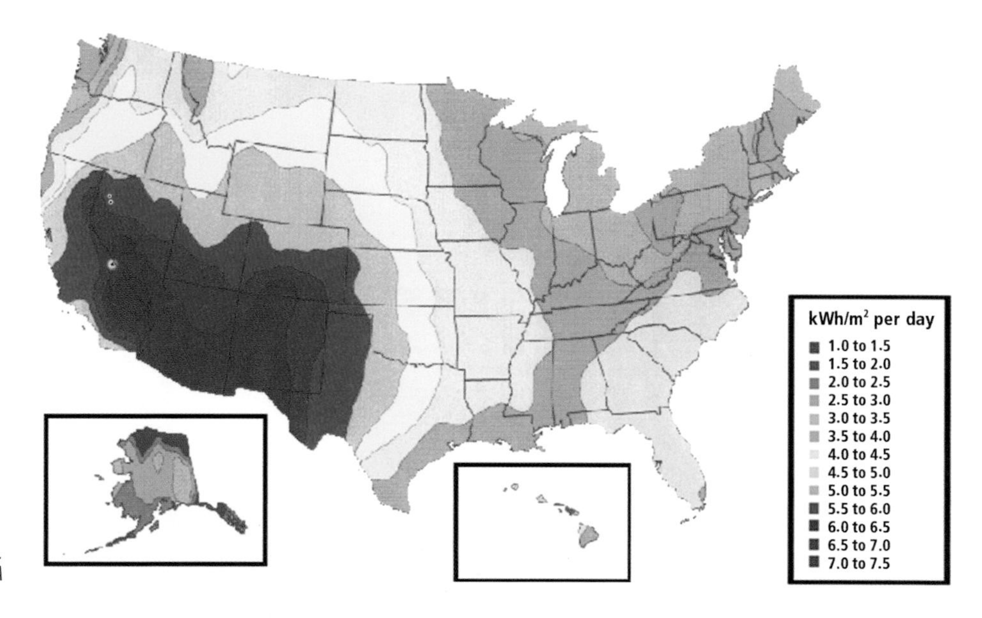

Figure 1.10b. USA

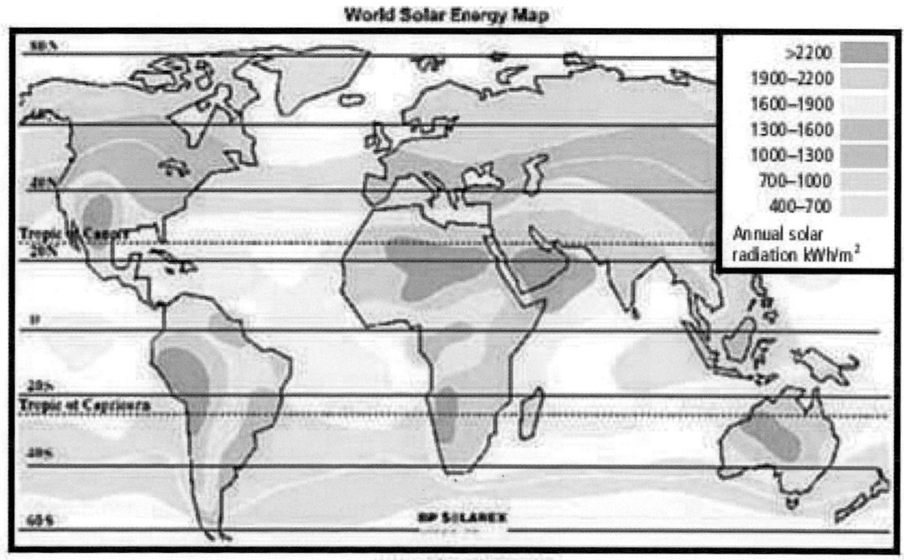

Figure 1.10c. The world

MEASURING SOLAR RADIATION

Devices that measure the global solar irradiance on a horizontal surface are called *pyranometers* (Figure 1.11). If these devices are screened from the sun's direct rays by a fixed ring that covers the whole path of the sun in the sky, then the device measures only the diffused radiation. The radiation receiver is seated beneath a spherical glass cover and consists of a star-shaped arrangement of black and white thermo-elements. These elements generate thermo-electromotive forces, depending on their temperature, which can be measured. Pyranometers are relative measuring instruments that have to be calibrated. Other global solar irradiance measuring devices that are available on the market and are cheaper than pyranometers possess a *solar cell* as a receiver, as in the MacSolar (Figure 1.12), for example.

The simplest and most commonly used device for measuring the sunshine duration is the Campbell–Stokes sunshine recorder (Figure 1.13). This consists of a solid glass sphere, which generates a focal point on the side that is turned away from the sun and which is always at the same distance. A correspondingly curved flameproof paper strip is placed around the sphere. A track is burned on the paper strip. When clouds cover the sun, the burnt track is interrupted.

Figure 1.11. Pyranometer made by Kipp & Zonen

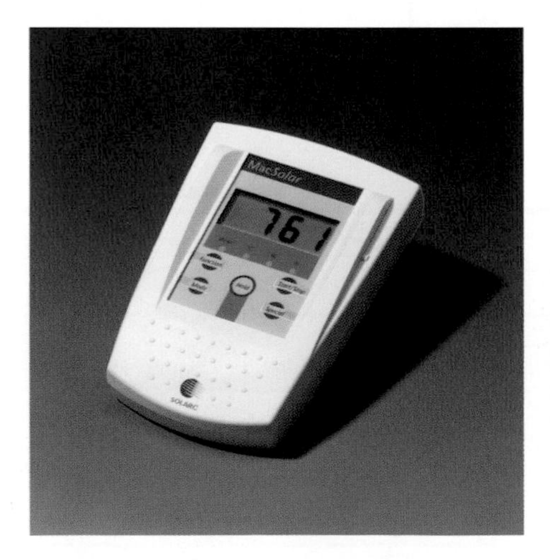

Figure 1.12 (left). MacSolar

Figure 1.13 (right). Campbell–Stokes sunshine recorder

1.1.3 Influence of orientation and tilt angle

The variables or figures that have been given so far referred to a horizontal receiving surface, such as a flat roof. Because the angle of incidence of the sun varies over the course of the year, the maximum radiation yield can be obtained only if the receiving surface is inclined at an angle to the horizontal. The optimum angle of inclination is larger in the low-radiation months than in the summer because of the low elevation of the sun.

In solar technology the angle descriptions listed in Table 1.3, and shown in Figure 1.14, are normally used. However, note that sometimes in architecture and building the following values are used as angle descriptions for the solar azimuth: North = 0°, East = 90°, South = 180°, West = 270°.

Table 1.3. Angle descriptions in solar technology

Sun's height	γ_s	Horizon = 0°	Zenith = 90°	
Solar azimuth	α_s	South = 0°	East = –90°	West = +90°
Surface inclination	β	Horizontal = 0°	Vertical = 90°	
Surface azimuth	α	South = 0°	East = –90°	West = +90°

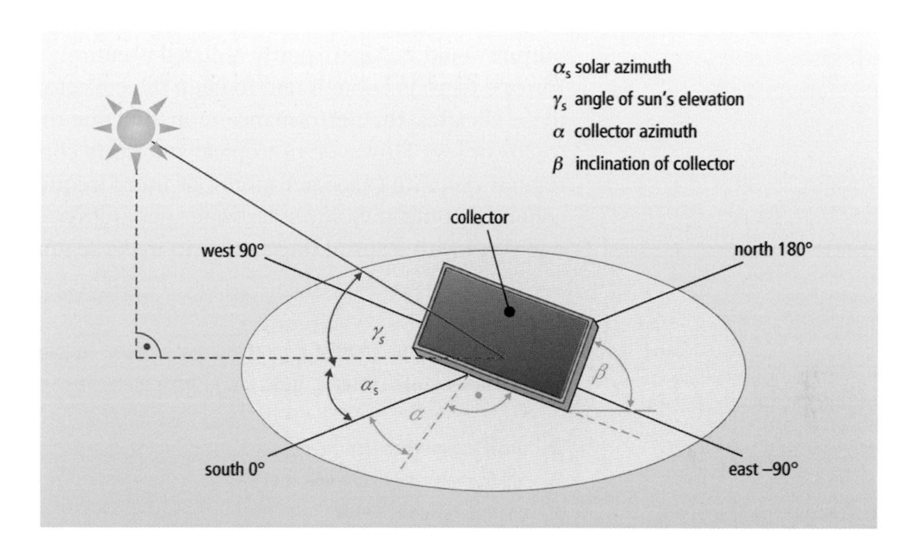

Figure 1.14. Angle descriptions in solar technology (northern hemisphere)

What about the influence of roof alignment and inclination on the insolation (incoming solar radiation)? Figure 1.15 shows the values measured in central Europe (Berlin) for the calculated average annual totals of global solar irradiance for differently oriented surfaces. Lines of equal radiation totals are shown in kWh/m² per year. On the horizontal axis the alignment can be read off and on the vertical axis the angle of inclination can be seen.

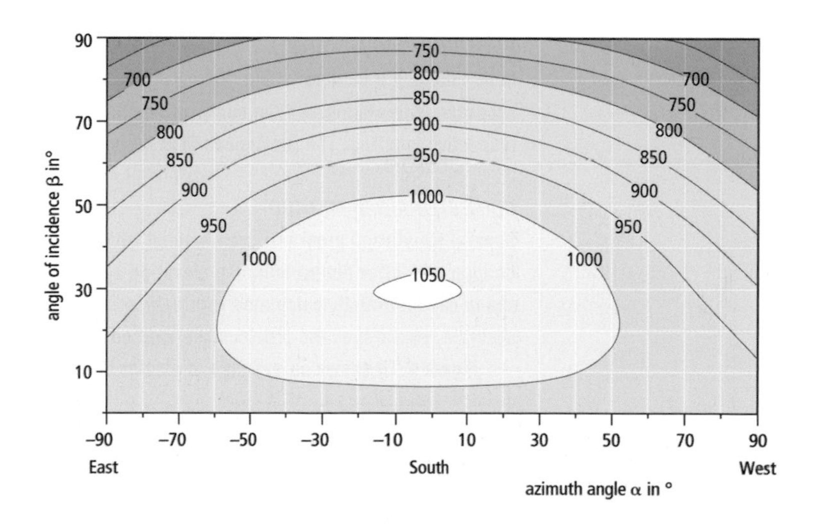

Figure 1.15. Annual total global solar irradiance for differently oriented receiving surfaces (central Europe)

According to the annual average, optimum irradiance occurs with a southern alignment ($\alpha = 0°$) and an inclination of $\beta = 30°$. The graph also shows that a deviation from the optimum alignment can be tolerated over wide ranges, as no significant losses of radiation are involved. Roughly speaking, for central European regions (latitudes of about 50°N) any collector angle between 30° and 60° in combination with an orientation between south-east and south-west will give almost optimal irradiance.

A general, worldwide rule of thumb is that a solar collector should roughly face the Equator and the optimal tilt angle should be close to 0.7 times the latitude, but always at least 10° or the minimum tilt angle specified for the solar collector (many solar collectors require a minimum tilt angle of 15°). For instance, for a latitude of 40°S, the collector tilt angle should be about 30° facing north.

However, deviations from that optimum will in general give only a slight reduction of the irradiation. The optimum angle of inclination for the winter months (October to March) is 50° but, in general, orientations between south-east and south-west and inclination angles between 30° and 60° will lead to a loss of solar radiation of less than 15% compared with the optimum angles.

1.1.4 Collection of dust, and need for cleaning

Whether dust accumulation will take place to such a degree that the solar collector's output would be significantly reduced is entirely dependent on the climate. In central Europe there is enough rain to keep the collector clean. Measurements show that, in these climates, the performance reduction due to dust on the collectors will be limited to 2% or less. However, in tropical and dusty climates this can be very different, and regular cleaning (once per month or more frequently) is needed to maintain the collector's output. Some solar water heating systems in tropical countries are even equipped with a sprinkling system to make regular cleaning easier.

SHADING

Shading reduces the yield of a solar thermal system. To take account of shading of the receiving surface by the surroundings (houses, trees etc.), three methods can be used:

- **graphical method (indicative)**
- **photographic method (indicative)**
- **computer-aided method.**

GRAPHICAL METHOD

This method requires a scale drawing of the layout of the surroundings, details of the height of each object that could shade the potential collector position, and a *solar altitude diagram* for the latitude at which the collector is to be located. First, the elevation and azimuth angles of the relevant objects must be established, and then the shade silhouette must be plotted in the solar altitude diagram. If large areas of shade arise in time periods with high radiation, then the expected radiation received must be reduced correspondingly.

PHOTOGRAPHIC METHOD

In this method a camera with a 'fish-eye' lens is used in connection with special solar-geometrical accessories to photograph the surrounding silhouette while blending in the solar altitude diagram of the respective location. The results can then be read directly off the photograph.

COMPUTER-AIDED METHOD

Several simulation programs are provided with shade simulators (see for instance TRNSYS and Sundi in Chapter 10). After determining the elevation and azimuth angles of important objects, the influence of shade can be directly calculated within the scope of the system simulation. This method yields more accurate results than the previous two methods.

Figure 1.16 shows an example of a solar altitude diagram with surrounding silhouette.

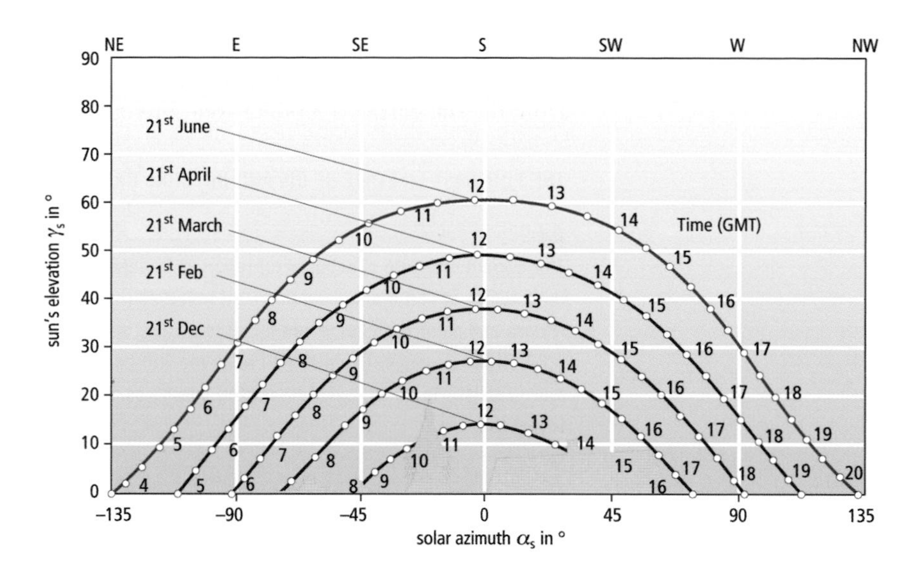

Figure 1.16. Solar altitude diagram with example silhouettes of objects (for latitude of about 50°)

1.2 The finiteness of energy resources

The available resources of fossil fuels (coal, oil, natural gas and uranium) are being consumed at an ever-increasing rate to meet the growing energy requirement on our planet. Because stocks are finite, this process will inevitably lead us into a cul-de-sac. The key that leads us out of this dilemma is to save energy, to use energy rationally, and to use renewable energy sources: sun, wind, water and biomass.

Figure 1.17 emphasises the relationship between fossil fuel reserves, energy requirement and the radiation supplied by the sun.

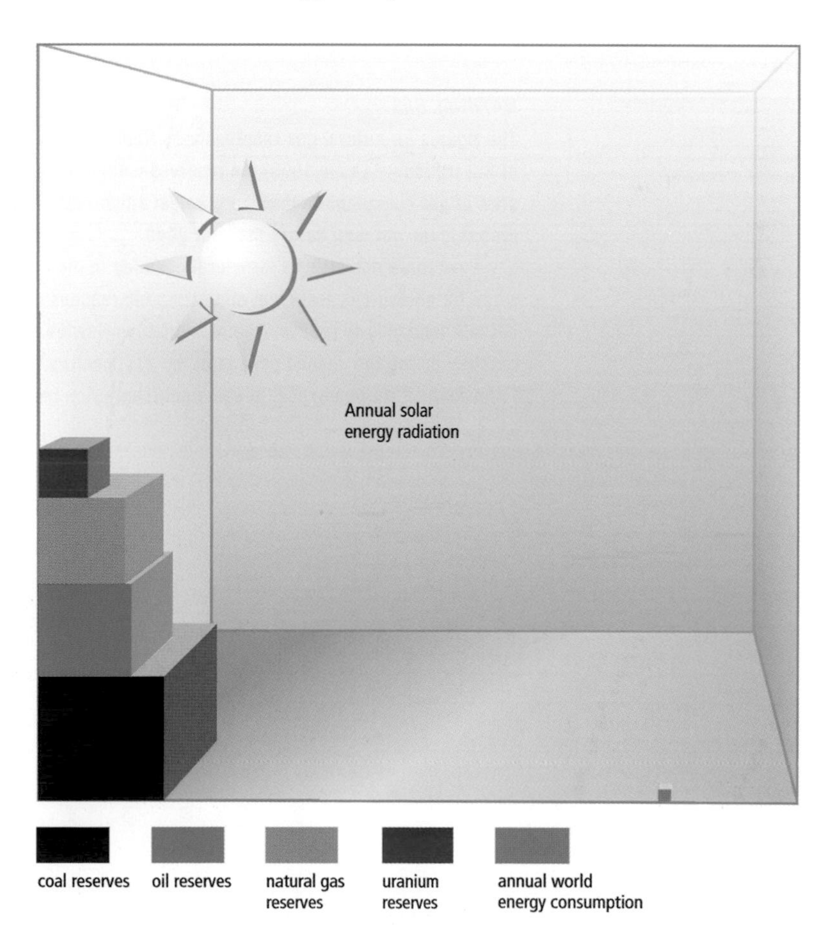

Figure 1.17. The energy cube

Each year the sun provides a multiple of the world energy consumption, and indeed even a multiple of all the known fossil fuel reserves. Put into numbers, it is 1.5×10^{18}

kWh/a, or 1500 million billion kilowatt-hours per annum. This is more than 10,000 times the energy that the human race needs at present. Moreover, the radiation supply from the sun carries a 5 billion-year guarantee.

THE FORECAST LIFETIME OF OIL AND NATURAL GAS RESERVES

OIL

The current estimate of secured oil reserves is between 118 and 180 billion tonnes: the latter figure includes so-called *non-conventional* oil reserves (heavy oils, tar sand, oil shale, and oil deposits in deep waters and polar regions). From this it is evident that even with the same oil extraction rate as in 1995 of 3.32 billion tonnes per year the oil reserves will be exhausted by about 2050 (Figure 1.18). However, it is more realistic to assume that increased energy consumption will result in a faster rate of use of the reserves.

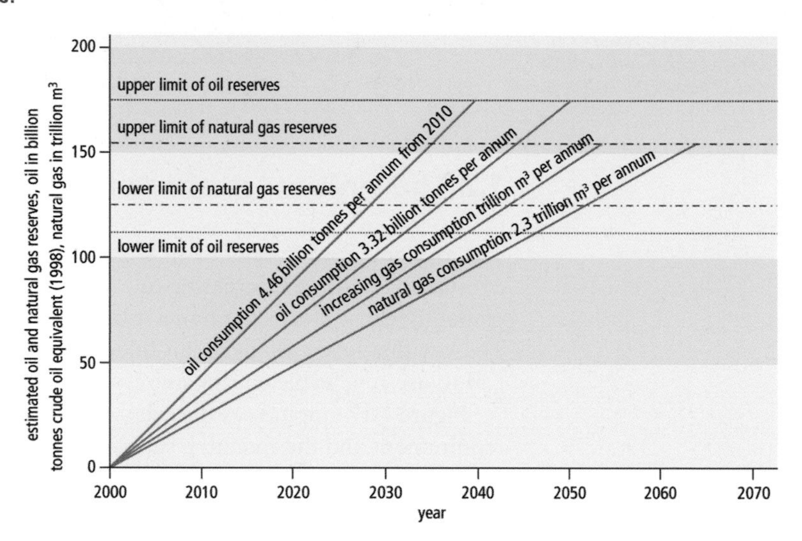

Figure 1.18. Reserves and forecast lifetime of oil and natural gas[4]

NATURAL GAS

The figures for natural gas reserves vary from 131 to 153 trillion m³. With a steady annual extraction rate of 2.3 trillion m³ (as in 1995) the reserves would be exhausted after 57–65 years. But it is exactly in this area of gas consumption that the greatest annual rate of increase is found, so that the reserves will very probably run out well before the year 2040.

Even more decisive for structural changes in the energy supply is the question of the point in time when for geological, technical and economic reasons oil and gas production can no longer be increased, but will tend only to reduce. The maximum worldwide extraction rate (the big rollover) is expected to be reached during the second decade of the 21st century: that is, between 2010 and 2020 (Figure 1.19). From then on we can expect to see significant price increases.[3]

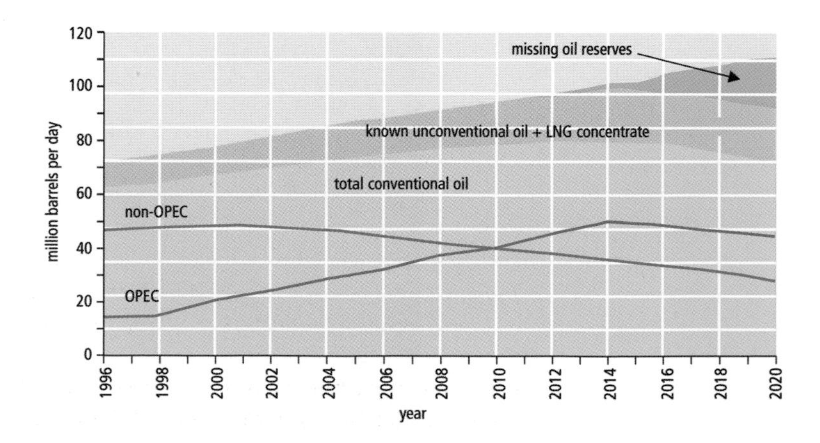

Figure 1.19. The OPEC proportion constantly increases from 27% today to over 60% in the year 2020. In 2020 an under-supply of liquid hydrocarbons of about 17% – or a quarter of today's world oil production – is expected.[5]

Based on experience, we know that in the area of technical innovations the changeover period will last for several decades. This means that immediate action is required. Thermal solar technology uses the unlimited supply offered by the sun and makes it possible for every solar plant operator to make an active contribution to mitigating the increasing dangers connected with the shortage of resources.

1.3 Climate change and its consequences

Based on the apparent finite energy resources, the environment and the climate are being dramatically changed and damaged to an ever-greater extent by the burning of fossil fuels.

The cause of this is the emission of hazardous substances such as sulphur dioxide, oxides of nitrogen and carbon dioxide connected with the incineration process. Sulphur dioxide and oxides of nitrogen are among those hazardous substances that play a significant part in causing acid rain. Carbon dioxide (CO_2) is the greenhouse gas that is mainly responsible for the heating up of the earth's atmosphere. At present CO_2 concentrations in the earth's atmosphere are increasing at an ever-increasing rate.

GREENHOUSE EFFECT

The earth–atmosphere system absorbs the visible, short-wave radiation from the sun in the wavelength range of approximately 0.3–3.0 μm. The result of this is heating up of the earth's surface and the atmospheric layers. In turn, each heated body radiates according to its temperature. However, this heat emission takes place in a longer wavelength range of between 3.0 and 30 μm.

CO_2 molecules can retain part of this heat energy radiated back from the earth's surface and the atmosphere. This process is called the *greenhouse effect*, as the CO_2 layer in the atmosphere can be compared with the glass panes in a greenhouse, which let the light in but keep the heat from getting out (Figure 1.20). Through the natural content of CO_2 in the earth's atmosphere the temperature of the earth is currently +15°C on average. Without this natural content it would be –15°C (natural greenhouse effect[6]). Through the burning of coal for industrial production and the generation of electricity as well as the use of oil products (petrol, heating etc.) the amount of CO_2 released by mankind from the 19th century onwards has become so great that nature can no longer compensate for this increase. The consequence is an additional greenhouse effect that causes the earth's temperature to increase continuously.

4 Production of additional greenhouse gases, mainly carbon dioxide (CO_2) increases the greenhouse effect
3 Part of the heat radiation remains in the earth atmosphere system and increases its temperature (greenhouse effect)
2 Long-wave heat radiation can partially escape into space
4
2
3
1
1 Short-wave solar radiation is absorbed by the atmosphere and the earth's surface

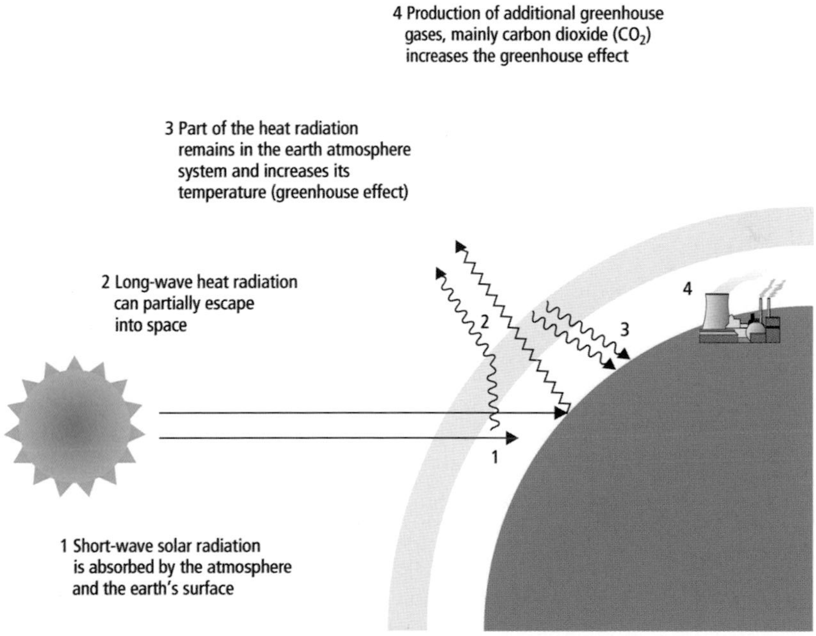

Figure 1.20. The greenhouse effect

Most scientists now agree on the potentially catastrophic effects of an increase in the annual average global temperature, some of which have already become reality:

- **heating of the oceans, and melting of the glaciers, leading to an increase in the sea level, with the result that some coastal regions have been flooded and building land has been lost (more than one third of the world's population live in coastal regions)**
- **displacement of the vegetation zones, resulting in drastic changes in food production, and a dramatic reduction in the variety of species**
- **the release of huge amounts of CO_2 and methane from the thawing ground of the tundra, which further increases the effect**

- **the 'Mediterraneanization' of the temperate latitudes, with hot, dry summers and milder but wet winters, resulting in shortages of water in summer, floods in winter, and a high load on already severely damaged forests**
- **an increase in tropical cyclones, the creation of which depends on the heating of the oceans**
- **intensification of known climate phenomena, such as El Niño, with an increase in the frequency of torrential storms in the otherwise dry areas of South America).**

It is only a reduction in energy consumption, more efficient use of the still-available energy resources, and the increased use of renewable energies that can reduce CO_2 emissions and protect the world's resources. The use of solar energy plays a central role in this, as the sun's energy is inexhaustible, abundantly available, and its use does not harm the environment.

1.4 Good arguments for solar systems

Every year the number of freshly installed collector surfaces increases, with an annually increasing rate of growth (the collector surface installed in 2003 in Europe was around 1,450,000 m^2). The European market has grown on average by 18% per year in the period 1994 to 1999.[7]

Every new square metre of installed solar collector surface increases the trend and is an active contribution to the protection of the climate:

- Solar system owners do not wait for political decisions or global changes, but simply act.
- The solar system is an easily visible sign of a high level of responsibility, environmental awareness and commitment.
- Solar system owners are pleased with every ray of the sun, and experience their environment with more awareness.
- They enjoy bathing, showering or washing their clothes using water heated by the sun – particularly in the summer, when the heating boiler can remain switched off.
- The solar system makes them slightly less dependent on increases in energy prices.
- Solar system operators enjoy tax advantages and government funding in many countries (see Chapter 11).
- Solar systems increase both the value of the property and its image. 'Solar houses' can be sold or rented out more easily.
- Thermal solar systems for the provision of hot water are technically mature and have a service life of about 20 years.
- A standard solar system covers between 50% (in northern latitudes) and 90% (in subtropical and tropical climates) of the yearly energy required for the provision of hot water. Even in northern latitudes, 90% of the energy demand between May and September can be covered.
- Solar systems for swimming pool water heating are economical to install, and their cost can be amortized over a very short period of time.
- Within the course of its life a solar system supplies about 13 times more energy than was used to make it.
- Solar systems require very little maintenance, and their energy is permanently available.
- By taking up solar technology the trade gains new areas of work, which are secure for the future.
- Solar technology creates lasting employment in production, installation and servicing.

2 Components of solar thermal systems

2.1 How does a solar thermal system work?

The solar collector mounted on the roof converts the light that penetrates its glass panes (short-wave radiation) into heat. The collector is therefore the link between the sun and the hot water user. The heat is created by the absorption of the sun's rays through a dark-coated, usually metal, plate – the *absorber*. This is the most important part of the collector. In the absorber is a system of pipes filled with a *heat transfer medium* (usually water or an antifreeze mixture). This takes up the generated heat. Collected together into a pipe it flows to the *hot water store*. In most solar water heating systems – by far the most commonly used type of solar thermal systems – the heat is then transferred to the domestic water by means of a *heat exchanger*. The cooled medium then flows via a second pipeline back to the collector while the heated domestic water rises upwards in the store. According to its density and temperature, a *stratified system* is set up in the store: the warmest water is at the top (from where it leaves the tank when the taps are turned on) and the coldest is at the bottom (where cold water is fed in).

In central and northern Europe, as well as in the USA, Canada and other countries, thermal solar systems operate with a water–glycol mixture that is circulated in a closed circuit (forced circulation). This system, which has a solar circuit separated from the domestic water circuit, is called an *indirect system* (see Figure 2.1). In some countries systems also exist with pure water as the heat transfer medium (for instance the so-called *drainback systems*) or with direct circulation of the domestic water through the collector.

The controller will only start the solar circuit pump when the temperature in the collector is a few degrees above the temperature in the lower area of the store. In this way the heat transfer liquid in the collector – having been warmed by the sun – is

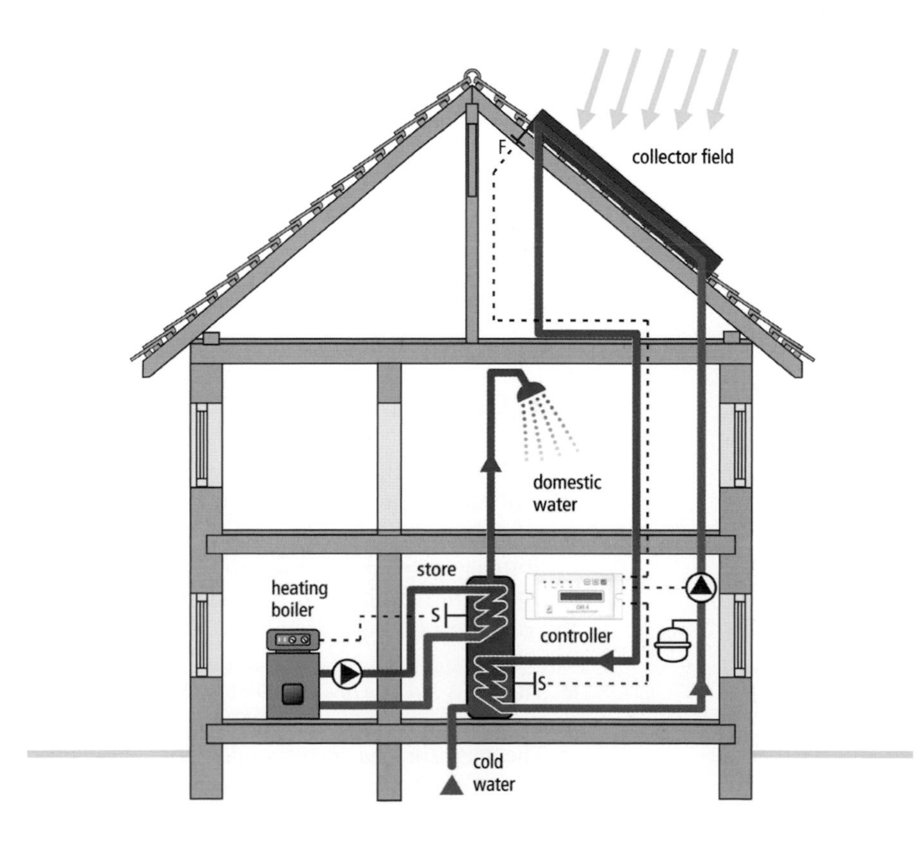

Figure 2.1. Standard solar water heating system with heating boiler for additional heating (S = temperature sensor)

transferred into the lower heat exchanger, where the heat is transferred to the stored domestic water.

In Australia, Israel and other Mediterranean countries, as well as many other countries, especially with tropical or subtropical climates, systems are designed based on the principle that hot water rises. These are called *thermosyphon* systems, and the storage tank is almost always located outdoors, directly on top of the solar collector; see Figure 2.2.

Figure 2.2. Standard thermosyphon solar water heater with outdoor tank Source: Solahart

For temperate climates, in a solar system for one- and two-family homes with dimensions of about 0.6–1.0 m^2 of collector surface per person and approximately 40–60 l of storage volume per person, the water is mostly heated by the solar system in the summer. This provides an annual degree of coverage (proportion of solar energy to the total energy required for domestic water heating) of about 50–60%. The remaining 40–50% has to be covered by auxiliary heating. For pumped systems, this is often done by means of an extra heat exchanger in the top of the store. Other common solutions are to use the solar water heater as a preheater and connect the solar-heated water to a conventional boiler, or (mainly for sunny climates) to use an electrical element immersed in the store.

Another decisive factor in establishing the level of supplementary energy required is the target domestic water temperature on the boiler controller. The lower this is set, for example 45°C, the higher the coverage proportion of solar energy and correspondingly the lower the proportion of auxiliary energy, and vice versa. However, in some countries, domestic hot water regulations pose a lower limit on this temperature setting, of 60°C.

The individual components of thermal solar systems are introduced in the following sections.

2.2 Collectors

Collectors have the task of converting light as completely as possible into heat, and then of transferring this heat with low losses to the downstream system. There are many different types and designs for different applications, all with different costs and performances. See Figures 2.3 and 2.4.

Different definitions of area are used in the manufacturers' literature to describe the geometry of the collectors, and it is important not to confuse them:

- The *gross surface area* (collector area) is the product of the outside dimensions, and defines for example the minimum amount of roof area that is required for mounting.
- The *aperture area* corresponds to the light entry area of the collector – that is, the area through which the solar radiation passes to the collector itself.

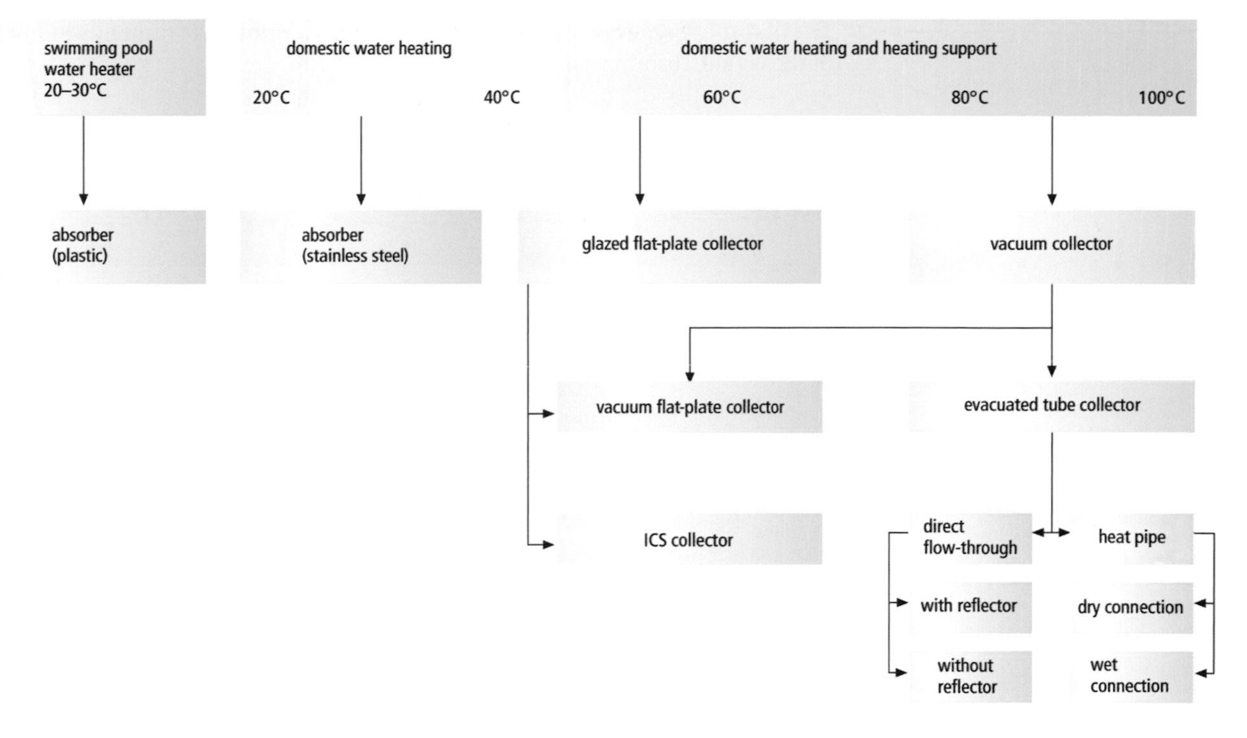

Figure 2.3.
Overview of types of collector

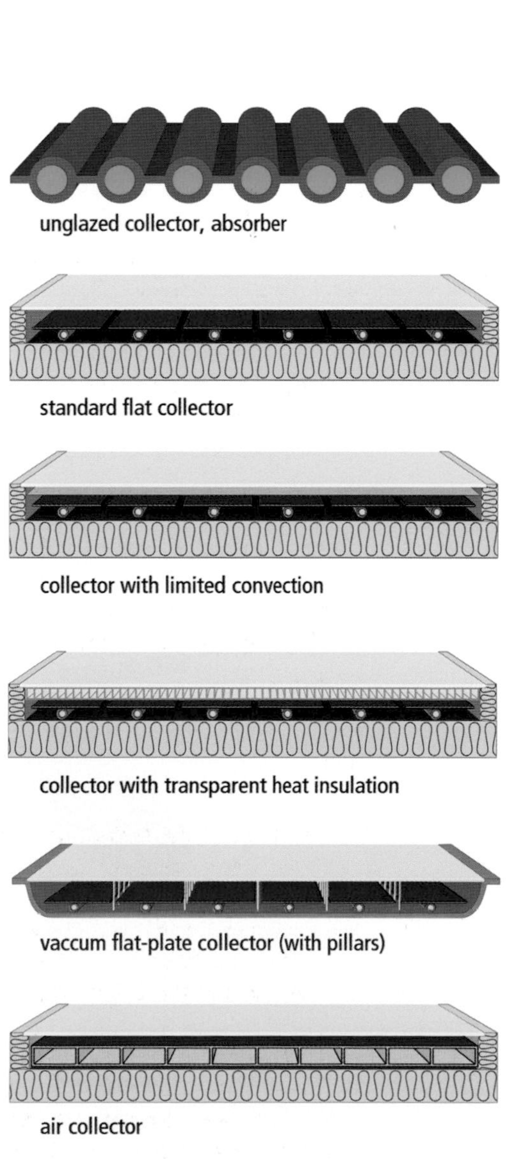

Figure 2.4.
Different collector designs

- The *absorber area* (also called the *effective collector area*) corresponds to the area of the actual absorber panel. See Figures 2.5–2.7.

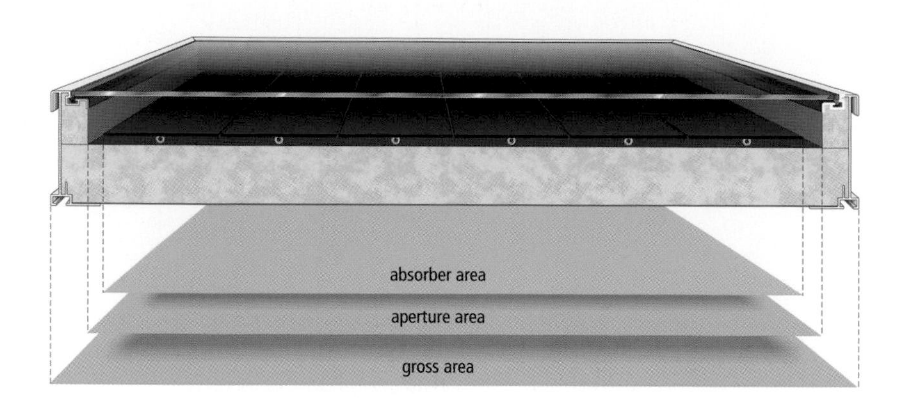

Figure 2.5.
Cross-section of a flat-plate collector with description of the different areas

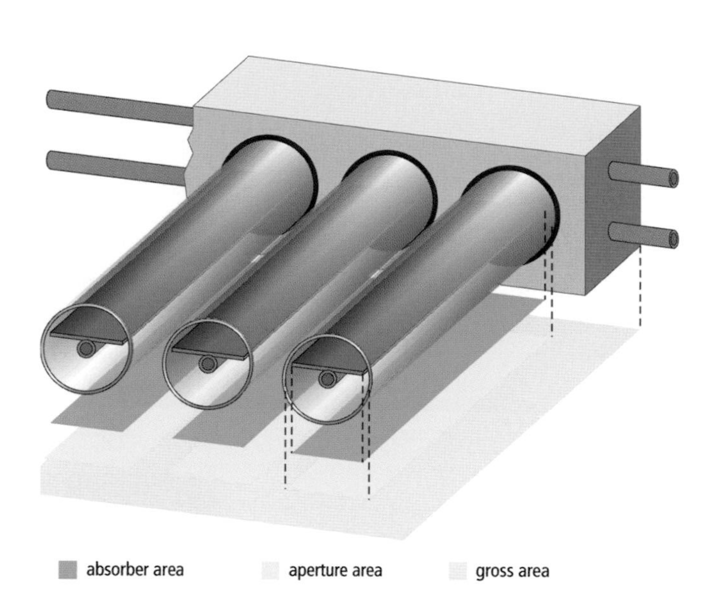

Figure 2.6.
Cross-section of a heat-pipe evacuated tube collector with description of the different surface areas

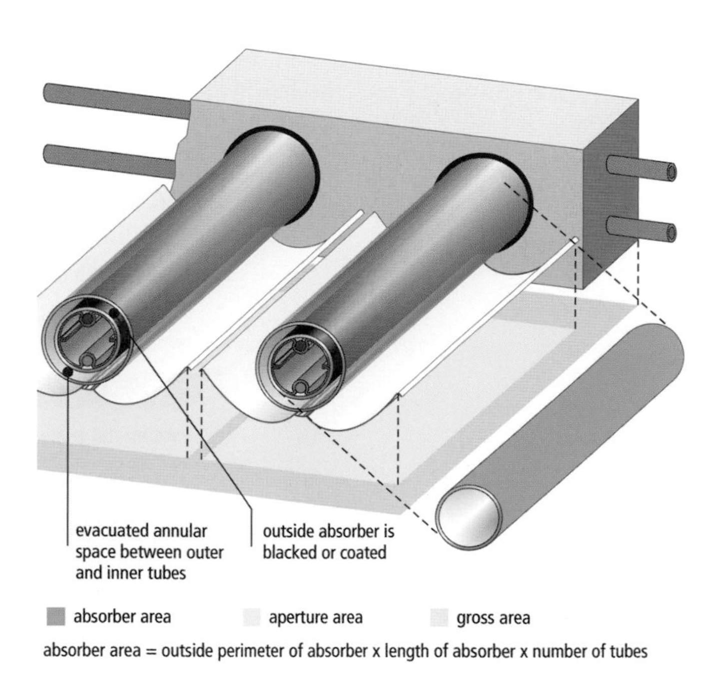

Figure 2.7.
Cross-section of a double evacuated tube collector ('Sydney tubes') with description of the different surface areas

When comparing collectors, the *reference area* is important – that is, the surface area from which the collector's characteristic values are drawn. In the collector test methods according to EN 12975 the reference area is equal to either the aperture area or the absorber area.

For the energy yield, it is not the collector (gross) area that is crucial but the absorber area. (The exception to this is evacuated tube collectors with reflectors (see section 2.2.3). In this case the receiving area – that is, the aperture area – is crucial. The radiation that impinges on this area is reflected to the absorber.)

2.2.1 Unglazed collectors

The simplest kind of solar collectors are unglazed collectors. These have no glazing or insulated collector box, so that they consist only of an absorber (see also section 2.2.2). Unglazed collectors can be found in various application areas, but they are used mainly as a plastic absorber for heating swimming pool water (see Chapter 7). They are also sometimes found as a selectively coated stainless steel absorber for preheating domestic water. This collector has a lower performance at equal operating temperature than a glazed flat-plate collector as it lacks the glass cover, housing and thermal insulation. It therefore has higher thermal losses and can be used only at very low operating temperatures, but because of its simple construction it is inexpensive. See Table 2.1.

Table 2.1. Comparison of yields and cost for stainless steel absorber and flat-plate collector[a]

	Yield (kWh/m²)	Cost (€ or US$ per m²)	Cost (£ per m²)
Unglazed stainless steel absorber	250–300	140–160	98–112
Flat-plate collector	350–500	200–350	140–245

[a]Not including fixing, mounting and VAT

Advantages of the unglazed collector:

- The absorber can replace the roof skin, saving a zinc sheeting, for example. This leads to better heat prices through reduced costs.
- It is suitable for a diversity of roof forms, including flat roofs, pitched roofs and vaulted roofs. It can easily be adapted to slight curves.
- It can be a more aesthetic solution for sheet metal roofs than glazed collectors.

Disadvantages:

- Because of the lower specific performance, it requires more surface area than a flat collector.
- Because of the higher heat losses, the temperature increase (above the air temperature) is limited.

2.2.2 Glazed flat-plate collectors

2.2.2.1 DESIGN

Almost all glazed flat-plate collectors currently available on the market consist of a metal absorber in a flat rectangular housing. The collector is thermally insulated on its back and edges, and is provided with a transparent cover on the upper surface. Two pipe connections for the supply and return of the heat transfer medium are fitted, usually to the side of the collector. See Figure 2.8.

1. frame
2. seal
3. transparent cover
4. frame – side-wall profile
5. thermal insulation
6. full-surface absorber
7. fluid channel
8. fixing slot
9. rear wall

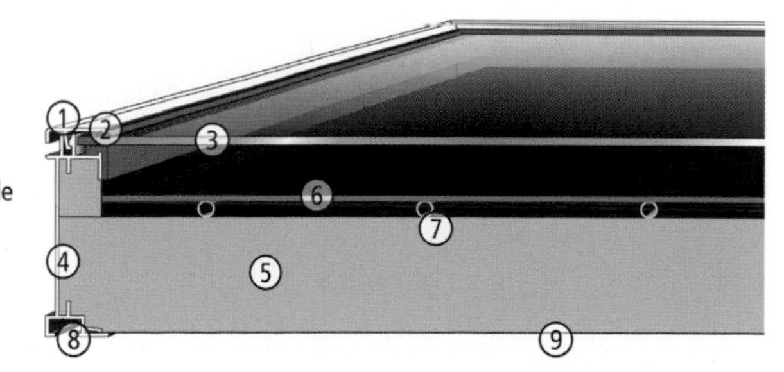

Figure 2.8. Section through a glazed flat-plate collector

Without the glass cover, glazed flat-plate collectors weigh between 8 and 12 kg per m^2 of collector area; the glass cover weighs between 15 and 20 kg/m^2. These collectors are made in various sizes from 1 m^2 to 12.5 m^2, or larger in some cases.

ABSORBER

The core piece of a glazed flat-plate collector is the *absorber*. This consists of a heat-conducting metal sheet (made of copper or aluminium for example, as a single surface or in strips) with a dark coating. The tubes for the heat transfer medium, which are usually made of copper, are connected conductively to the absorber. When the solar radiation hits the absorber it is mainly absorbed and partially reflected. Heat is created through the absorption and conducted in the metal sheet to the heat transfer medium tubes or channels. Through these tubes flows the liquid heat transfer medium, which absorbs the heat and transports it to the store. A variant is the so-called *cushion absorber*, which has full-surface flow-through.

The task of a solar collector is to achieve the highest possible thermal yield. The absorber is therefore provided with a high light-absorption capacity and the lowest possible thermal emissivity. This is achieved by using a *spectral-selective coating*. Unlike black paint, this has a layered structure, which optimizes the conversion of short-wave solar radiation into heat while keeping the thermal radiation as low as possible. See Figure 2.9.

Figure 2.9. Absorption and emission behaviour of different surfaces

Most spectral-selective layers have an absorption rate of 90–95%, and an emission rate of 5–15%. Commonly used selective coatings consist of black chrome or black nickel. See Table 2.2. However, the latest developments in selective coatings with improved optical characteristics currently offered on the market have been applied either in a vacuum process or by sputtering. These processes feature a significantly lower energy consumption and lower environmental load during manufacturing in comparison with black-nickel and black-chrome coatings, which are usually applied by electroplating. In addition, the energy gain of these absorbers is higher at higher temperatures or at low levels of solar irradiance than that of absorbers with black-chrome or black-nickel coatings.[8]

Table 2.2. Advantages and disadvantages of various absorber types

Type	Advantage	Disadvantage
Roll-bonded absorber	Good thermal properties, no mixed materials; simplifies subsequent recycling	Subject to corrosion of aluminium in connection with copper tube
Absorber strips with pressed-in copper tube	High flexibility in size; cheap because of greater volume of production	Many solder points
Absorber with tube system pressed in between metal sheets	No mixed materials; simplifies subsequent recycling	High production cost as connection possible only on plain metal sheet
Absorber with soldered-on tube system	Very flexible in size and flow rate	Heat transfer not optimal
Full flow-through stainless steel absorber	Good heat transfer to liquid	High weight, thermal inertia
Serpentine absorber	Only two solder points in tube system	Higher pressure loss than tube register
Tube register (full-surface absorber)	Lower pressure loss than serpentine absorber	Many solder points in tube system; expensive
Tube register (vane absorber)	Lower pressure loss than serpentine absorber	Many solder points in tube system

RADIATION AND INTERACTION WITH MATTER

When short-wave sunlight (wavelength 0.3–3.0 μm) hits an object, such as a solar cover, it is reflected more or less strongly according to the surface structure (material, roughness and colour). White surfaces reflect much more than dark surfaces. The proportion of reflected radiation (especially for glass panes) also depends on the angle of incidence of the radiation (Fresnel's law). The remaining portion is absorbed by the object or, for translucent material, is allowed to partially pass through. Finally, the absorbed portion is converted into long-wave thermal radiation (wavelengths 3.0–30 μm) and radiated according to the surface structure.

These processes are described physically as the degrees of reflection, absorption, transmission and emissivity of a body.

- ***degree of reflection***, ρ:

$$\rho = \frac{\text{Reflected radiation}}{\text{Incident radiation}}$$

- ***degree of absorption*** **(absorption coefficient),** α:

$$\alpha = \frac{\text{Absorbed radiation}}{\text{Incident radiation}}$$

- ***degree of transmission***, τ:

$$\tau = \frac{\text{Transmitted radiation}}{\text{Incident radiation}}$$

- ***emissivity***, ϵ:

$$\epsilon = \frac{\text{Emitted radiation}}{\text{Absorbed radiation}}$$

The variables ρ, α, τ **and** ϵ **are dependent on the material and wavelength. The sum of** ρ, α **and** τ **is equal to 1 (100%).**

For solar thermal technology the *Stefan–Boltzmann law* is significant. This says that a body emits radiation corresponding to the fourth power of its temperature.

$$\dot{Q} = \sigma T^4$$

where $\dot{Q}$ **is the emitted thermal radiation (W/m²),** σ **is the Stefan–Boltzmann constant =** 5.67×10^{-8} **(W/m²K⁴), and** T **is the absolute body temperature (K).**

In order to reduce the emissions and hence increase the efficiency of the collectors, new absorber coatings are being developed continuously.

As a material for the absorber plate, copper possesses the requisite good thermal conduction. The thermal transmission between absorber plate and tube takes place through the best possible heat-conducting connection.

A further factor for a large energy yield is a low heat capacity, which permits a fast reaction to the ever-changing level of solar radiation. For absorbers with flow channels this is lower (0.4–0.6 l of heat transfer liquid per m^2 of absorber surface) than for full-surface flow absorbers, such as for example cushion absorbers with 1–2 l/m^2.

INSULATION

To reduce heat losses to the environment by thermal conduction, the back and edges of the collector are heat insulated.

As maximum temperatures of 150–200°C (when idle) are possible, mineral fibre insulation is the most suitable here. It is necessary to take account of the adhesive used. This must not vaporize at the temperatures given, otherwise it could precipitate onto the glass pane and impair the light-transmitting capacity.

Some collectors are equipped with a barrier to reduce convection losses. This takes the form of a film of plastic, such as Teflon, between the absorber and the glass pane. In some countries collectors are offered with translucent heat insulation under the glass panel.

HOUSING AND GLASS PANEL

The absorber and thermal insulation are installed in a box and are enclosed on the top with a light-transmitting material for protection and to achieve the so-called greenhouse effect.

Glass or occasionally plastic is used for the cover. Low-ferrous glass (which is highly transparent) is mainly used, in the form of safety glass, 3–4 mm thick. The light-transmitting coefficient is maximally 91%.

Requirements for the transparent cover are as follows (see also Table 2.3):

- high light transmittance during the whole service life of the collector
- low reflection
- protection from the cooling effects of the wind and convection
- protection from moisture
- stability with regard to mechanical loads (hailstones, broken branches etc.).

The first products are now coming onto the international market with special coatings on the glazing to reduce reflections and thus increase the efficiency.

SEALS

Seals prevent the ingress of water, dust and insects. The seals between the glass panel and the housing consist of EPDM (ethylene propylene diene monomer) material or silicon rubber. The rear wall is sealed to the frame with silicon. For tube entry, seals made of silicon or fluorinated rubber are suitable (maximum application temperature 200°C).

Table 2.3. Characteristics of different cover and box materials

Cover	Glass	Plastic		
Transmission values	Long-term stability	Deterioration due to embrittlement, tarnishing, scratches		
Mechanical stability	Stable	Stable		
Cost	Higher	Lower		
Weight	Higher	Lower		
Housing	**Aluminium**	**Steel plate**	**Plastic**	**Wood, bonded waterproof**
Weight	Low	High	Medium	High
Processing	Easy	Easy	Medium	Difficult
Energy requirement	High	Low	Medium	Low
Cost	High	Low	Low	Medium
Other	Increase in energy recovery time, recyclable	Hardly ever used	Seldom used	Ecological material only for in-roof mounting

2.2.2.2 WORKING PRINCIPLE OF A GLAZED FLAT-PLATE COLLECTOR

See Figure 2.10.

The irradiance (G_0) hits the glass cover. Here, even before it enters the collector, a small part of the energy (G_1) is reflected at the outer and inner surfaces of the pane. The selectively coated surface of the absorber also reflects a small part of the light (G_2) and converts the remaining radiation into heat. With good thermal insulation on the rear and on the sides of the collector using standard, non-combustible insulating materials such as mineral wool and/or CFC (chlorofluorocarbon)-free polyurethane foam sheets, the energy losses through thermal conduction (Q_1) are reduced as much as possible.

The transparent cover on the front of the collector has the task of reducing losses from the absorber surface through thermal radiation and convection (Q_2). By this means only convection and radiation losses from the internally heated glass pane to the surroundings occur.

From the irradiated solar energy (G_0), because of the various energy losses G_1, G_2, Q_1 and Q_2, the remaining heat (Q_3) is finally usable.

COLLECTOR EFFICIENCY COEFFICIENT

The *efficiency*, η, of a collector is defined as the ratio of usable thermal power to the irradiated solar energy flux:

$$\eta = \frac{\dot{Q}_A}{G_O}$$

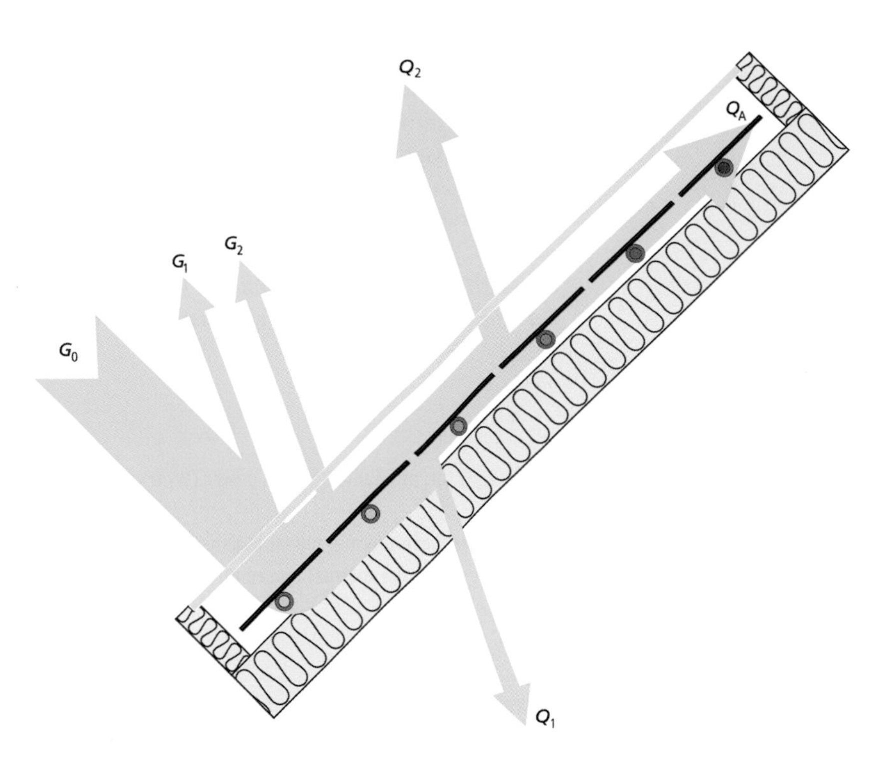

Figure 2.10.
Energy flows in the collector

irradiance (G_0) – reflection losses (G_1 and G_2) – thermal losses ($\dot{Q}_1$ and $\dot{Q}_2$) – available heat quantity ($\dot{Q}_A$)

The efficiency is influenced by the design of the collector: more specifically, it is influenced by the particular optical (G_1 and G_2) and thermal ($\dot{Q}_1$ and $\dot{Q}_2$) losses (see Figure 2.11).

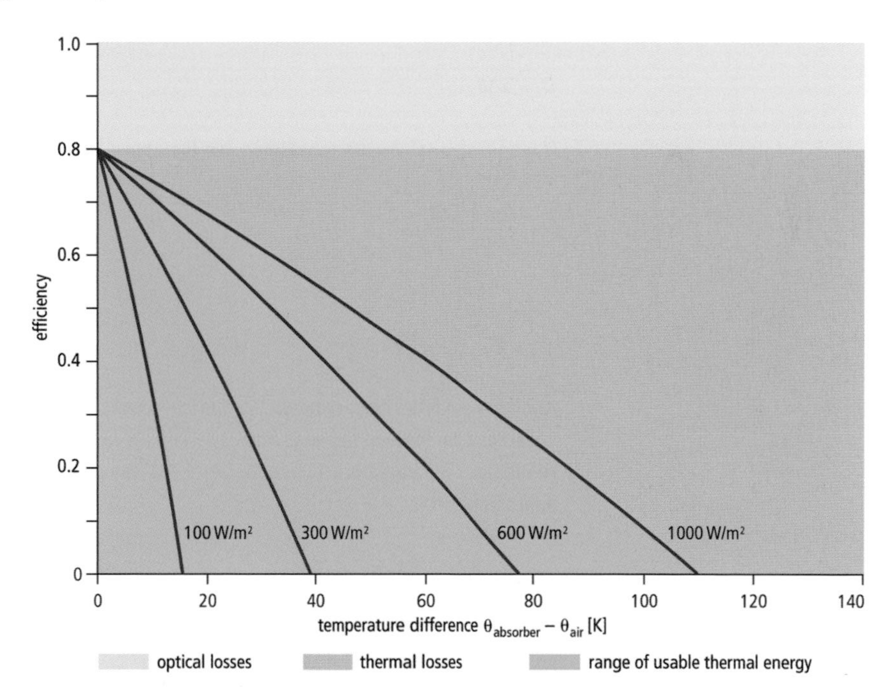

Figure 2.11.
Optical and thermal losses

optical losses thermal losses range of usable thermal energy

The optical losses describe the proportion of the solar irradiance that cannot be absorbed by the absorber. They are dependent on the transparency of the glass cover (degree of transmission, τ) and the absorption capacity of the absorber surface (degree of absorption, α) and are described by the *optical efficiency*:

$$\eta_0 = \tau\alpha$$

The thermal losses are dependent on the temperature difference between the absorber and the outside air, on the insolation, and on the construction of the collector. The influence of the construction is described by the *heat loss coefficient*, k (or k-value), which is measured in W/m^2K.

As the temperature difference between the absorber and the outside air increases, the heat losses increase for a constant irradiance, so that the efficiency reduces. It is therefore important for the yield of a thermal solar system to ensure a low return temperature and a high irradiance.

CHARACTERISTIC CURVE EQUATION AND THE THERMAL LOSS COEFFICIENT

The efficiency of a collector can in general be described by:

$$\eta = \frac{\dot{Q}_A}{G}$$

where Q_A is the available thermal power (W/m^2), and G is the irradiance incident on the glass pane (W/m^2).

The available thermal power is calculated from the available irradiance at the absorber, converted into heat, minus the thermal losses through convection, conduction and radiation:

$$\dot{Q}_A = G_A - \dot{Q}_L$$

where G_A is the available irradiance (W/m^2), and $\dot{Q}_L$ represents the thermal losses (W/m^2).

The available irradiance is obtained mathematically from the product of: the irradiance hitting the glass pane, G; the degree of transmission of the glass, τ; and the degree of absorption of the absorber, α:

$$G_A = G\tau\alpha$$

The thermal losses are dependent on the temperature difference between the absorber and the air, $\Delta\theta$. To a first approximation (for low absorber temperatures) this relationship is linear, and can be described by the heat loss coefficient, k (W/m^2K):

$$\dot{Q}_L = k\Delta\theta$$

If the various values are substituted into the above equation, we obtain for the collector efficiency:

$$\eta = \frac{G\tau\alpha - k\Delta\theta}{G}$$

$$= \frac{\eta_0 - k\Delta\theta}{G}$$

At higher absorber temperatures the thermal losses no longer increase linearly with the temperature difference but instead increase more strongly (by the power of 2) as a result of increasing thermal radiation. The characteristic line therefore has some curvature and the equation in a second order approximation is:

$$\eta = \eta_0 - \frac{k_1\Delta\theta}{G} - \frac{k_2\Delta\theta^2}{G}$$

where k_1 is the linear heat loss coefficient (W/m^2K), and k_2 is the quadratic heat loss coefficient (W/m^2K^2).

In the literature a k_{eff} value is also sometimes given. This is calculated from the k_1 and k_2 values:

$$k_{eff} = k_1 + k_2\Delta\theta$$

when k-values are discussed in the following sections the k_1 value is meant.

NUMERICAL VALUES

The characteristic numbers given are the criteria for comparing the qualities of different collectors. Good glazed flat-plate collectors with spectral-selective absorbers have an optical efficiency, η_0, greater than 0.8 and a k-value of less than 3.5 W/m^{2}K.

The average annual efficiency of a complete system with glazed flat-plate collectors is 35–40%. With an annual amount of solar radiation of 1000 kWh/m^2 (as in central Europe) this corresponds to an energy yield of 350–400 kWh/m^{2}a. In sunnier climates, the radiation may increase to over 2200 kWh/m^2 and the corresponding energy yield of the system may then surpass 770–880 kWh/m^{2}a. These yields assume a sensibly dimensioned system and corresponding consumption.

ADVANTAGES AND DISADVANTAGES OF A GLAZED FLAT-PLATE COLLECTOR

Advantages:

- It is cheaper than a vacuum collector (see section 2.2.3).
- It offers multiple mounting options (on-roof, integrated into the roof, façade mounting and free installation).
- It has a good price/performance ratio.
- It has good possibilities for do-it-yourself assembly (collector construction kits).

Disadvantages:

- It has a lower efficiency than vacuum collectors, because its k-value is higher.
- A supporting system is necessary for flat roof mounting (with anchoring or counterweights).
- It is not suitable for generating higher temperatures, as required for, say, steam generation, or for heat supplies to absorption-type refrigerating machines.
- It requires more roof space than vacuum collectors do.

2.2.2.3 SPECIAL DESIGNS

HYBRID COLLECTOR

A hybrid collector is a combination of a thermal glazed flat-plate collector with solar cells that convert the sunlight into electrical energy. The heat created is used, for instance, to heat domestic water. The solar cells are electrically isolated on the surface of a liquid-cooled absorber with which they are thermally conductively connected. The electrical yields are in the same range as those of conventional photovoltaic systems, and the thermal yields are in the range of collectors without selectively coated absorbers. These collectors are still being developed. Whether this will be a product that succeeds on the market, or whether thermal collectors and photovoltaic modules will in future be offered separately but possibly in a uniform grid dimension and roofing system, remains to be seen.

ICS (INTEGRAL COLLECTOR AND STORAGE) SYSTEM

Here the collector and water store form one structural unit. The following components are not required: heat exchanger, piping for the solar circuit, controller and recirculation pump. The ICS presents an interesting alternative to the standard glazed flat-plate collector from the points of view of efficiency and price/performance ratio. However, simple collectors, with only a glass pane covering them, are not frost-proof, and must therefore be emptied in winter in climates where freezing can occur. Some products use *translucent insulation material* (TIM) behind the glass pane to reduce heat losses, but there is still a risk of freezing of the supply and return lines. ICS systems are widely used in southern Europe.

2.2.3 Vacuum collectors

2.2.3.1 EVACUATED TUBE COLLECTORS

To reduce the thermal losses in a collector, glass cylinders (with internal absorbers) are evacuated in a similar way to Thermos flasks (Figure 2.12). In order to completely suppress thermal losses through convection, the volume enclosed in the glass tubes must be evacuated to less than 10^{-2} bar (1 kPa). Additional evacuation prevents losses

Figure 2.12. The principle of vacuum thermal insulation

through thermal conduction. The radiation losses cannot be reduced by creating a vacuum, as no medium is necessary for the transport of radiation. They are kept low, as in the case of glazed flat-plate collectors, by selective coatings (small ϵ value). The heat losses to the surrounding air are therefore significantly reduced. Even with an absorber temperature of 120°C or more the glass tube remains cold on the outside. Most vacuum tubes are evacuated down to 10^{-5} bar.

ABSORBER, GLASS TUBE, COLLECTOR AND DISTRIBUTOR BOXES

For evacuated tube collectors, the absorber is installed as either flat or upward-vaulted metal strips or as a coating applied to an internal glass bulb in an evacuated glass tube. The forces arising from the vacuum in the tube are very easily taken up by the high compression strength of the tubular form.

An evacuated tube collector consists of a number of tubes that are connected together and which are linked at the top by an insulated distributor or collector box, in which the feed or return lines run. At the base the tubes are fitted to a rail with tube holders. There are two main sorts of evacuated tube collector: the direct flow-through type and the heat-pipe type.

DIRECT FLOW-THROUGH EVACUATED TUBE COLLECTORS

In this design (Figure 2.13) the heat transfer medium is either led via a tube-in-tube system (coaxial tube) to the base of the glass bulb, where it flows back in the return flow and thereby takes up the heat from the highly spectral-selective absorber, or it flows through a U-shaped tube.

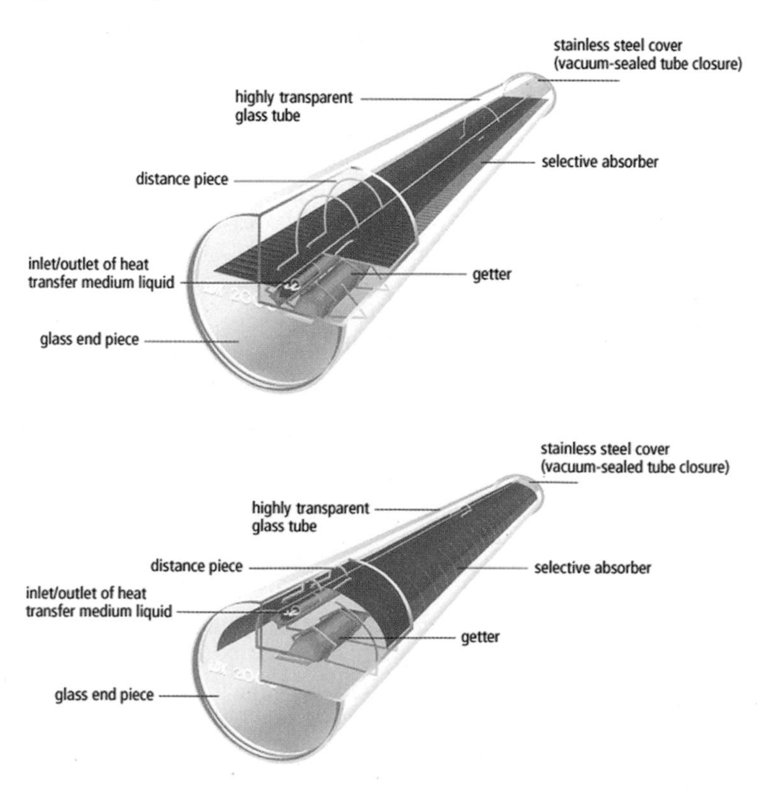

Figure 2.13. Cross-sectional view of a direct flow-through evacuated tube collector (Lux 2000 and 2000R, ThermoLUX, Kempten)

Direct flow-through evacuated tube collectors can be oriented towards the south, but they can also be mounted horizontally on a flat roof.

A particular design of direct flow-through evacuated tube collector marketed in some countries is the *Sydney collector* (Figure 2.14). The collector tube consists of a vacuum-sealed double tube. The inner glass bulb is provided with a selective coating of a metal carbon compound on a copper base. Into this evacuated double tube is plugged a thermal conducting plate in connection with a U-tube to which the heat is transferred. Several tubes are combined into one module (between 6 and 21, according to supplier). To increase the radiation gain the collector is fitted with external reflectors in the sloping roof version. The flat roof version requires a light background such as gravel or reflective foil, as it does not have reflectors.

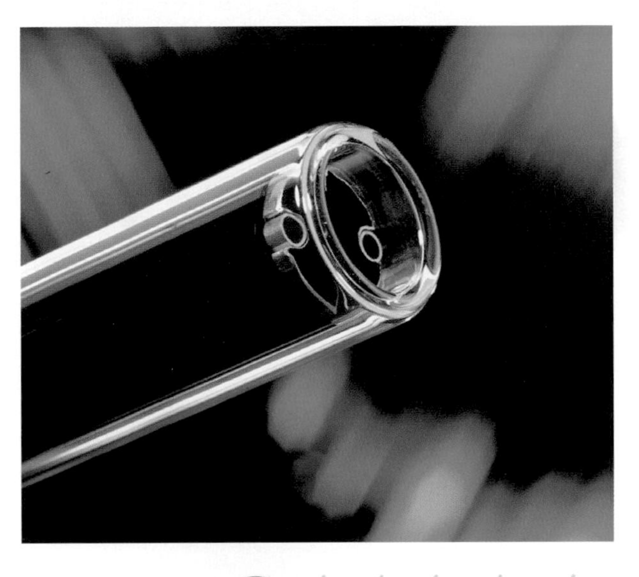

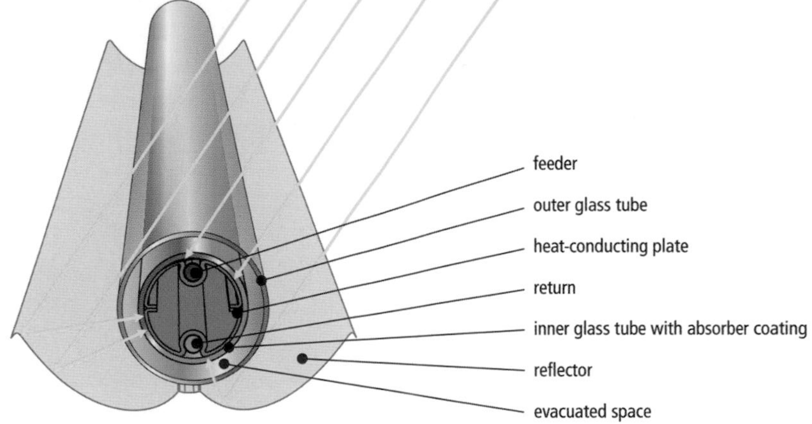

Figure 2.14. Sydney tube (microtherm, Lods)

HEAT-PIPE EVACUATED TUBE COLLECTORS

In this type of collector a selectively coated absorber strip, which is metallically bonded to a heat pipe, is plugged into the evacuated glass tube. The heat pipe is filled with alcohol or water in a vacuum, which evaporates at temperatures as low as 25°C. The vapour thus occurring rises upwards. At the upper end of the heat pipe the heat released by condensation of the vapour is transferred via a heat exchanger (condenser) to the heat transfer medium as it flows by. The condensate flows back down into the heat pipe to take up the heat again. For appropriate functioning of the tubes they must be installed at a minimum slope of 25°.

Heat pipe evacuated tube collectors are offered in two versions, one with a dry and one with a wet connection. For the dry type (Figure 2.15), the condenser completely surrounds the collector, and provides a good heat-conducting link to a double tube heat exchanger: The heat transfer takes place from the condenser via the tube wall to the heat transfer medium. This permits defective tubes to be exchanged without emptying the solar circuit.

For the wet connection (Figure 2.16) the condenser is immersed in the heat transfer medium. If tubes have to be exchanged, it is necessary to at least empty the collection device at the top.

The UK manufacturer Thermomax has designed a stagnation temperature limiting system, in which a *memory spring* is used to close the condenser from a temperature of 130°C and hence thermally decouple the system from the collection device.

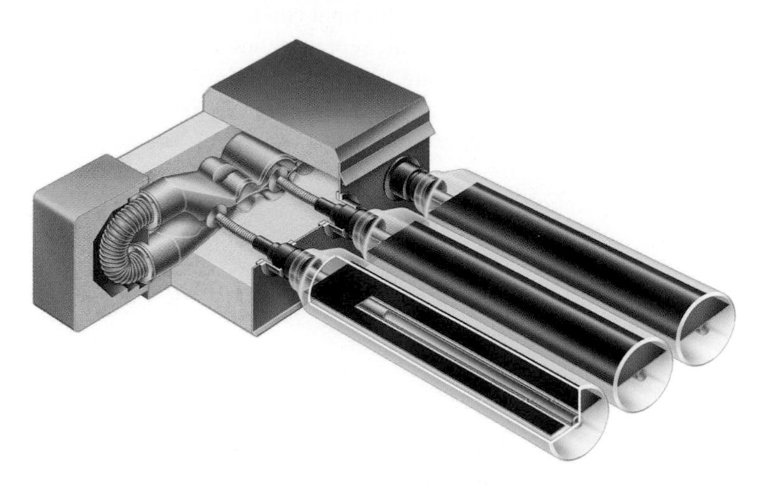

Figure 2.15. Cross-section of heat-pipe evacuated tube collector with dry connection. Source: Viessmann, Allendorf

Figure 2.16. Cross-section of a heat-pipe evacuated tube collector and condenser with memory spring and wet connection. Source: Elco-Klöckner, Hechingen

NUMERICAL VALUES

The optical efficiency of evacuated tube collectors is somewhat lower than that of glazed flat-plate collectors because of the tube form (η_0 = 0.6–0.8), but because of the better heat insulation the *k*-value is below 1.5 W/m^{2}K.

The average annual efficiency of a complete system with evacuated tube collectors is between 45% and 50%. With an annual solar irradiance of 1000 kWh/m^2 (as in central Europe) this corresponds to an energy yield of 450–500 kWh/m^{2}a. These yields assume appropriate dimensioning of the system.

ADVANTAGES AND DISADVANTAGES OF AN EVACUATED TUBE COLLECTOR

Advantages:

- It achieves a high efficiency even with large temperature differences between absorber and surroundings.
- It achieves a high efficiency with low radiation.
- It supports space heating applications more effectively than do glazed flat-plate collectors.
- It achieves high temperatures, for example for steam generation or air-conditioning.
- It can be easily transported to any installation location because of its low weight; sometimes the collector is assembled at the installation site.
- By turning the absorber strips (in the factory or during assembly) it can be aligned towards the sun (only relevant for certain products).
- In the form of direct through-flow tubes it can be mounted horizontally on a flat roof, hence providing less wind load and lower installation costs. In this way penetration of the roof skin is avoided.

Disadvantages:

- It is more expensive than a glazed flat-plate collector.
- It cannot be used for in-roof installation.
- It cannot be used for horizontal installation for heat pipe systems (inclination must be at least 25°).

2.2.4 Collector accessories

Collectors cannot be installed without additional materials. They are:

- for on-roof installation: roof hooks, special bricks, rails, vent tiles
- for in-roof installation: covering frames
- for flat roof or free installation: bracing, counterweights, bearers.

For more information see Chapter 4.

2.2.5 Collector characteristic curves and applications

Figure 2.17 shows typical efficiency curves and areas of application with the same global solar irradiance for the following collector types: swimming pool absorber, glazed flat-plate collector, and evacuated tube collector. At $\Delta\theta = 0$ each collector type is at its highest efficiency (η_0). At its maximum temperature – that is, when it has reached its stagnation temperature – the efficiency equals zero.

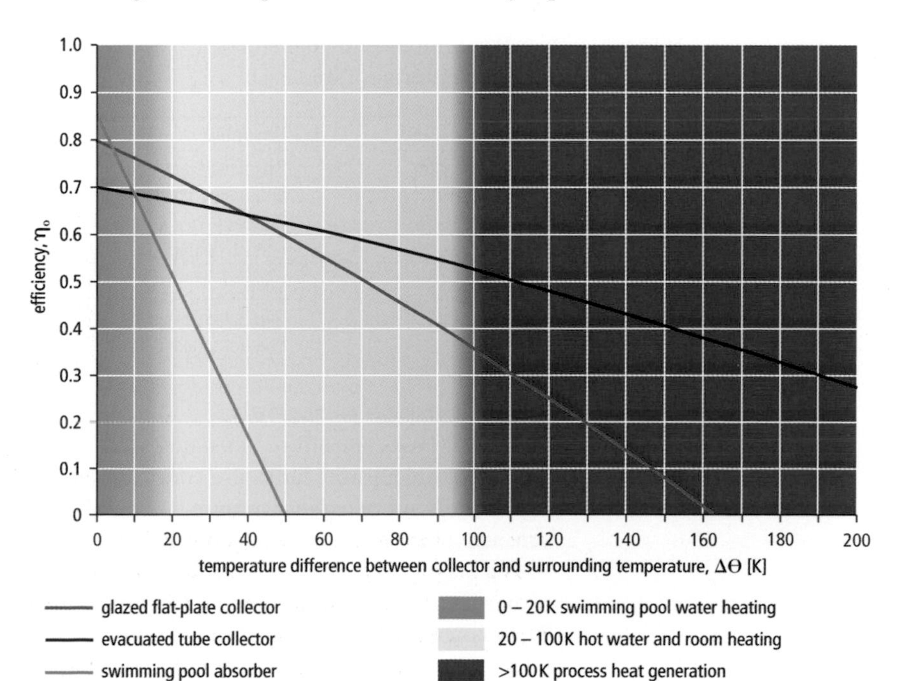

Figure 2.17. Efficiency characteristic curves for different types of collector and their areas of application (at irradiation of 1000W/m²K)

2.2.6 Stagnation temperature

If the circulation pump fails in the event of strong solar irradiance, or if – for example in holiday times – no hot water is used so that the store is hot (60–90°C) and the system switches off, no more heat is drawn from the collector. In this situation the absorber heats up until the heat losses through convection, heat radiation and heat conduction reach the thermal output of the absorber. The greater the insolation, the higher the stagnation temperature. Well-insulated glazed flat-plate collectors achieve maximum stagnation temperatures of 160–200°C, evacuated tube collectors 200–300°C, or – with a reflector – as much as 350°C.

STEAM GENERATION WITH FLOW-THROUGH EVACUATED TUBE COLLECTORS

The generation of steam in connecting pipes is considerably more pronounced for evacuated tube collectors than for glazed flat-plate collectors. In the connecting pipes, convective heat transport processes occur in the liquid–steam phase. Therefore the thermal transfer medium is converted to the steam condition in a large section of the connecting pipeline. This condition, which can be particularly significant for short pipes, such as in central roof heating units, has the following consequences.

All materials in the solar circuit must be checked with respect to the increased stagnation condition requirements (insulation materials, solders, temperature sensors). The connecting technology must be matched to the temperatures in the solar circuit, 160°C or more. Soft solders cannot therefore be used. Expansion vessels must never be installed in the vicinity of tube collectors: longer and non-insulated supply lines must be provided to the expansion vessel. The additional volume of steam in other parts of the solar circuit must be taken into account when dimensioning the vessel.

For heat-pipe tubes the thermal load is significantly lower, as thermal decoupling between tube and collector takes place at temperatures of no more than 150°C.The whole content then has vaporised, and no further heat transmission takes place.

2.3 Heat stores

The energy supplied by the sun cannot be influenced, and rarely matches the times when heat is required. Therefore the generated solar heat must be stored. It would be ideal if this heat could be saved from the summer to the winter (seasonal store) so that it could be used for heating. In some countries, such as Switzerland, this has already been done for several years in low-energy houses with hot water stores of several m^3 in volume and collector surface areas of several tens of square metres. There are stores that store heat chemically, currently available as prototypes, which should be available on the market in the near future. Even for short-term storage over one or two days, to bridge over weather variations, developments are still taking place.

We differentiate heat stores according to the application, the compression strength and the material (Table 2.4).

Table 2.4. Classification of solar stores

Type	Unvented	Vented
Secondary water	Stainless steel Enamelled steel Copper Plastic-coated steel	 Copper Stainless steel
Buffer	Steel	Plastic
Thermal with domestic hot water	Steel/stainless steel Steel/enamelled steel	

2.3.1 Storage materials

Unvented tanks are offered in stainless steel, enamelled or plastic-coated steel or copper. Stainless steel tanks are comparatively light and maintenance-free, but significantly more expensive than enamelled steel tanks. Also, stainless steel is more sensitive to water with a high chloride content. Enamelled tanks must be equipped with a magnesium or external galvanic anode for corrosion protection reasons (voids or hairline cracks in the enamel). Cheaper plastic-coated steel tanks are also offered. Their coating must be free of pores, and they are sensitive to temperatures above 80°C. Vented plastic tanks are similarly sensitive to higher temperatures. In the UK, the predominant material used for tanks is copper. The advantages of copper are its lightness and the ease of fabricating it in different sizes.

2.3.2 Domestic hot water stores

Figure 2.18 shows an example of an unvented solar store, as often used in temperate climates. It has the following features:

- two heat exchangers for two heat sources (bivalent): a solar heat exchanger and an additional heat exchanger for a heating boiler
- direct connection to the cold water supply
- pressure tank 4–6 bar operating pressure.

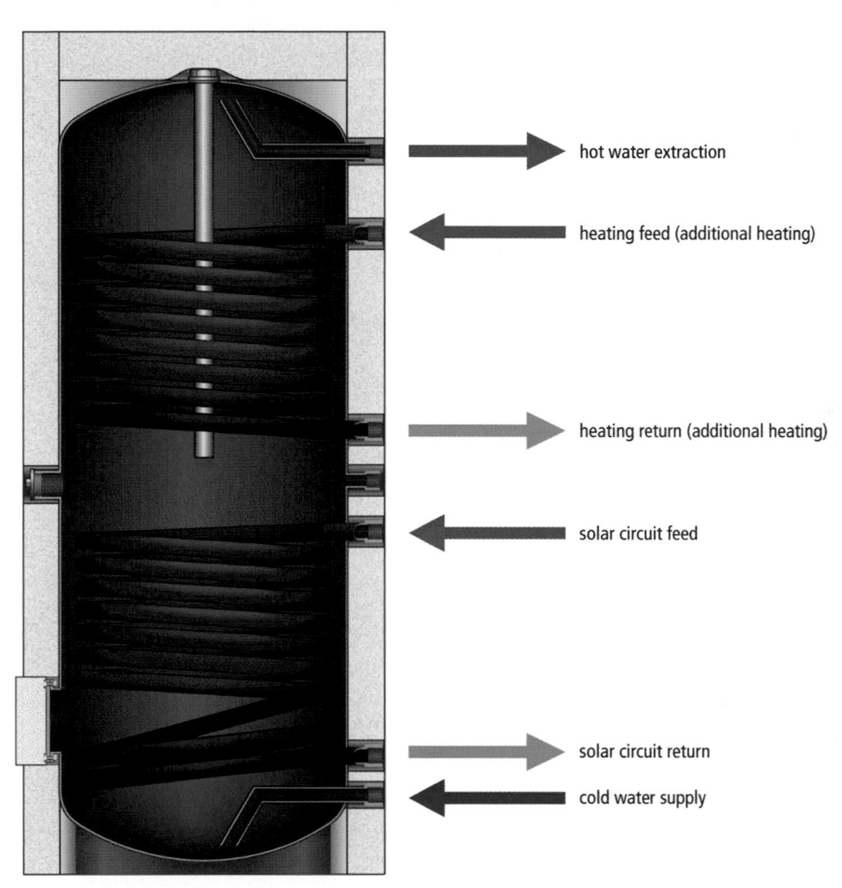

Figure 2.18. Upright model solar tank with auxiliary heat exchanger

Examples of typical store volumes are:

- UK: 150–200 l
- Germany: 300–500 l in the one- and two-family home segment of the market
- USA: 50–100 gal (approx 200–400 l)
- Australia: 300–400 l for 3–4 persons.

See also Chapter 3, section 3.5.2.4. Larger stores can take up greater quantities of energy, but for constant collector surface areas can lead to increased switch-on frequencies of the heating boiler when the store is not stratified because the temperature level in the tank is lower than for a smaller tank.

When the store contains water for consumption, one should be wary of limescale precipitation at temperatures above 60°C, which can block the heat exchanger surface (see section 2.3.4). Moreover, lime residues are deposited gradually on the base of the store.

There are various design features that have a significant impact on the suitability of a solar heat store. These are described below.

2.3.2.1 NARROWNESS OF THE HOT WATER STORE

Because of short-term variations in irradiation, the solar store should be able to contain the hot water consumption of about one to two days. Every time a tap is turned on, cold water flows into the lower area of the store, so that cold, warm and hot water are found in the one store at the same time. Because of the different densities, a temperature stratification effect forms. The 'lighter' hot water collects at the top, the 'heavier' cold water in the lower tank area. This stratification effect has a positive effect on the efficiency of a solar system.

As soon as hot water is drawn off, cold water flows in: this should not mix with the hot water. The slimmer and taller the store, the more pronounced the temperature stratification will be. Upright stores have the best stratification; the recommended height–diameter ratio for optimum stratification is at least 2.5:1. The coldest possible lower zone ensures that, even with low irradiance, the solar system can still operate with high efficiency at a low temperature level (see Figure 2.17).

When installing slim stores, narrow doors seldom cause problems; however, there may however be problems with the overall height (the tilted dimension should be noted). What is important is the free height at the installation location, as heating pipes, water drainage pipes and similar frequently run below the roof.

Not all products follow the above reasoning. In many countries, solar water heaters are sold that do use horizontal stores. These stores feature less stratification. Some such systems are sold with immersed electrical elements as backup heater. Such electrical elements are sensible only in hot climates where they are used for, at most, one or two months per year. In these systems the whole tank can heat up when the electrical element is activated, destroying the solar efficiency.

THE TEMPERATURE AND ENERGY CONTENTS OF A STORE

Let us consider a 300 l solar store in which, after charging and hot water consumption, the temperature layer system shown in Figure 2.19 has built up as a result of its slender construction. The energy content of this tank at a cold water temperature of 15°C is:

$$Q = mc_w\Delta\theta$$

where Q is the heat quantity (Wh), m is the mass (kg), c_w is the specific heat capacity of water = 1.16 Wh/kgK, and $\Delta\theta$ is the temperature difference (K).

$$Q = 100\text{ kg} \times 1.16\text{ Wh/kgK} \times 0\text{ K} + 100\text{ kg} \times 1.16\text{ kWh/kgK} \times 15\text{ K} + 100\text{ kg} \times 1.16\text{ Wh/kgK} \times 45\text{ K}$$
$$= 6960\text{ Wh}$$

If the same energy content is charged into a store that cannot form layers there will be a uniform mixed temperature of:

$$\theta_m = \frac{6960\text{ Wh}}{300\text{ kg} \times 1.16\text{ Wh/kgK}} + 15°\text{C}$$
$$= 35°\text{C}$$

In this case additional heating is required in the standby area of the store (Figure 2.20); however, in the first instance 100 l of 60°C or 150 l of 45°C water can be drawn off. The calculation of the mixed temperatures is possible using the following equation:

$$\theta_m = \frac{m_1\theta_1 + m_2\theta_2}{m_1 + m_2}$$

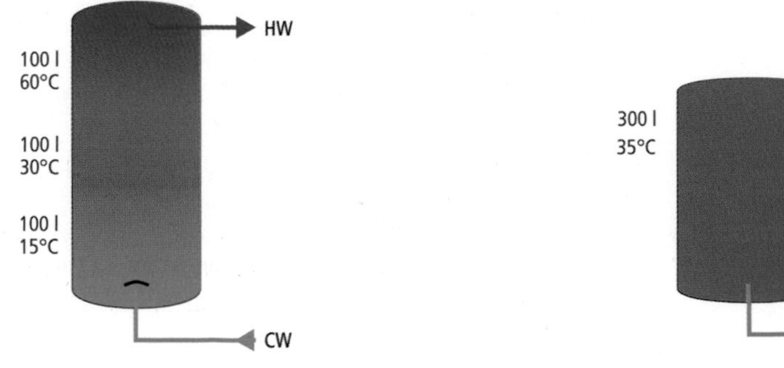

Figure 2.19 (left). Usable stratification

Figure 2.20 (right). Additional heating required

2.3.2.2 BAFFLE PLATE AT COLD WATER INPUT

When a tap is opened, cold water will flow into the tank. To prevent mixing between the cold water and the warm water still in the tank, a baffle plate is often used.

2.3.2.3 HOT WATER EXTRACTION

With conventional unvented stores, the hot water is often taken out from the top. After the water has been taken out, the warm water standing in the pipes cools down again. The cooled water falls down into the upper layers of the tank (single tube circulation). This causes destruction of the stratification effect, and heat losses in the order of magnitude of up to 15% of the total store losses (Figure 2.21). A better system is one in which the pipe for the hot water outlet is led inside the store from the top to the bottom through a bottom flange or outside the store downwards within the heat insulation. In this way heat losses via the disturbed heat insulation in the hottest zone are avoided. At the very least the hot water outflow – if it is led to the side through the heat insulation – should be provided with a 180° pipe bend (syphon).

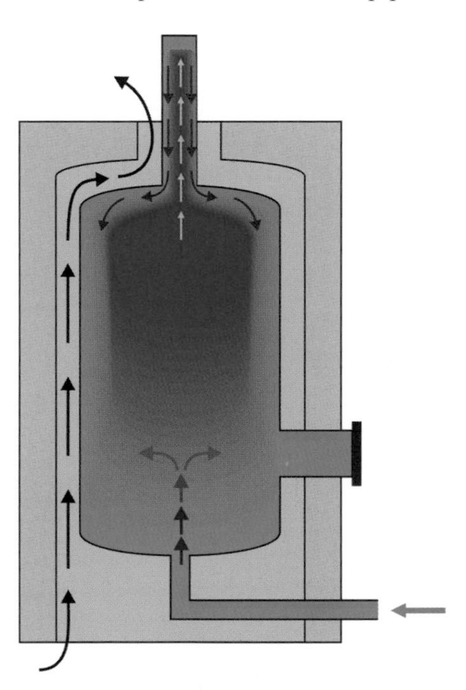

Figure 2.21. Avoidable heat losses through design deficiencies[9]

2.3.2.4 HEAT EXCHANGERS AND THEIR CONNECTIONS

Ideally the heat exchanger connections for the solar and additional heating circuit should both be installed with two 180° pipe bends so that heat losses via the connections are reduced. The position of the additional heat exchanger in the upper area of the store guarantees rapid heating up of the standby volume (daily requirement), without removing the possibility of being able to store even small amounts of the sun's energy effectively in the cooler area of the store. The solar circuit heat exchanger should reach as far as possible down into the store, so that the tank's contents can be heated right down to the base of the tank.

2.3.2.5 INSULATION OF THE STORE

Good heat insulation is essential for an efficient solar store, and this includes the floor of the tank. It should fit well all round (otherwise there would be losses through convection), and consist of CFC- and PVC-free materials (melamine resin foam and polyethylene (PE) jacket, for example), with a thermal conductivity, $\lambda < 0.035$ W/mK. As for a collector, a k-value (W/m²K) can also be defined for a store as the ratio of the thermal conductivity to insulation thickness ($k = \lambda/D$). The product of the k-value with the store surface area, A, gives the heat loss rate, kA (W/K), of the tank.

In total, the heat loss rate should be less than 2 W/K. A very good store with a heat loss rate of, say, 1.5 W/K rather than 3 W/K, at a temperature difference of 35 K loses about 450 kWh less energy per annum. This corresponds to an additional collector surface area of about 0.5–1 m² or additional costs of €200–400 ($200–400, or £140–280).

The following insulation options are offered on the solar market: flexible foam; rigid foam envelope, which can be retrofitted; or in-situ foam-treated stores with plastic or metal jacket. It is particularly important to ensure that the insulation is closed tightly round the pipe connections without any gaps, and that flanges, plugs etc. are also insulated.

2.3.2.6 STORE TEMPERATURE SENSOR: SOLAR CIRCUIT

This sensor should always be fitted in the lower third area of the store, in the middle of the solar circuit heat exchanger, because in this way automatic switching of the pump takes place earlier than if the sensor is fitted higher. This means that, even where small amounts are drawn off, the system has the chance of recharging with solar energy. If no submerged sleeve is available at the corresponding height on the store, a contact sensor can also be used. This causes no problems if a clamping block for the sensor is provided (outside the store, passing from the top to the bottom). The height can be selected in this case within certain limits. If no sensor-clamping block is provided there may be fixing problems. As it is important to have good thermal contacts for control of the system (firm seating and heat conducting paste), poor thermal contact may lead to malfunctioning of the complete system.

2.3.2.7 STORE TEMPERATURE SENSOR: ADDITIONAL HEATING

If an auxiliary heating source is connected to keep the upper part of the store at a set temperature, this sensor provides the information for the control of the additional heating and signals the information for the beginning and end of the auxiliary heating process. It should be fitted at the same height as the auxiliary heat exchanger, or higher, but never below it. It is part of the boiler controller.

2.3.3 Buffer store

Buffer stores (or thermal stores) can be vented or unvented and filled with heating water. The heat stored in them can be either directly fed into the heating system (heating support) or transferred via a heat exchanger to the domestic hot water. Buffer storage is nothing new for plumbers. Buffer storage is used to avoid the cycling of heating boilers: the boiler first charges the store, which leads to longer burner running times (longer service life, lower hazardous substance emissions). The radiators are then supplied from the store.

2.3.4 Combination stores and innovative store designs

The combi store is a combination of a buffer and a domestic hot water store. A smaller domestic hot water store section is installed in the upper warm area of a buffer store, whose upper surface acts as a heat exchanger.

It is suitable for use in solar systems for domestic hot water heating, with or without space heating. Because of the feed system, the pipework is simple and the control system is uncomplicated. All heat generators (solar collectors, heating boilers) and all heat consumers (hot water, heating) work on the same buffer. The heating system is connected in the upper area to the buffer store, where it heats the domestic water. The middle area can be used for increasing the temperature of the return of the heating water. In the lower area there is the heat exchanger for the solar energy feed. The inner domestic water store is heated through its wall.

In several European countries, including Germany and Austria, special constructions can be found in the store to optimize the stratification.

Innovative store concepts are in various stages of development. These include integration of a gas-burning boiler into the tank unit (Denmark, the Netherlands, Germany, Switzerland) and stores based on the sorption (adsorption or absorption) principle (Germany).

2.3.5 The thermostatic mixing valve

The thermostatic mixing (blending) valve controls and limits the temperature of the hot water. If the hot water taken from the solar store is too hot, the thermostatic mixing valve mixes cold water into it. Independently of the design and temperature of the water heater, it ensures a constant temperature at the withdrawal point and therefore protects the users from scalding. European standard EN 12976-1 states that a thermostatic mixing valve should be installed if the domestic hot water temperature can reach over 60°C in the hot water system. In the case of domestic hot water stores this occurs if the store tank temperature has not been already limited to 60°C in order to reduce lime formation.

A thermostatic mixing valve must be installed immediately at the hot water outlet of the solar store (without blocking safety vents). Most models can be set variably to temperatures between 30°C and 70°C (up to 10 bar). There must be a filter in the cold water pipe to protect the mixer from contamination and hence to avoid malfunctions.

2.3.6 Connection of washing machines and dishwashers

Washing machines and dishwashers require the largest part of their power consumption for heating the water. The connection of a solar system offers the opportunity of replacing expensive electrical power with the sun's energy, and the opportunity of increasing the efficiency of the system.

In several European countries *hot fill* dishwashers and washing machines are on the market, which are suitable for connecting to the hot water pipe system. These can be directly connected to the solar system to make immediate use of the sun's energy.

Most washing machines do not have a hot water connection; for these machines special control units are available in many countries that can regulate the cold and hot water flows to the machine according to the desired washing programme.

2.3.7 Legionella contamination

After several severe outbreaks of Legionella (including an outbreak in the Netherlands, where, in 1999, 29 people died after a Legionella infection caused by an infected whirlpool at a demonstration during a flower exhibition), strict legal requirements were developed for official public buildings in several countries including the Netherlands and Germany. These requirements can significantly affect the design of solar water heaters. For instance, in the Netherlands there is a duty on the building owner to keep a log book (with information on, for example, prevention measures taken, temperature measurements taken, and planned modifications to the installation etc.). Within the scope of the prevention measures, the cold water temperature must never exceed 25°C and the hot water temperature must never fall below 60°C. This means that all solar water heaters must have an auxiliary heater always on and set to at least 60°C.[10] In Germany, strict requirements are put on hot water installations with a volume exceeding 400 l, which has caused many solar water heaters to be designed in such a way (often using tank-in-tank constructions) that this volume is not reached.

At present methods such as pasteurization are being tested in practice in order to overcome these problems. To what extent these will be implemented or specified at the European level with a harmonized procedure remains to be seen.

LEGIONELLA

***Legionella pneumophila*, a rod-shaped bacterium, is naturally found in water. The concentration of Legionella bacteria is very low in cold natural water, and hence normally there is no danger to health. Legionella bacteria multiply most quickly at temperatures between 30°C and 45°C. They are destroyed at temperatures above 50°C, and this process is significantly accelerated with increasing temperatures.**

Crucial factors for the risk of an infection are:

- ***The way the bacteria are absorbed.* As long as they are absorbed via the gastro-intestinal tract – that is, if water is swallowed during swimming or when cleaning teeth – they are harmless. A danger of infection occurs only if they enter the lungs, for example by inhaling finely dispersed water droplets in the air (aerosol), such as during showering.**
- ***The concentration of bacteria.* Legionella can occur in dangerous concentrations in air-conditioning plants with air humidifiers, in whirlpools, and in large hot water systems where hot water stands for long times, such as in hotels or hospitals with large stores and long pipelines.**
- ***The time span spent in a contaminated environment.***
- ***Personal resistance to illness.***

2.4 Solar circuit

The heat generated in the collector is transported to the solar store by means of the solar circuit. This consists of the following elements:

- the pipelines, which connect the collectors on the roof and the stores
- the solar liquid or transport medium, which transports the heat from the collector to the store
- the solar pump, which circulates the solar liquid in the solar circuit (thermosyphon and ICS systems do not have a pump)
- the solar circuit heat exchanger, which transfers the heat gained to the domestic hot water in the store
- the fittings and equipment for filling, emptying and bleeding
- the safety equipment. The expansion vessel and safety valve protect the system from damage (leakage) by volume expansion or high pressures.

2.4.1 Pipelines

Copper is the most frequently used material for the heat transport pipelines between collector and store. Many types of fittings made of copper, red bronze or brass are available for Cu/Cu connections and the transitions to other system components with threaded connections. If the installation conditions are restricted in the area of the collector connections, or in the case of self-assembly, flexible stainless steel corrugated pipes are used (this means solder-free pipe connections). Steel pipes are also often used, mostly for larger systems. Using galvanized pipes together with antifreeze fluids is not recommended, as this leads to corrosion problems.

INSULATION OF PIPELINES

In conventional heat installations, insufficient attention is often given to the insulation. But what is the use of the best heating technology if a large and unnecessary amount of heat is lost on the transport routes and in the store?

According to European Standard EN 12976 (see Chapter 12), for pipes up to 22 mm external diameter the thickness of the layer of insulation should be at least 20 ± 2 mm. For pipes with external diameters between 28 and 42 mm, the insulation thickness should be 30 ± 2 mm. The insulation material used should have a thermal conductivity of $\lambda \leq 0.035$ W/mK.

It is important to insulate the complete piping without gaps and open spots. That means that even the fittings, valves, store connections, plugs, flanges and similar must be well insulated. In the solar circuit a temperature-resistant insulation material must be used, such as Aeroflex or Armaflex HT, up to 150–170°C (see Figure 2.22). External pipelines must be UV and weather resistant as well as being waterproof, and should offer protection from animal damage, for example by installing them in a metal jacket.

Figure 2.22. Temperature-resistant thermal insulation (Armstrong Insulation, Pfaffnau)

Apart from the traditional separately installed combination of pipe, thermal insulation and electric cable, prefabricated pipeline units, consisting of copper pipe or corrugated hose of stainless steel including heat insulation and electric cable, from the roll are offered on the market for installation as the solar feed and return and the collector sensor.

2.4.2 Solar liquid

The solar liquid transports the heat produced in the collector to the solar store. Water is the most suitable medium for this, as it has some very good properties:

- high thermal capacity
- high thermal conductivity
- low viscosity.

Moreover, water is:

- non-combustible
- non-toxic
- cheap.

As the operating temperatures in collectors can be between –15°C and +350°C, if we use water as the heat transfer medium we shall have problems with both frost and evaporation.

In fact, water:

- freezes (at 0°C)
- evaporates (at 100°C).

Through the addition of 40% propylene glycol, which is predominantly used at present, frost protection down to –23°C is achieved, together with an increase in the boiling point to 150°C or more, according to the pressure.

Water is highly corrosive. This is further increased by the propylene glycol. For this reason a whole range of inhibitors are used, each inhibitor offering corrosion protection for a specific material.

Other effects of the addition of glycol are:

- reduced thermal capacity
- reduced thermal conductivity
- increased viscosity
- increased creep capacity.

Apart from a sufficiently high thermal capacity and conductivity, together with good frost and corrosion protection, another requirement for the solar liquid is its biological degradability. It must be non-toxic and non-irritant. All the requirements given are best fulfilled at present by a water–glycol mixture with liquid inhibitors.

In Figures 2.23–2.26 some characteristics of various concentrations of mixtures of water and propylene glycol are shown.

SOLAR LIQUIDS IN EVACUATED TUBE COLLECTORS

Stagnation temperatures of up 350°C in evacuated tube collectors demand excellent thermal stability of the heat transfer medium. At these temperatures, boiling of the liquid cannot be prevented. Heat transfer media are required that evaporate without leaving any residue.

Decisive factors in the corrosion of metal materials in solar energy systems by liquids are the oxygen content, the presence of dissolved solids, and their pH values. In order to reduce the significantly higher corrosive nature of mixtures with propylene glycol in comparison with water, inhibitors are added. These form a protective film on the metal surface. In the heat transfer media that are normally used for temperatures of up to 200°C, dissolved substances have been used. When the liquid evaporates, solid, non-vaporizable residues remain on the pipe walls, which in the most unfavourable case can no longer be taken up by the re-condensing liquid. Insufficient or lack of re-solubility in these coatings can lead to

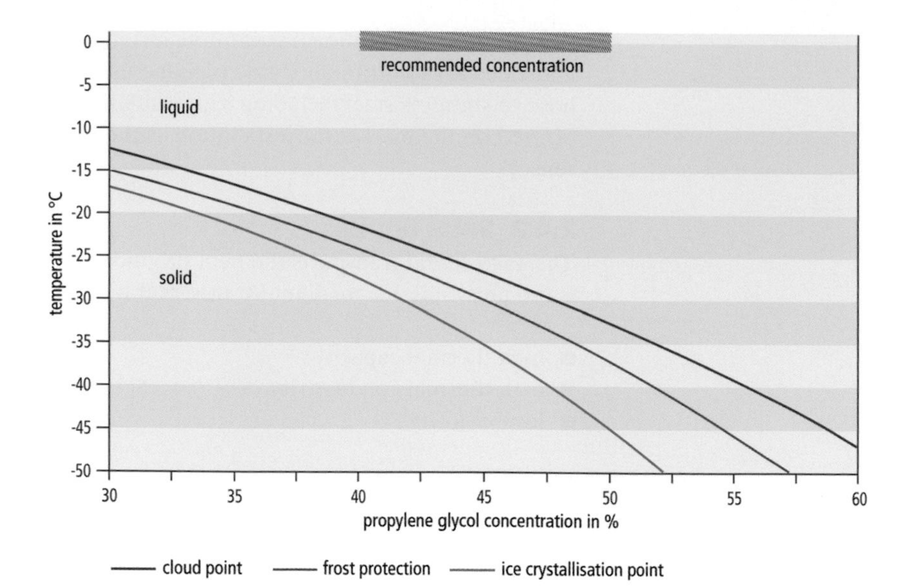

Figure 2.23. Frost protection from water/propylene glycol mixtures, as a function of concentration. The pour point temperatures are determined according to DIN 51 583, and the ice flocculation point temperatures to ASTM D 1177[11]

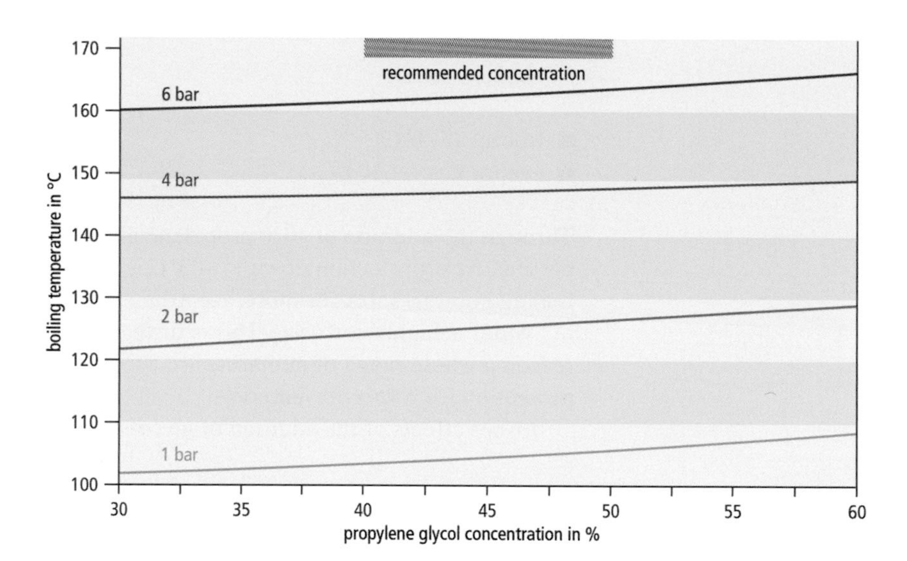

Figure 2.24. Boiling temperature in °C of water/propylene glycol mixtures as a function of operating pressure[11]

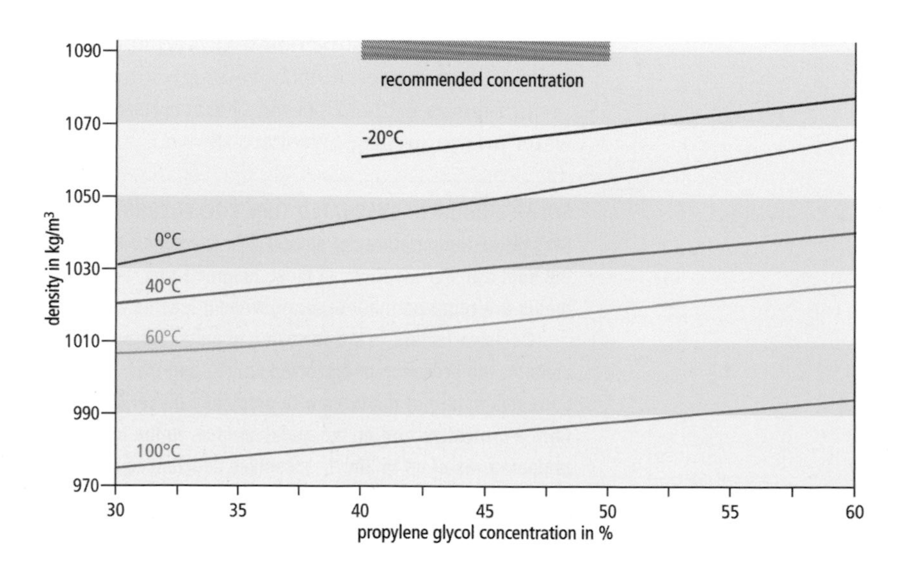

Figure 2.25. Density of water/propylene glycol mixtures as a function of temperature[11]

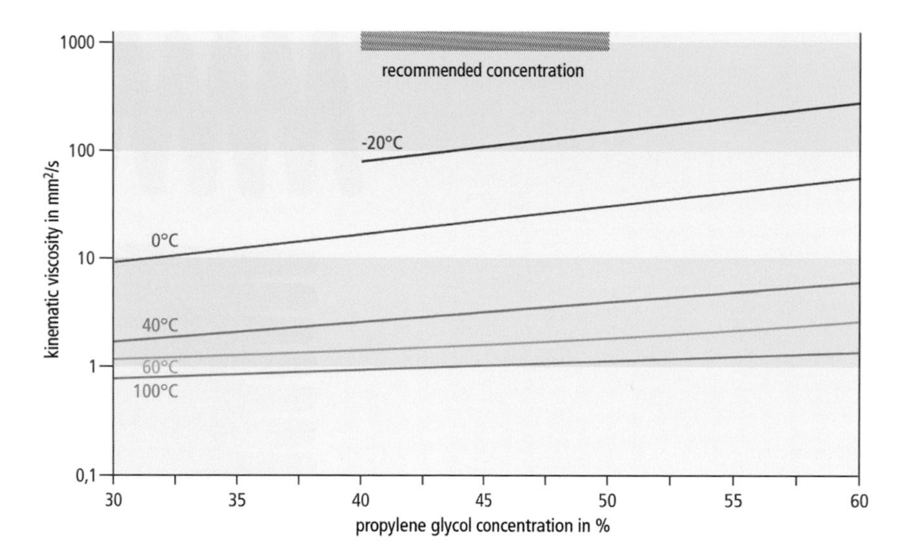

Figure 2.26. Kinematic viscosity (in mm²/s) of water/propylene glycol mixtures as a function of temperature[11]

restrictions and even blockages in the flow channels over time, particularly for very small cross-sections such as the U-pipe in the Sydney collector.

By the use of new liquid inhibitors, the formation of this residue can be effectively prevented as, in the stagnation situation, the medium – that is, propylene glycol, water and inhibitors – evaporates and can re-condense.

2.4.3 Solar pumps

For pumped systems, the use of electricity for pumps should be kept as low as possible, and therefore over-dimensioning of the power of the pump should be avoided. For systems in one- and two-family homes an accurate calculation of the pressure losses in the solar circuit is not necessary if nominal widths for pipelines are selected according to Table 3.3 in Chapter 3. The pumps offered by system manufacturers within the framework of complete systems possess a suitable range of powers. Through the use of three to four switchable power settings, the volumetric flow can be selected so that, at the maximum performance of the collector (with high irradiance), a temperature difference of between 8 and 12 K is produced between the feed and return lines. This situation should occur at a medium power setting so that the performance of the pump can be increased or reduced as required. For large collector surface areas, however, a calculation of the pressure losses in the pipe network must be carried out in order to determine the pump capacity (see example calculation in Chapter 3, section 3.5.2.6)

Heating pumps are actually designed for a high volumetric flow with low delivery heights: that is, for other conditions than those that apply to solar energy systems. In fact they operate here with very low efficiencies of between 2% and 7%. Several specialised pumps are on the market that have been optimized for application in solar energy systems, with low power consumption and sufficient head.

2.4.4 Solar heat exchanger (heat transfer unit)

For the transfer of the heat gained from the sun to the domestic hot water, a heat exchanger (heat transfer unit) is required in twin-circuit systems. We can differentiate between *internal* and *external* heat exchangers.

2.4.4.1 INTERNAL HEAT EXCHANGERS

As internal heat exchangers, *finned tube* and *plain tube* types are available (Figure 2.27). The plain tube heat exchanger possesses a greater heat exchange capacity per square metre of exchanger face. Compared with the finned tube heat exchanger a multiple of the pipe length is required. Plain tube heat exchangers are installed in the factory, whereas finned tube heat exchangers, because of their more compact design, can be installed into the store on site by means of special flanges and seals. However,

Figure 2.27.
Top: Finned tube heat exchanger
Bottom: Plain tube heat exchanger

the effectiveness of heat exchangers can be reduced by the build-up of limescale. Even a layer 2 mm thick reduces the heat transfer capacity of a heat exchanger by 20%; a 5 mm thickness reduces it by more than 40%.

The vertical installation of heat exchangers is more favourable, and should be preferred, because it promotes the stratification effect in the store system.

The connections to the solar circuit should be made so that the exchanger flow is from top to bottom, so that as much heat as possible can be transferred.

For thermosyphon systems, the proper circulation of the fluid inside the heat exchanger also needs to be ensured.

2.4.4.2 EXTERNAL HEAT EXCHANGERS

External heat exchangers are manufactured as *plate* (Figure 2.28) or *tubular* (Figure 2.29) heat exchangers. The heat transfer medium and domestic or heating water to be

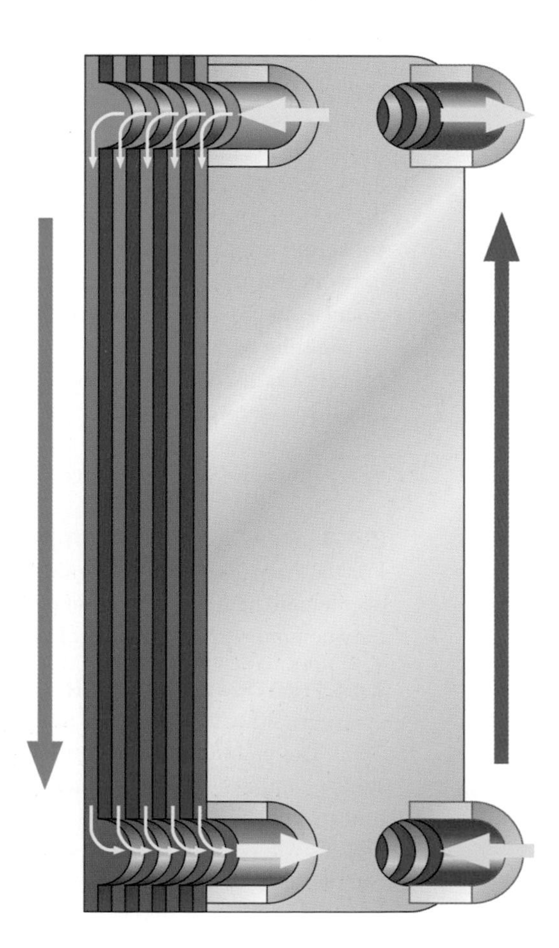

Figure 2.28.
Plate heat exchanger

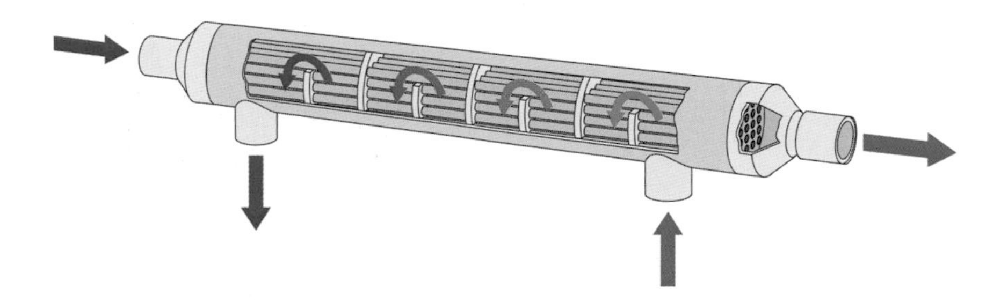

Figure 2.29. Tubular heat exchanger

heated flow past one another in a counter-current. External heat exchangers are insulated, with prefabricated heat insulating shells.
Advantages of external heat exchangers:

- The heat transfer capacity is higher than for internal heat exchangers.
- There is hardly any reduction in performance from limescale build-up.
- Several stores can be charged from one single heat exchanger.

Disadvantages of external heat exchangers:

- They are more expensive than internal heat exchangers.
- In most cases an additional pump is required on the secondary side of the heat exchanger. Therefore they are not an option for thermosyphon systems.

External heat exchangers* are mostly used in large systems. In such systems a heat exchanger can charge several stores, by which means the costs are lower compared with the installation of several internal heat exchangers.

2.4.5 Return-flow prevention

It is important to install a check valve or gravity brake in the return flow between the pump and collector (Figure 2.30). This should be dimensioned so that the lifting force of the hot heat transfer medium is not sufficient alone to open the valve. Thus the store is prevented from being cooled down by the collector when the pump is not working (mostly at night). To prevent in-tube circulation in tubes larger than 15 mm × 1 mm, a further return-flow preventer could be installed in the flow pipe.

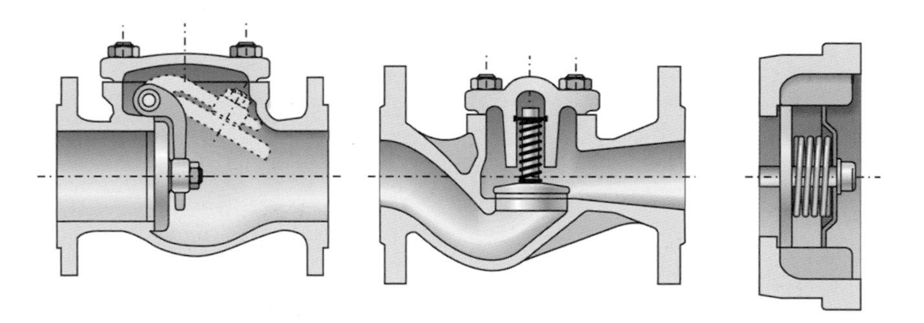

Figure 2.30. Return-flow preventers

In some systems, return flow is prevented by letting the hot water pipe from the store make a downward bend at least 30 cm deep. In thermosyphon systems return-flow prevention is more difficult, as the check valve may not disturb the flow.

2.4.6 Rapid air bleeders

At the highest point of every solar energy system, an automatic air bleeder with stop valve or a manual air bleeder must be provided (Figure 2.31). Only in the case of special concepts with a high flow rate, or drainback systems, is it possible to have the air bleeding in the cellar by the use of corresponding pumps. The air bleeder must be

*Plate heat exchangers are referred to here; tubular heat exchangers are used mainly in swimming pool systems.

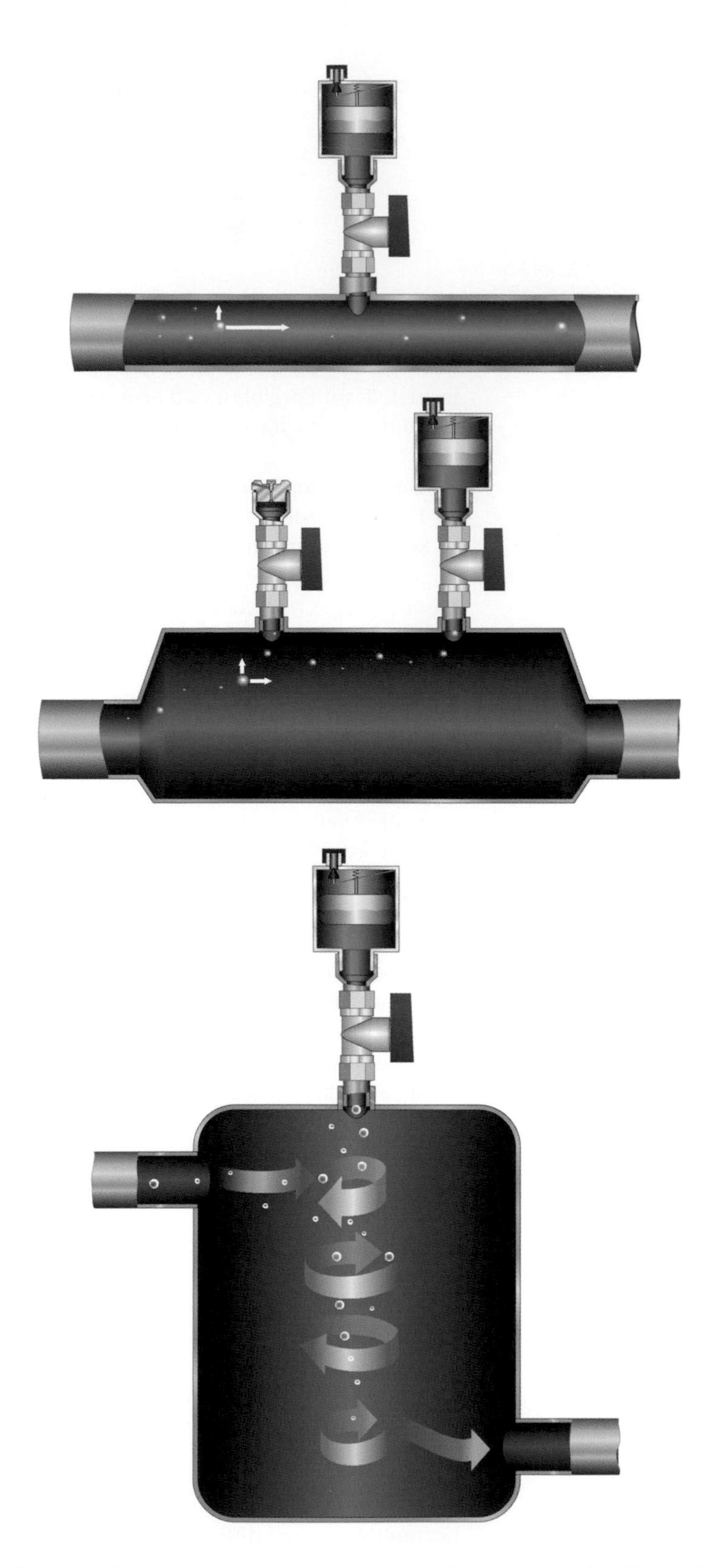

Figure 2.31.
Rapid air bleeder, manual air bleeder, check valve, air separator. Top: poor air bleeding. Middle and bottom: good to very good air bleeding

glycol-resistant and have a temperature resistance of at least 150°C (all-metal air bleeder). This serves to bleed the air from the solar circuit after filling with the heat transfer medium and during operation as necessary. The valve must be closed, as otherwise there is the risk that, during a stagnation situation, evaporated heat transfer medium will escape via the automatic air bleeder.

2.4.7 Flowmeters

Small mechanical volumetric flow display units permit control of the volumetric flow. With certain flowmeters (for example Taco-Setter; see Figure 2.32) the volumetric

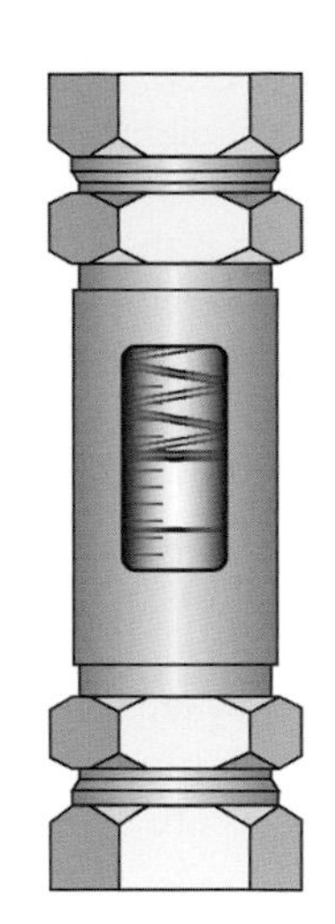

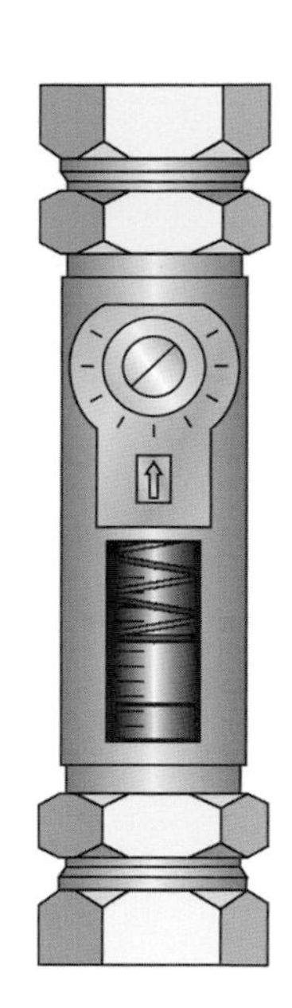

Figure 2.32. Taco-Setter. Manufacturer: Taconova GmbH, Singen

flow can be reduced within certain limits by means of a flow control valve. It is, however, better to reduce the volumetric flow by means of the pump power settings because in this way it is also possible to save power. It is very easy to see with the flowmeter whether a flow exists or not.

2.4.8 Safety devices in the solar circuit

2.4.8.1 SAFETY VALVES

According to European standard EN 12976, each section of the collector field that can be shut off must be fitted with at least one safety valve. ICS systems must be fitted with at least one safety valve, which may be integrated with an inlet combination. The safety valve should resist the temperature conditions it is exposed to, especially the highest temperature that can occur. The safety valve should resist the heat transfer medium. The safety valve should be dimensioned so that it can release the highest flow of hot water or steam that can occur. The dimensions of the safety valve(s) should be well designed.

2.4.8.2 MEMBRANE EXPANSION VESSELS (MEV)

The expansion vessel is an enclosed metal container. In the middle of the tank an expandable membrane separates the two media flexibly: on one side there is nitrogen,

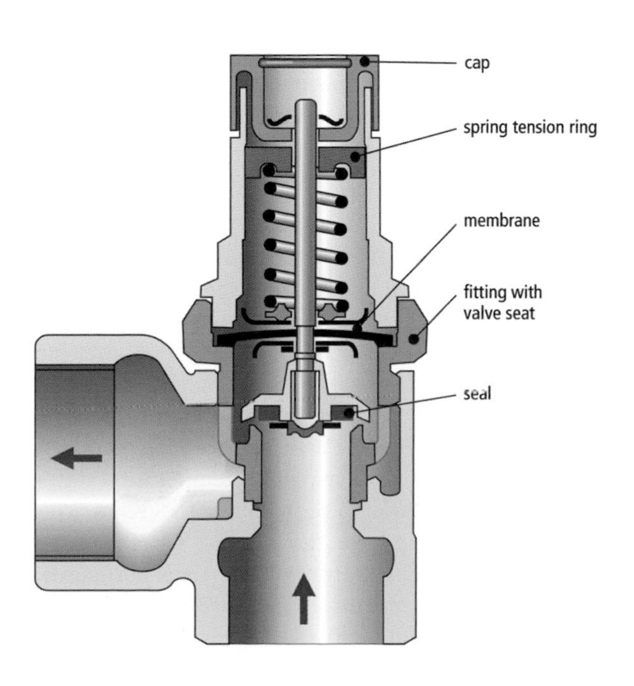

Figure 2.33. Safety valve

which is under an initial pressure, and on the other side there is the solar liquid, which enters the MEV when heated and after a pressure rise (see Figure 2.34).

The MEV is connected to the collector piping system so that it cannot be cut off and thus absorbs the temperature-related volume changes in the solar liquid. The size of the expansion vessel should be sufficient for the liquid content of the collector field. If the heat cannot be taken away by the solar liquid (for constant heat absorption in the solar collectors, and if the solar pump fails) the liquid evaporates when it reaches the evaporation temperature (this is pressure dependent).

If the expansion vessel is designed for additional take-up of the collector content, the solar liquid can be displaced into the vessel so that the maximum permitted operating pressure is not reached and the safety valve does not respond (intrinsically safe system).

Expansion vessels are available in the standard sizes 10 l, 12 l, 18 l, 25 l, 35 l and 50 l.

The glycol resistance of the membrane (made of, for example, EPDM) is important. As this is not always true for membrane expansion vessels for heat installations, special expansion vessels are used in solar thermal systems.

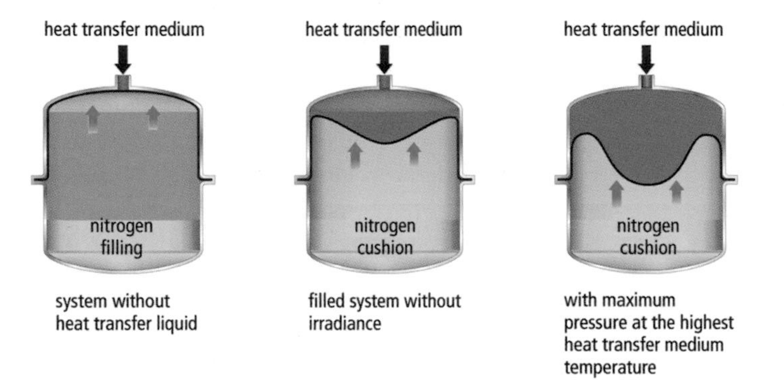

Figure 2.34. Different operating conditions in membrane expansion vessels

2.4.9 Solar station

In several systems on the market, the functions of pump, control unit, valves and connections have been integrated into a single unit, a so-called *solar station*.

2.5 Controller

The controller of a solar thermal system has the task of controlling the circulating pump so as to harvest the sun's energy in the optimum way. In most cases this entails simple electronic temperature difference regulation.

Increasingly, controllers are coming onto the market that can control different system circuits as one single device, and in addition are equipped with functions such as heat measurement, data logging and error diagnostics.

Thermosyphon systems do not have a controller.

2.5.1 Control principles for temperature difference control

Two temperature sensors are required for standard temperature difference control. One measures the temperature at the hottest part of the solar circuit before the collector output (flow); the other measures the temperature in the store at the height of the solar circuit heat exchanger. The temperature signals from the sensors (resistance values) are compared in a control unit. The pump is switched on via a relay when the switch-on temperature is reached (see Figure 2.35).

The switch-on temperature difference depends on various factors. Standard settings are from 5 K to 8 K. In principle, the longer the pipeline from the collector to the store, the greater the temperature difference that should be set. The switch-off temperature difference is normally around 3 K. A third sensor can be connected for temperature measurement in the upper area of the store, which permits the draw-off temperature to be read.

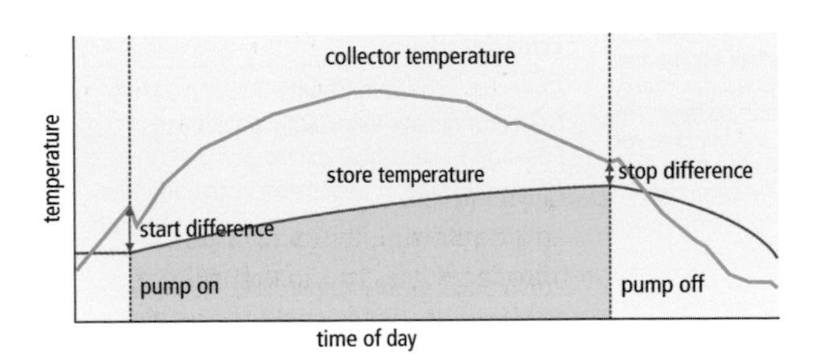

Figure 2.35. Function of a temperature difference controller shown as the daily progression of the collector and store temperature (schematic)[12]

An additional function is switching off the system when the maximum store temperature has been reached, as a means of overheating protection.

Frost protection is effected either by adding antifreeze to the collector fluid (as already discussed), or by using the drainback system. In the latter system, the collector circuit is only partly filled with water, and when the pump is off the collector is completely dry. This obviously places special requirements on the design of the collector and the piping. Drainback systems, when well designed so that no water is left in the collector or any piping that could freeze when the pump is switched off, automatically work together correctly with a temperature differential controller. When the danger of freezing occurs, the pump will be switched off because the store will then be always warmer than the collector.

2.5.2 Digital controller with special functions

Among the functions of modern controllers are operating hours and heat quantity measurement, remote displays, remote diagnostics and PC interfaces.

The measurement and reporting of the running time solar circuit pump aids function control and provides initial information on the system. With the help of heat quantity measurement the function and yield of a solar energy system can be monitored and seen more accurately. A heat quantity meter consists of:

- a volumetric flowmeter (turbine or impeller wheel counter installed in the return line close to the store, or sometimes magnetic induction)
- one temperature sensor in each of the feed and return lines
- suitable electronics for calculating the heat quantity.

The presetting of the water–glycol mixing ratio at the heat quantity meter is important for correct heat quantity recording.

With the help of heat quantity measurements it is possible to check the performance of a solar energy system, and to calculate the saved CO_2 emissions. As the additional heating system always ensures sufficient supplies of hot water, a fault in the solar energy system is often not found until fairly late or not at all. Error signals are therefore necessary to detect faults in time. There are several options for this:

- *Remote display of system parameters.* With a remote display it is possible to bring the system parameters into the living room. In this way really easy monitoring is possible, as well as awareness of the daily energy production performance.
- *Controller with error signals.* This controller possesses an automatic system for controlling particular variables (such as resistance of the collector sensor), which in the case of a fault, such as a short circuit in the sensor line, issues an error code to the display, which is explained by means of a table. More advanced controllers operate with an integrated error diagnosis that provides information on particularly frequently occurring errors via a cause–effect reference (see Table 2.5).[13]
- *System monitoring by continuous evaluation of system data.* A data logger records all the system values (for example temperatures, volumetric flow) measured by the solar control system: these can then be evaluated via a suitable PC interface. For larger solar energy systems, central recording of the operating data is possible by means of remote data transmission via the telephone network, which regularly checks the system parameters. In this way, better monitoring of the system is possible as well as rapid error rectification.

Table 2.5. Example of cause–effect assignment of errors in a thermal solar energy system for water heating. The temperature difference ($\Delta\theta$) refers to the temperature in the feed and return lines of the solar circuit[14]

Error description	Resulting effect
Collector connection faulty	Pump cycles
Collector sensor incorrectly positioned	Pump cycles
Leakage in solar heat exchanger	System pressure too high
Initial pressure of expansion vessel too low	System pressure occasionally too high
Initial pressure of expansion vessel too high	System pressure occasionally too low
Shut-off valve in solar circuit closed	Temperature difference ($\Delta\theta$) too high
Cable break between controller and pump	Temperature difference ($\Delta\theta$) too high
Collector temperature sensor defective/inaccurate	Pump runs at wrong times
Store temperature sensor defective/inaccurate	Pump runs at wrong times
Inputs on the controller defective	Temperature difference ($\Delta\theta$) too high; pump runs at night
Outputs on the controller defective	Temperature difference ($\Delta\theta$) too high; pump runs at night
Controller software defective	Temperature difference ($\Delta\theta$) too high
False setting of volumetric flow	Temperature difference ($\Delta\theta$) too high
Air bleeding incorrect	Temperature difference ($\Delta\theta$) too high
Gravity brake open	Pump also runs at night
Gravity brake contaminated	Pump also runs at night
$\Delta\theta$ setting unfavourable	Temperature difference ($\Delta\theta$) too high
Incorrectly programmed timer	Pump runs at night
Power failure	System pressure too high
Anode in store defective	Store corroded

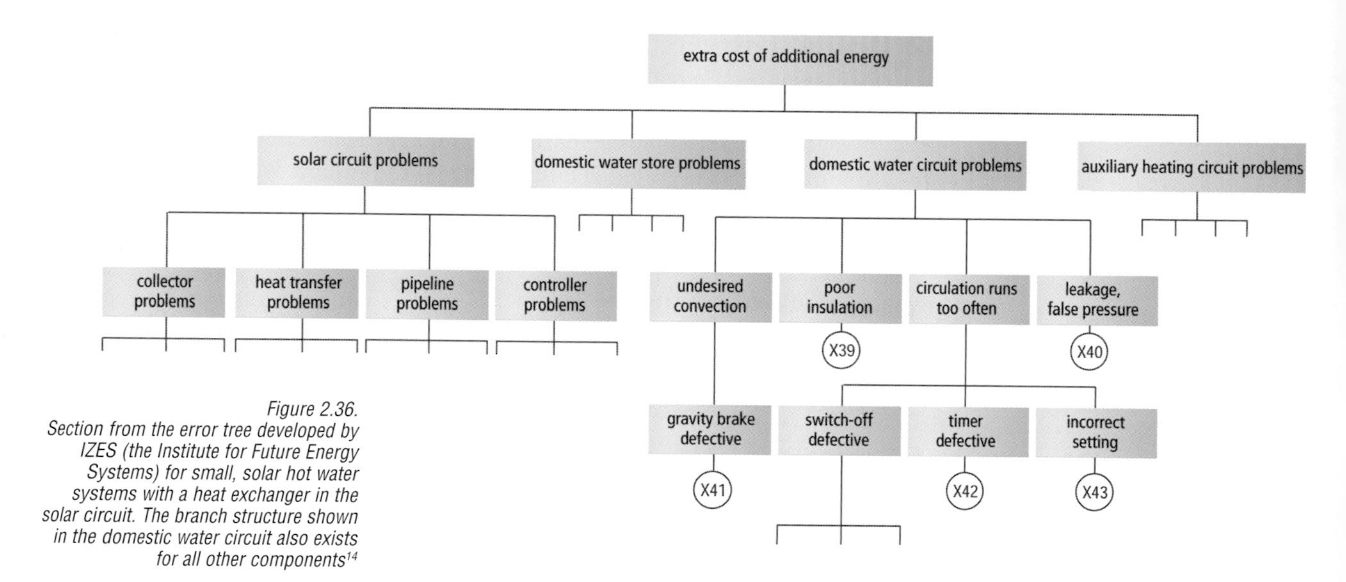

Figure 2.36. Section from the error tree developed by IZES (the Institute for Future Energy Systems) for small, solar hot water systems with a heat exchanger in the solar circuit. The branch structure shown in the domestic water circuit also exists for all other components[14]

2.5.3 Temperature sensors

The effectiveness of the controller is heavily dependent on the correct mounting and function of the temperature sensors. The collector sensor can be fixed as a contact sensor to the feed-collecting pipe of the absorber, or fixed directly to the absorber in front of the collector output, or used there as a submerged sensor (preferably always inside the collector).

The measuring point of the collector sensor should not be shaded.

In systems with a bypass circuit, the radiation from the sun and the temperature of the water flowing into the store (instead of the collector temperature) are measured in front of the three-way valve. The solar store sensor should be mounted at the height of the lower half of the solar circuit heat exchanger as a submerged sensor or contact sensor.

As temperature sensors and controllers are often matched to one another, it is normally not possible to exchange temperature sensors from different manufacturers.

The requirements with respect to temperature resistance are extremely high, in view of the high stagnation temperatures, especially in evacuated tube collector systems, and also with respect to the accuracy, for example during heat quantity recording.

2.5.3.1 SENSOR ELEMENTS

THERMISTOR

This widespread sensor type comprises elements that have an increasing electrical line resistance with increasing temperature; they can normally be used up to a maximum temperature of 150°C. In the normal measuring ranges for solar energy systems, their accuracy is ± 1.5 K. When used in solar controllers, they are connected to the controller by two wires.

When used for measuring collector temperatures, they can become damaged or even destroyed during long stagnation times in the collector. They are unsuitable for temperature measurement in vacuum tubes.

PT100, PT1000

Both of these elements are made from platinum wires, and they are suitable for temperatures from –200°C to +850°C. Pt100 sensors are most frequently found in *four-wire circuits*. The Pt1000 sensors found on the market are equipped with two wires. Qualities A and B are differentiated according to the application temperature and measuring accuracy. Quality A is mostly used for accurate measurements in combination with *paired sensors*. Paired sensors (each with two sensors) are subject to a calibration procedure and, for example, for the calculation of heat quantities, are very accurate, as they behave consistently over the whole measuring range during temperature measurement. Class B elements are mostly used for the control of solar energy systems: their accuracy in system operation is $\pm(0.3 + 0.005\theta)$°C, where θ is the temperature to be measured (in °C).

Because of their high temperature resistance, platinum elements are also used for collectors with very high stagnation temperatures.

PV CELLS FOR MEASURING RADIATION

For controllers whose programmes are dependent on measuring solar radiation, *encapsulated PV cells* are generally used. They are suitable for simple radiation measurement, as is required here. Their resistance, which changes with the intensity of the radiation, acts as an input variable for the solar energy system controller. It must be ensured that the materials used for the cells and the housing do not change their properties in the course of time, which over the years would lead to permanent error measurements (for example do not use light-sensitive plastics, and protect the sensors from condensed water).

2.5.3.2 SENSOR HOUSING AND CONNECTING CABLES

The sensor housing protects the sensor element from damage and the influences of weather, and must be temperature- and corrosion-resistant. The materials used are therefore mostly tinned brass or stainless steel. In order to seal the housing from moisture the sleeves for the sensor cable are frequently fitted with a plastic that is rolled-in or pressed. The quality of this seal is of great importance for the permanent use of the sensor. If the seal is not permanent, or if it is damaged by high temperatures in the collector, the consequence may be incorrect temperature displays or even a short circuit in the sensor.

The connecting cables to the sensor are usually PVC- or silicon-sheathed cables. In the high temperature area PTFE cables are also used. In order to prevent damage to the cables of the collector sensor, protective pipes made of stainless steel or similar material are used in the external area.

PVC cables should not be used because of the potential environmental problems. The only permanently safe alternative for collector sensors is silicon cables with PTFE sheathing on the cores. These cables can resist temperatures of up to 230°C.

Unsuitable sensors can lead to faults or even to the complete failure of the solar energy system. Because there is little difference between the prices for high-quality platinum sensors with suitable housings and connecting cables and those for lower-quality sensors, there is no justification for the use of the latter.

Then we shall list the essential variants of each and consider the most important systems, which can be combined from these. *Note that, for clarity, the schematic diagrams in this chapter do not show safety valves or vents.*

3.2 Systems for charging/discharging the store

3.2.1 Charging by means of solar energy

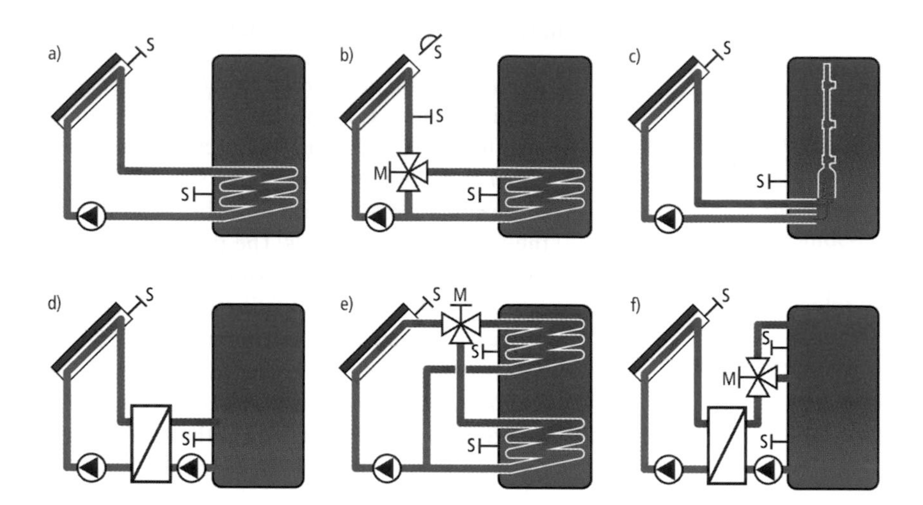

Figure 3.2.
Types of store charging with solar energy (S = sensor)

(a) *Internal solar heat exchanger.* The heat exchanger is normally designed as a finned or plain tube coil, which is arranged in the lower area of the store. Thermal transmission to the domestic water takes place through thermal conduction, and as a result of this convection takes place: that is, the heated water rises as a result of its lower density.

(b) *Internal solar heat exchanger with bypass circuit.* This is a variant of system (a) for larger systems. A radiation sensor measures the solar radiation. At a threshold value of, for example, 200 W/m^2 the controller switches on the solar pump, and the three-way valve initially bypasses the heat exchanger. The solar circuit heats up. When the set temperature difference with the store has been reached at the flow sensor, the controller switches over the valve and the store is charged with heat.

(c) *Stratum charging with self-regulating stratified charger.* The central core of this store-charging method is a riser pipe with two or more admission ports at different heights and a heat exchanger installed below. The heat exchanger heats up the water surrounding it in the riser pipe, and the water rises up. This causes a pronounced temperature-stratification effect, and in the upper area a useful temperature is very soon achieved.

(d) *External heat exchanger.* The solar liquid flows through the primary side of an external plate heat exchanger. For charging the store a second circulating pump draws cold water from the bottom of the store. This flows through the secondary side of the heat exchanger in a counter-flow and then flows back into the middle of the store. An external heat exchanger has better thermal transfer properties than an internal type. Stores without internal fittings can also be used. The bypass circuit (b) can be implemented without the three-way valve by controlling the two pumps separately. For this purpose sensor arrangement (b) should be selected.

(e) *Stratified charging system* with two internal heat exchangers and feed via a three-way valve at two different heights.

(f) As (e) but with an *external heat exchanger.*

The advantage of stratified charging systems (rapid reaching of useful temperatures in the standby area) is greatest in connection with the *low-flow* concept. In low-flow systems, sometimes also called *matched-flow* or *single-pass* systems, only about 25 l of heat transfer fluid circulates per square metre of collector area. The effect of this is that the efficiency of the collectors is increased slightly, and that higher temperature differences occur between the collector and water tank. This can increase the system efficiency, because less pump energy is needed, and after only a few hours a layer of hot water may accumulate at the top of the tank. If hot water is needed at that moment, it can be supplied entirely or almost entirely from the solar heat, and auxiliary energy is saved. Most solar water heaters can be used as low-flow systems, but there are also specially designed systems with optimized water tanks and heat exchangers. The advantage of low-flow systems is that thinner, and hence cheaper, standpipes can be used. These also lose less heat and, together with the minimal amount of heat transfer fluid and a smaller pump, help to reduce the total cost of the solar heating system.

3.2.2 Charging by means of auxiliary heating

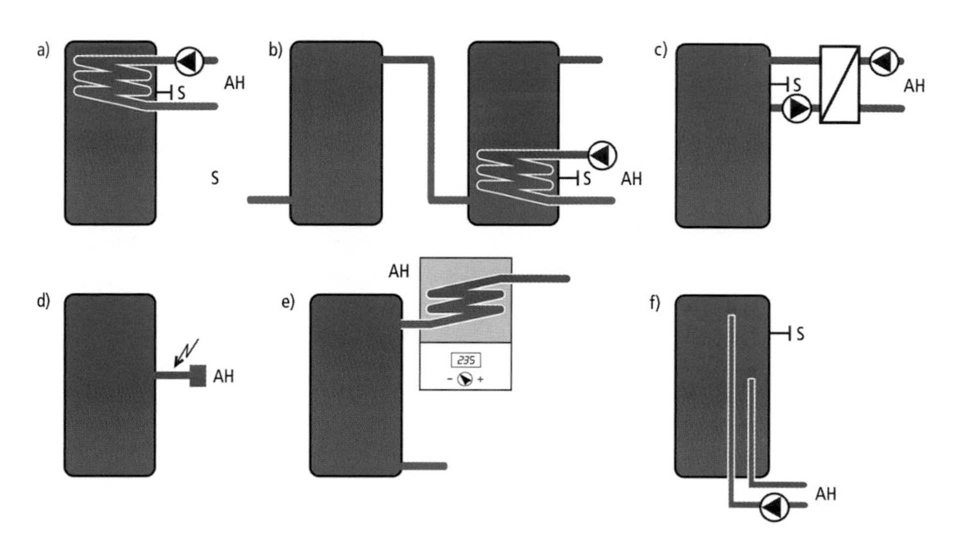

Figure 3.3. Types of store charging with auxiliary heating

(a) *Domestic water store.* The standby volume is heated as necessary via the upper heat exchanger, using an oil or gas boiler with store priority switching. Recently, in several countries, wood has started to be used again (for example in the form of pellets) as a growing fuel resource for heating purposes that is CO_2-neutral. In the case of the cascade connection of two stores the additional heat exchanger can be located in the top or bottom of the second store according to the size of the required standby volume.

(b) *Domestic water store.* Auxiliary heating of the standby volume by an external heat exchanger.

(c) *Domestic water store.* Electrical auxiliary heating of the standby area. Electrical power is acceptable for use as heating only in exceptional cases, as electricity generation involves high losses. (Electricity is generated in a conventional power station with an efficiency of about 30–50%, and on top of this loss are the line losses to the consumer.)

(d) *Domestic water store.* Top-up heating by a downstream instantaneous heater (gas or electricity). The device must be thermally controlled and designed to accept preheated water: that is, it only heats up as much as is required to achieve a set exit temperature. Advantage: the whole store is available for solar energy, and there are lower store losses compared with top-up heating inside the store. Disadvantage: in the case of electricity, similar to (c). In the case of gas or oil, there are very few compatible appliances.

(e) *Buffer (thermal) store.* Top-up heating via admission pipe. The store-charging pump draws the water to be heated from the middle of the store. This is heated by the boiler and fed into the top of the store via an admission pipe.

(f) *Buffer (thermal) store.* Direct top-up heating of a separate buffer store. Warm heating water is withdrawn from the middle of the store and hot water is fed in again at the top, as in (e), but in this case the pipes are installed at a corresponding height at the side.

3.2.3 Store discharge

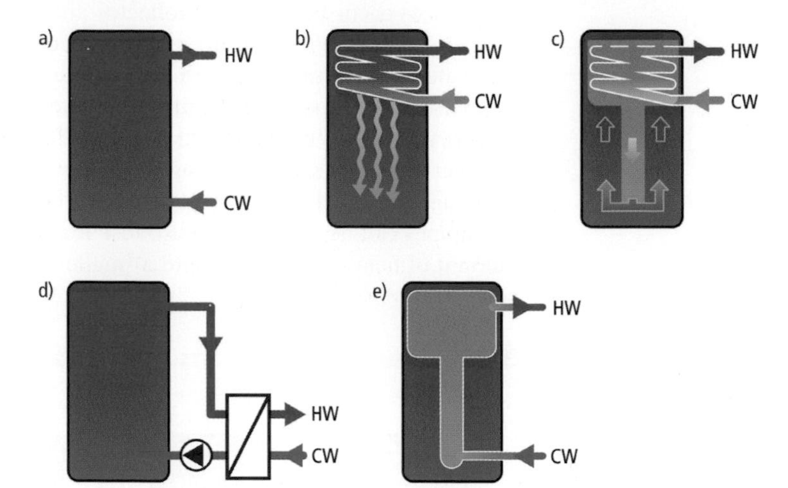

Figure 3.4.
Methods of store discharge

(a) *Domestic water store*. The water is withdrawn from the top in the hottest part of the store (standby area). Cold water flows in at the bottom in a corresponding amount. Almost the whole amount of stored hot water can be withdrawn.
(b) *Buffer store*. Discharge takes place in the upper area via an internal heat exchanger. The disadvantage is that the upper storage area is significantly cooled down by the cold water (8–12°C) flowing into the heat exchanger. This causes a significant amount of circulation, which causes mixing of the waters: that is, the store temperature layers are destroyed.
(c) *Improvement on (b)*. The cooled store water descends within a downpipe and pushes the warmer water uniformly upwards. In this way the hottest water is always available at the top, and the stratification is hardly disturbed.
(d) *Buffer store*. Discharging is through an external heat exchanger.
(e) *Combined store*. Cold water enters the domestic water store right at the bottom. It heats up according to the layers in the buffer store and is removed from the hot area at the top.

Two groups of systems can be combined from the modules described above:

- systems for domestic water heating
- systems for domestic water heating and heating support.

These are described in the next two sections.

3.3 Systems for heating domestic water

The standard system (Figure 3.5) has been widely accepted for use in small systems, and is offered by many manufacturers. It is a twin circuit (indirect) system with an internal heat exchanger for solar heat feed and a second one for top-up heating by a heating boiler. In the store there is domestic water, which can be limited to a set

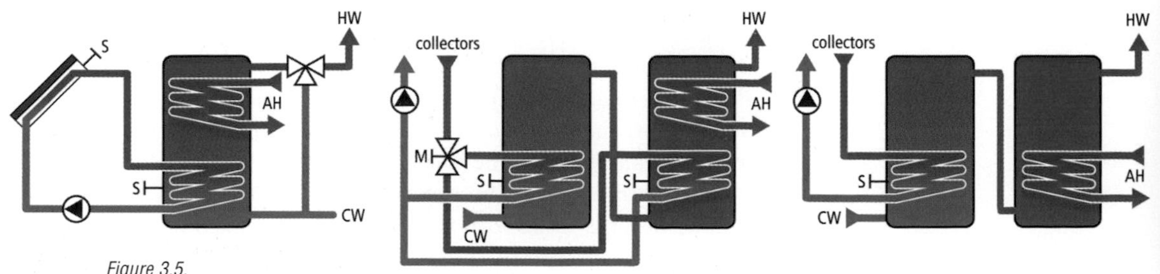

Figure 3.5.
Standard solar energy system

maximum draw-off temperature by means of a thermostatic three-way blending valve (see Chapter 2, section 2.3.6).

The circuitry is comparatively easy to implement, as well-tried control principles are used. The solar circuit pump is switched on as soon as the temperature in the collector is 5–8 K higher than in the lower store area. When the temperature on the boiler controller for the standby volume falls below a set temperature, the boiler provides the necessary top-up heating. During this time, the space heating circuit pump (if connected) is normally switched off (domestic hot water priority switching). In the case of the cascade connection of two stores, either both stores can be heated by solar energy where the draw-off store is charged as a priority, or only the preheating store is charged by solar energy and the draw-off store is top-up heated as necessary.

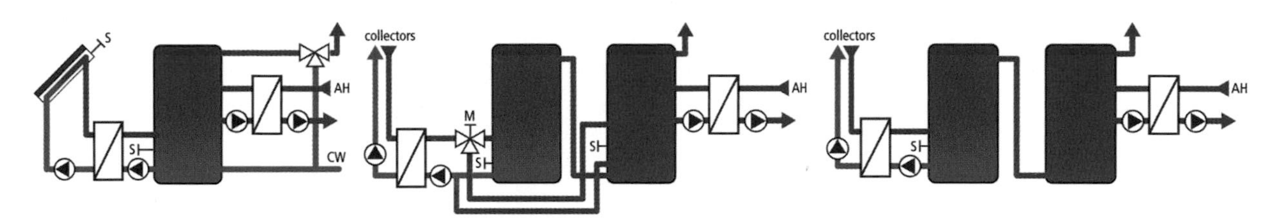

Figure 3.6. The same system configuration as Figure 3.5 but with external heat exchangers

Through the use of a special *stratified store* either as a domestic hot water store or as a buffer store (Figure 3.7), which is used only for domestic water heating for hygiene reasons, the heat from the solar collector is specifically fed into the matching temperature layer in the store. The significantly reduced mixing process leads to a usable temperature level much faster than for all the other systems described here, and the frequency of auxiliary heating is reduced. When the stratified store operates with buffer water, the heat for the domestic water is discharged by means of an external once-through heat exchanger. Also important for the performance of this system is good matching of the discharge control system to the different tapping rates.

Figure 3.8 shows a *buffer store* with external charging and an internal output heat exchanger, which includes an internal downpipe and direct auxiliary heating by the boiler (for hygiene reasons it is exclusively used here for domestic water heating).

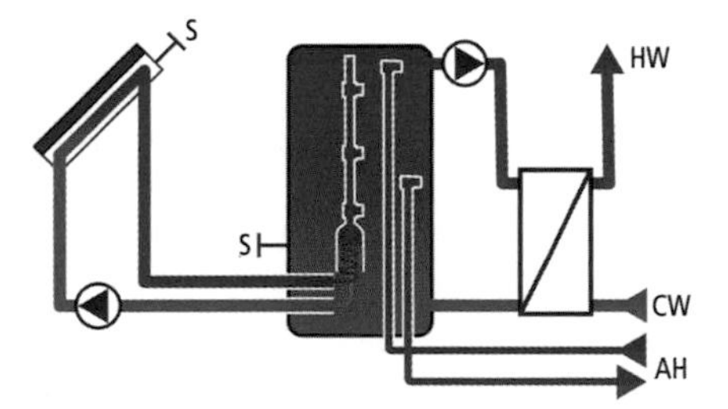

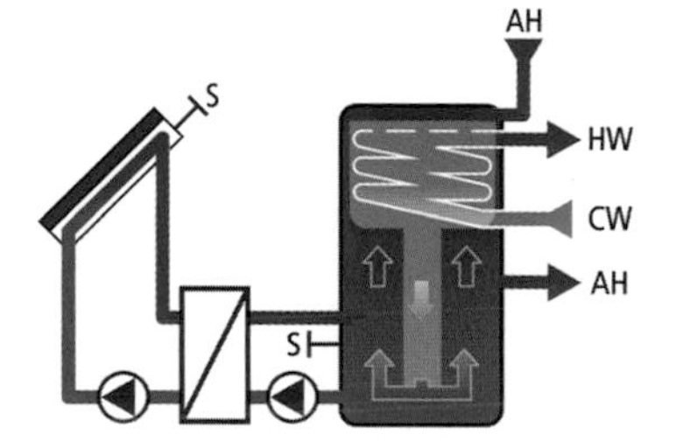

Figure 3.7 (left). Stratified store as buffer storage

Figure 3.8 (right). Buffer store with external charging and internal output heat exchanger

In a *twin-store system* (see Figure 3.9) the solar circuit charges a buffer store via an internal or external heat exchanger, from which in turn a downstream domestic water store is supplied with heat. This then receives auxiliary heating in the upper area. This variant has proved to be more favourable than auxiliary heating in the buffer store. [15] The temperature in the buffer store is thus determined entirely by the solar radiation.

Lower energy losses occur compared with auxiliary heating of the buffer store. In larger systems the store volume is divided into buffer and domestic water areas for water hygiene reasons (Legionella problems) and for energy saving.

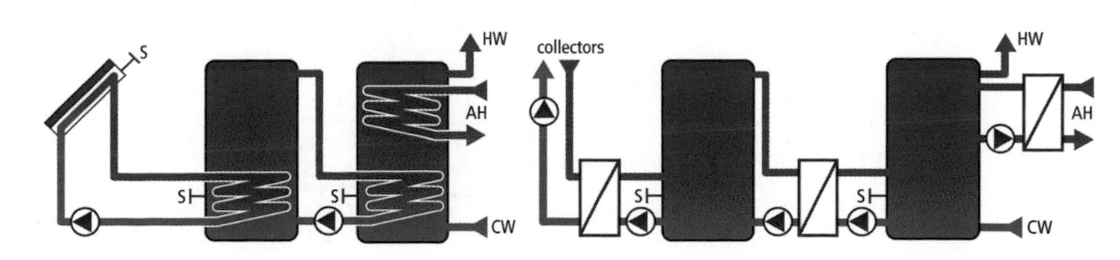

Figure 3.9. Twin-store systems

Figure 3.10 shows a *combined store system*: for hygienic reasons, this is used here exclusively for domestic water heating. In comparison with standard systems, this store contains a smaller domestic water volume. Hence the water dwell time in the store is shorter. The surrounding buffer (thermal storage) water is used only as intermediate storage for the heat.

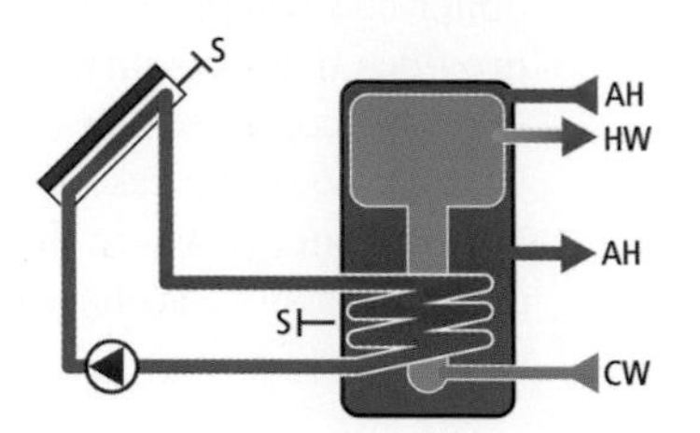

Figure 3.10.
Combined store system

3.4 Systems for heating domestic water and space heating

If a solar energy system is considered in the planning stage of the whole heating system, and the house has a central space heating installation, it is possible to use solar heating to augment the space heating. The reduced heat requirements of low-energy houses and the higher performance of modern solar energy systems encourage the trend to install solar systems with space heating support in countries with cold or temperate climates. In spite of the comparatively low solar fraction proportion for room heating (typically 10–20%), combined systems for domestic water heating and space heating support have a higher primary energy substitution than pure domestic water systems.

The space heating support is implemented in such a way that either the boiler operates on the store and this then operates on the space heating circuit, or the temperature of the heating return flow is raised by means of the solar system. The heating boiler then only has to supply little or no heat.

By means of a larger collector surface area, these systems are over-dimensioned in the seasons in which space heating is not necessary. For the removal of excessive heat see Chapter 2, section 2.5.4.

3.4.1 Combined store system (store-in-store system)

See Figure 3.11. A small domestic water store is permanently fitted in the buffer store. The solar circuit is led via an internal heat exchanger, and the auxiliary heating is charged directly into the upper area. The heating circuit draws heating water from halfway up the store and feeds the cooled heating circuit return flow back into the bottom of the buffer store. The cold water that flows back into the domestic water store does, however, impair the heat layers in the event of high draw-off rates. The system requires no expensive control technology. It is mostly found in central Europe, particularly in Switzerland.

3.4.2 System with buffer store, internal heat exchanger for heat removal and downpipe

See Figure 3.12. The store can be charged by the solar circuit via an external heat exchanger at two levels. The corresponding level is selected according to the temperature. Domestic hot water is removed via a special internal heat exchanger, which is fitted above a downpipe. If the heat exchanger cools down because of the inflow of cold water, a downward flow is set up within the pipe and then upwards within the store. In this way the heat exchanger always has sufficient flow rate. The

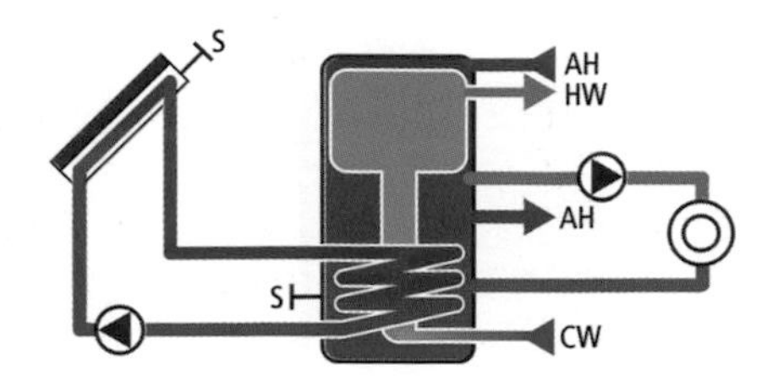

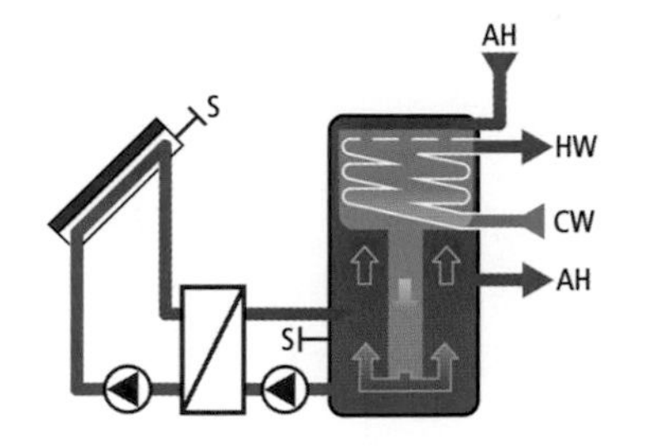

Figure 3.11 (left).
Combined store system for domestic water heating and space heating

Figure 3.12 (right).
Buffer store system

auxiliary heating heats up the upper area of the store. The heating return line is led either to the solar return level via the store or, if the store is too cold, directly to the heating boiler, bypassing the store. The system does not require an expensive control system, but temperature stratification is not optimally achieved.

3.4.3 Stratified store with hot water heating in once-through flow and heating support

Stratified stores, which can be used for heating domestic water only, are also designed for space heating support. The two-layer chargers shown in Figure 3.13 ensure that the solar heat and the heating return pipes, at different temperatures, are brought in at the matching temperature layers of the store. The auxiliary heating takes place in the upper area of the store. Removal of the heat for domestic water heating takes place via an external heat exchanger with a speed-controlled pump. The system operates very efficiently, but special attention must be paid to the control of the discharge pump. For prefabricated systems, however, this does not pose any practical restrictions.

3.4.4 Twin store system

See Figure 3.14. Here we have the classical separation between domestic water and heating buffer stores. The solar circuit charges both stores in the lower area, but the domestic water store has the priority. The auxiliary heating charges both stores in the upper area. Each store supplies the system that follows it. This type of system with additional equipment can be used for solar heating support, making use of the existing solar store.

Disadvantages, however, in comparison with the previously mentioned systems are a larger physical space requirement, higher pipework costs and higher store losses.

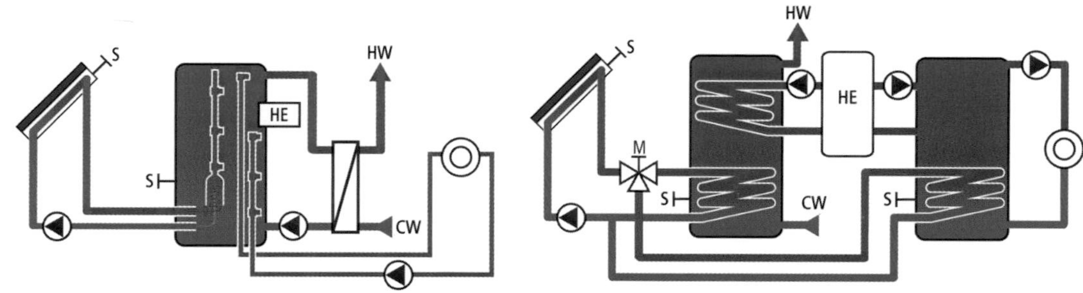

Figure 3.13 (left). Stratified store system

Figure 3.14 (right). Twin-store system

3.5 Planning and dimensioning

Right at the start of system planning it is important to record accurately the conditions at the site. This includes clarifying all the details that are important for planning and installation, and obtaining data about the building, the hot water consumption and, if necessary, the heat requirement for the house. It is also necessary to sketch all the important details that are required for preparing the quotation.

3.5.1 Important features for preparing the quotation

- Is the collector shaded by trees, parts of the building or other buildings (present/future depending on the time of year; see section 1.1.4)?
- Is the collector temperature sensor shaded?
- What is the most favourable alignment of the collector surface (see Figure 1.15)?
- What is the future accessibility of the collector for maintenance? Access to any chimney must always be guaranteed.
- Do not install collectors beneath aerials and similar equipment, because of bird droppings.
- What is the shortest possible path to the store (< 20 m)?
- Are there any requirements with respect to listed buildings?
- Use and complete checklists (see below).

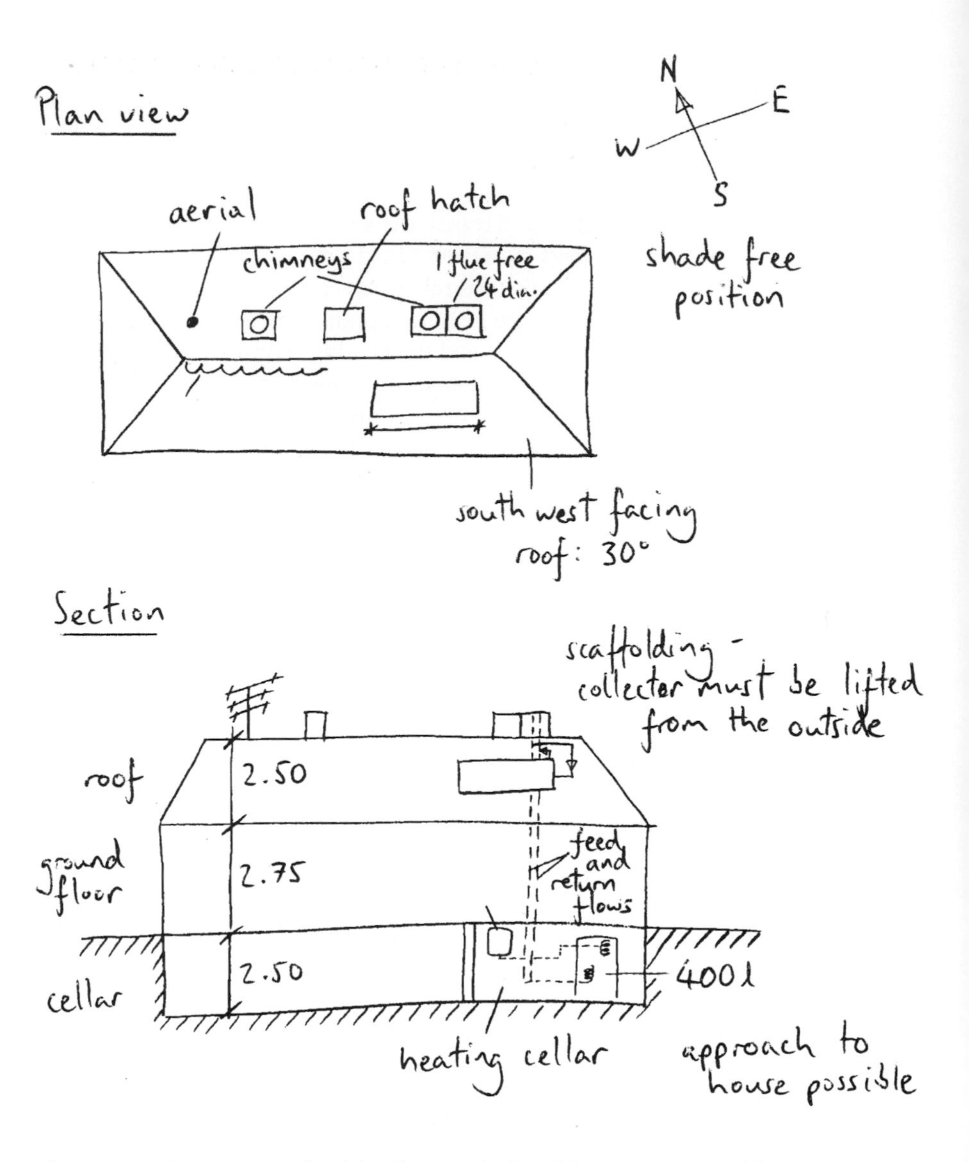

Figure 3.15.
Preparation of an initial system sketch[16]

- Is extra equipment required (such as an inclined hoist, or a crane)?
- Are any safety measures necessary (for example safety equipment, safety belts)?
- What type of collector installation is possible/desirable (on-roof, in-roof)?
- Who will install the collector fixing brackets (roofer)?
- Are roof tiles laid in a mortar bed (increased cost of installation)?
- Can the roof be walked on (fragile tiles)?
- Flat roof: bearing load of roof skin (fragile roof)?
- Note minimum distance from mortared ridge tiles.
- What is the best way to fit the piping with good insulation and make the roof penetrations?
- Does central hot water heating already exist?
- How will the solar store be transported to reach its installed location? For example, a 400 l store weighs 145 kg, is 1.7 m high and 0.62 m diameter.
- How does the auxiliary heating of the store take place (integration into the existing heating system)?
- How, and by whom, is the electrical work to be done (mains connection, lightning protection/earthing, control system)?
- Is a waste water connection available in the store installation room?
- If a solar energy system cannot be considered at present for cost reasons, at least the installation of collector pipes and sensor cables and possibly even the installation of a solar store should take place for later retrofitting.

DATA FOR A THERMAL SOLAR ENERGY SYSTEM

Name:
Road:
Telephone:
Property address:
Customer special requirements:

House type: ◯ One/two-family home ◯ Multiple-family home
No. of living units: Year of construction:

Use of sun's energy for: ◯ Hot water (DHW) ◯ Heating swimming pool

No. of people:
Estimated/measured hot water consumption (litres per person/day)
◯ Low (30 l) ◯ Average (50 l) ◯ High (80 l)
Other hot water requirements: ◯ Washing machine ◯ Dishwasher

Type of installation: ◯ In-roof ◯ On-roof ◯ Free-standing
◯ Flat roof ◯ Façade/wall/balustrade
Roof covering: Description, type: Colour:
◯ Tiles
◯ Stone tiles
◯ Slates
◯ Other
Pitch Alignment
Height from ground to eaves:
Usable installation area Height: Width:
Shade:

Height of solar store installation space (m):
Minimum door width (m):
Single length of connecting pipe from store to collector (m):
Work required for connection pipe from store to collector:
Work required for connection pipe from store to hot water distribution:

If store already exists: Type:
Make:
Volume: (l)
Heating boiler: Type:
Make:
Power (kW):
Circulation line: ◯ Yes ◯ No
Thermal insulation: ◯ Good ◯ Poor
Running time of circulation pump: ◯ Continuous ◯ From To

Only for swimming pool water heating

Pool surface area (m^2):

Covered: ○ Yes ○ No

Required water temperature (°C):

Swimming season From: To:

Only for heating support

The living floor area to be heated (m^2):

The space heating feed and return flow temperature (°C):

Fuel:

Fuel consumption (l/year, m^3/year):

Notes:

Place:

Date:

Filled in by:

3.5.2 The dimensioning of systems for domestic water heating

3.5.2.1 DESIGN OBJECTIVES

In temperate climates, a common design objective is that a solar water heater for one-family homes should cover about 90% of the hot water demand during the summer months by means of solar energy. This will have the effect that the heating boiler can remain dormant during most of this period (where drinking water regulations allow this). In this way not only is the environment protected, but also money is saved and the life of the heating boiler is extended.

In the remaining months, when the heating boiler has to operate anyway, the auxiliary heating system supplies the necessary heat not otherwise provided by the sun.

In sunny climates, the common design goal is either to supply the hot water consumption fully from solar energy, or to supply full coverage for most of the year and use a back-up heater (often an electrical element immersed in the solar tank) for only a few weeks or months per year.

3.5.2.2 SOLAR FRACTION

The solar fraction is described as the ratio of solar heat yield to the total energy requirement for hot water heating:

$$SF = \frac{Q_S}{Q_S + Q_{aux}} \times 100$$

where SF is the solar fraction (%), Q_S is the solar heat yield (kWh), and Q_{aux} is the auxiliary heating requirement (kWh). The higher the solar fraction in a solar energy system, the lower the amount of fossil energy required for auxiliary heating: in the extreme case (SF = 100%) none at all.

Figure 3.16a shows the monthly solar fractions for a solar thermal system in a temperate climate. It can be seen that in the case of 100% solar fraction during the radiation-rich months from May to August, an annual fraction of 60% is achieved.

When a system is properly designed, for instance to cover almost the complete demand in the summer months in central Europe, the addition of extra collector area would not lead to correspondingly higher output. In periods with high irradiation, the system would produce excess heat, which would lead not only to frequent high

(a)

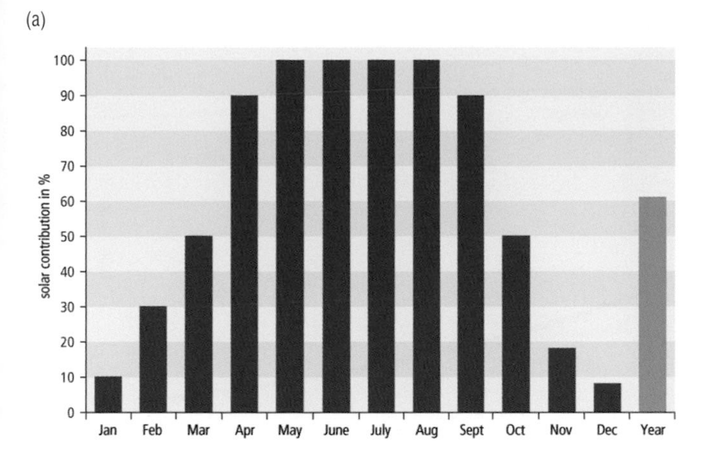

(b)

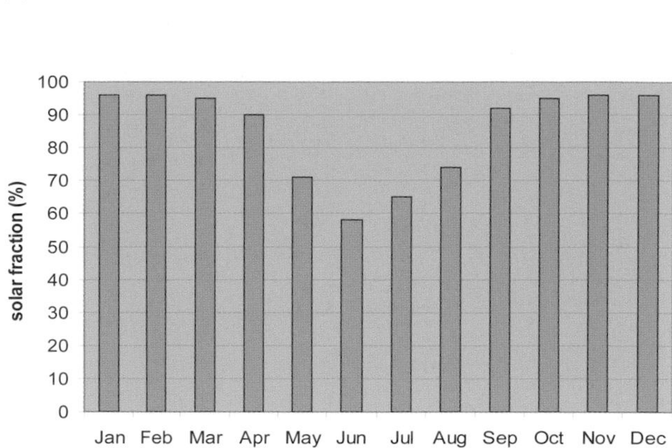

Figure 3.16.
Solar fraction for each month for: (a) a system in Northern Europe (55°N), designed to cover the hot water demand fully in the summer months; (b) a system in Sydney, designed to cover 85% of the hot water demand over the year

thermal loads on the collectors (stagnation), but also to a lower efficiency (additional costs are higher than additional yield). In periods with lower irradiation, the output would be higher but the total annual output per square metre of collector would be lower than with the original system.

In systems with an auxiliary heat exchanger inside the same tank as the solar-heated water, the tank heat losses result from both the solar heat and the auxiliary heat brought into the tank. Depending on how the heat losses from the tank are attributed to the solar heat and/or the auxiliary heat, the solar fraction may then be defined in different ways.

3.5.2.3 System efficiency

The system efficiency gives the ratio of solar heat yield to the global solar irradiance on the absorber surface with respect to a given period of time, for example one year:

$$SE = \frac{\dot{Q}_S}{E_G A} \times 100$$

where SE is the system efficiency (%), $\dot{Q}_S$ is the solar heat yield (kWh/a), E_G is the total yearly solar irradiance (kWh/m^{2}a), and A is the absorber surface area (m^2).

If the absorber surface area and the irradiance are known, and if the solar heat yield is measured (heat meter), the system efficiency can be determined:

Example:

$A = 6$ m^2
$E_G = 1000$ kWh/m^{2}a (central Europe)
$\dot{Q}_s = 2100$ kWh/a

Then the system efficiency is given by

$$SE = \frac{2100 \text{ kWh} \times \text{m}^2 \times \text{a}}{1000 \text{ kWh} \times \text{a} \times 6\text{m}^2} \times 100 = 35\%$$

The system efficiency is strongly dependent on the solar fraction. It is higher at lower solar fractions (when the solar water heater size is small compared with the hot water demand). If the solar fraction is increased by increasing the collector area, the system efficiency is reduced, and every further kilowatt-hour that is gained becomes more expensive. This counter-effect of the two variables can be seen in Figure 3.17.

3.5.2.4 Step 1: Determination of hot water consumption

The hot water consumption, V_{HW}, of those living in the house is a key variable for system planning, and if it cannot be measured, it should be estimated as closely as possible. When determining the requirements, a check should be made on the

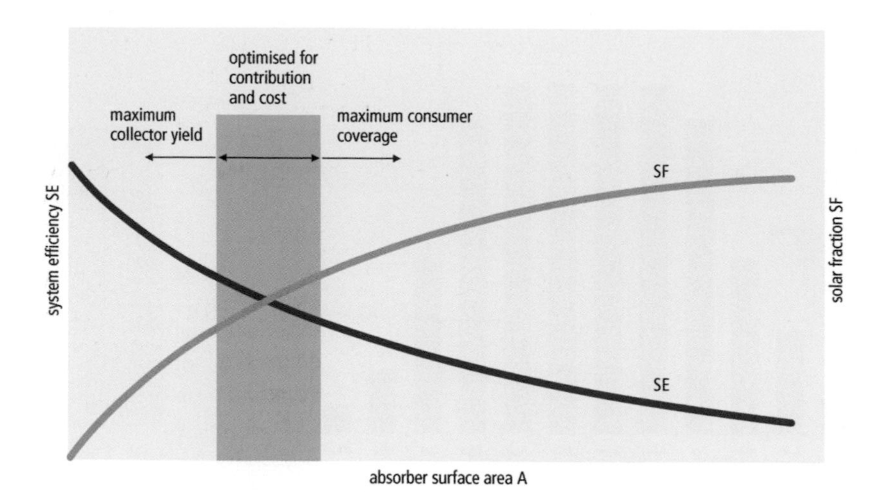

Figure 3.17. Solar fraction and system efficiency[17]

possibilities of saving domestic water (for example by the use of water- and energy-saving fittings). A lower water consumption means a smaller solar energy system and hence a lower investment.

However, it is not possible to estimate the hot water consumption of a household accurately, as individual differences are enormous. Of two similar families living in two identical neighbouring houses, one family might use twice as much hot water as the other. For large solar installations, the hot water consumption may be measured separately before designing a solar water heating installation.

During the design of solar energy systems for one- and two-family houses, the following average values can be used for estimating the hot water consumption:[18]

1 × hand washing (40°C) 3 l
1 × showering (40°C) 35 l
1 × bathing (40°C) 120 l
1 × hair washing 9 l
Cleaning 3 l per person per day
Cooking 2 l per person per day
1 × dishwashing (50°C) 20 l
1 × washing machine (50°C) 30 l

Depending on the fittings in the household, the following average consumption values per person per day can be calculated (usage temperature of hot water approximately 45°C):

- low consumption 20–30 l
- average consumption 30–50 l
- high consumption 50–70 l.

In the following, all the components for a thermal solar energy system for heating the domestic water for a four-person household in the UK will be dimensioned: collector surface area, domestic water store volume, solar circuit pipework, heat exchanger, circulating pump, expansion vessel and safety valve.

We assume an average hot water consumption of 50 l per person per day (45°C), and a requirement to supply the dishwasher and the washing machine with solar-heated water (either special machines or by using an adapter; see Chapter 2 section 2.3.7 on this subject). According to the information from the user the dishwasher and washing machine operate on average twice per week.*

Taking into account the different hot water temperatures, the daily hot water consumption is then calculated as follows:

V_{HW} = 4 persons × 50 l (45°C) + 16 l (45°C) = 216 l (45°C) per day

*2 × 20 l (50°C) + 2 × 30 l (50°C)</7 days ≈ 16 l (45°C) for dishwashing and the washing machine.

3.5.2.5 Step 2: Hot water heat requirement

The heat requirement, Q_{HW}, can be determined from the hot water consumption according to the following equation:

$$Q_{HW} = V_{HW} c_W \Delta\theta$$

where V_{HW} is the average hot water quantity (l or kg), c_W is the specific heat capacity of water (= 1.16 Wh/kgK), and $\Delta\theta$ is the temperature difference between hot and cold water (K).

In our example the necessary daily heat requirement for heating 216 l of water from 10°C (we assume this to be the cold water temperature for this example) to 45°C is given by:

$$\begin{aligned} Q_{HW} &= 216 \text{ kg} \times 1.16 \text{ Wh/kg K} \times (45 - 10) \text{ K} \\ &= 8770 \text{ Wh} \\ &= 8.77 \text{ kWh per day} \end{aligned}$$

Note that, depending on the domestic/drinking water regulations, the actual temperature of the hot water delivered should in some countries be higher, for instance 60°C. In such cases, the water will be mixed at the tapping point. The hot water supply system will have to deliver a smaller amount of water at the high temperature, which will however have the same energy content. This amount is calculated as follows:

$$V_{\theta 2} = \frac{\theta_1 - \theta_c}{\theta_2 - \theta_c} \times V_{\theta 1}$$

where θ_1 is the old temperature level, θ_2 is the new temperature level, θ_c is the cold water temperature, $V_{\theta 1}$ is the volume of water at the old temperature level, and $V_{\theta 2}$ is the volume of water at the new temperature level.

For example, the amount of water at 60°C equivalent to the above-mentioned 216 l at 45°C, using a cold water temperature of 10°C, is calculated as follows:

$$V_{60} = \frac{45 - 10}{60 - 10} \times 216 = 151 \text{ l}$$

HEAT LOSSES IN PIPING AND STORES

The significance of thermal insulation is often underestimated. In the following, estimates are made of the possible thermal losses from the solar circuit, the circulation lines and the solar store.

HEAT LOSSES IN INSULATED PIPES

It is possible to make a relatively good estimate of the losses if we consider only the heat conduction through the thermal insulation.[19]

The heat losses can be formulated as follows:

$$Q_{pipe} = \frac{2\pi\lambda\Delta\theta}{\ln (D_{wd}/D_{pipe})} \quad \text{(W/m)}$$

where λ is the thermal conductivity of the insulating material (W/mK), D_{wd} is the outside dimension of the insulated pipe (mm), D_{pipe} is the outside diameter of the pipe (mm), ln is natural logarithm, and $\Delta\theta$ is the difference between the temperature in the pipe, θ_{pipe}, and the ambient air temperature, θ_{air} (K).

Example:

λ = 0.04 W/mK (mineral wool)
D_{wd} = 54 mm
D_{pipe} = 18 mm
$\Delta\theta$ = 30 K
In this way Q_{pipe} is calculated as

$$Q_{pipe} = \frac{2\pi \times 0.04 \text{ W} \times 30 \text{ K}}{\ln (54 \text{ mm}/18\text{mm})\text{mK}} = 6.9 \text{ W/m}$$

With a total solar circuit length of 20 m and approximately 2000 operating hours per annum, heat losses of $\dot{Q}_{pipe}$ = 6.9 W/m × 20 m × 2000 h/a = 276 kWh/a. This corresponds to an approximate annual yield for a solar energy plant with 5 m² of glazed flat-plate collectors of 15% ($\dot{Q}_S$ = 5 m² × 1000 kWh/m²a × 0.35 = 1750 kWh/a).

If a circulation pipe is installed in the building, additional heat losses or an increased heat requirement will have to be allowed for. With a circulation line of 15 m and a running time for the circulation pumps of 2 h per day the following heat losses can be calculated:

$$\dot{Q}_{circ} = 6.9\ \text{W/m} \times 15\ \text{m} \times 2\ \text{h/day} \approx 76\ \text{kWh/a}$$

Therefore the resulting heat losses Q_G are

$$\dot{Q}_G = \dot{Q}_{pipe} + \dot{Q}_{circ}$$
$$= 276\ \text{kWh/a} + 76\ \text{kWhA}$$
$$= 352\ \text{kWh/a}$$

This corresponds to an annual heat gain for 1 m² of glazed flat-plate collector surface area.

HEAT LOSSES FROM NON-INSULATED PIPES

It is also possible to estimate the heat losses for non-insulated pipes. However, as very complicated relationships exist here between convection and heat radiation, a calculation is offered for simplification purposes that uses a very much simplified equation using factors.[19] This is valid only for piping diameters below 100 mm and air temperatures around the pipes of θ_A = 20°C:

$$\dot{Q}_{pipe} = D_{pipe}\ (29.85 + 0.027\theta_{pipe}\sqrt[3]{\theta_{pipe}})\Delta\theta\ \ (\text{W/m})$$

The pipe diameter must be inserted here in m.

Example:

D_{pipe} = 0.018 m
θ_{pipe} = 50°C
$\Delta\theta$ = 30 K

$$\dot{Q}_R = 0.018\ \text{m}(29.85 + 0.027 \times 50°\text{C}^3\ \sqrt{50°\text{C}}) \times 30\text{K} \approx 19\ \text{W/m}$$

Therefore, when the solar circuit is not thermally insulated, heat losses arise that are almost three times higher than those for insulated piping. These heat losses will be comparable to the yield of several square metres of solar collectors, and will greatly reduce the yield of the solar system.

HEAT LOSSES FROM STORES

The heat losses from a solar store increase in proportion to the area of its upper area, *A*, and the temperature difference between the store and the surroundings, $\Delta\theta$:

$$\dot{Q}_{St} \approx A\Delta\theta$$

With the help of the heat loss coefficient k in W/m²K, the following equation is derived:

$$\dot{Q}_{St} = kA\Delta\theta\ (\text{W})$$

For stores the *kA* value is normally given in W/K.

Example:

kA value = 1.6 W/K
$\Delta\theta$ = 30 K

$$\dot{Q}_{St} = 1.6\ \text{W/K} \times 30\ \text{K} = 48\ \text{W}$$

Over the course of a year this store has heat losses of

$$\dot{Q}_{St} = 48\ \text{W} \times 24\ \text{h/d} \times 365\ \text{days/a} \approx 420\ \text{kWh/a}$$

In the above case the heat losses in the store thereby correspond with the solar gain of about 1.2 m² of collector surface.

3.5.2.6 Step 3: Design and dimensioning of system components

There are four different methods:

- rough determination of size with an approximation formula
- detailed calculation of the individual components
- graphical design with nomographs
- computer-aided design with simulation programs.

ROUGH DETERMINATION OF SIZE WITH AN APPROXIMATION FORMULA

COLLECTOR SURFACE AREA

For a system designed for a *temperate climate*, a rough estimate for the essential system components can be made under the following assumptions:

- Average hot water requirement, V_{HW} = 35–65 litres (45°C) per person per day.
- Yearly average solar fraction = approximately 60% (covering almost the complete hot water demand in the summer months).
- Collector at optimal or almost optimal orientation and tilt angle (see section 1.1.3).
- No or little shading.
- Design goal is to cover the load almost completely in the months with high irradiation.

The rule of thumb for such situations is:

1–1.5 m^2 of glazed flat-plate collector area per person (E_G = 1000 kWh/m^2a)
0.7–1 m^2 of evacuated tube collector surface per person (E_G = 1000 kWh/m^2a)

In our example this approximation formula leads to a required glazed flat-plate collector surface area of 4–6 m^2.

For different irradiation levels (see Chapter 1), these values may be scaled. For instance, when the irradiance is 1300 kWh/m^2a, the collector area may be 1.3 times smaller.

For a *tropical climate*, the collector area can be estimated under the following assumptions:

- Average hot water requirement, V_{HW} = 35–65 litres (45°C) per person per day.
- Favourable solar irradiance conditions.
- Yearly average solar fraction = approximately 80% (auxiliary heating is necessary only in a few months).
- Collector at optimal or almost optimal orientation and tilt angle (see section 1.1.3).
- No or little shading.

The rule of thumb for such situations is:

0.35–0.6 m^2 of glazed flat-plate collector area per person (E_G = 2000 kWh/m^2a)

In our example this approximation formula leads to a required glazed flat-plate collector surface area of 1.4–2.4 m^2.

For different irradiation levels (see Chapter 1), these values may be scaled. For instance, when the irradiance is 1700 kWh/m^2a, the collector area must be 1.2 times larger.

DOMESTIC WATER STORE VOLUMES AND HEAT EXCHANGERS

In general, in order to bridge over a few sunless days without any auxiliary heating, the store volume should be designed to be 1–2 times the daily hot water consumption. In our example of a consumption of 216 l per day, this leads to a store volume of 200–400 l. For the dimensioning of internal heat exchangers the following approximation formulae apply:

- Finned tube heat exchanger: 0.35 m^2 exchanger surface area per m^2 of collector surface area.

Example:

Volumetric flow, $\dot{m}$= 240 l/h (6 m² × 40 l/m²h)
Flow speed, v = 0.7 m/s

$$D = \sqrt{\frac{4\frac{240\,\text{ls}}{0.7\,\text{hm}}}{3.1416}}$$

$$= \sqrt{\frac{4\frac{240\,\text{dm}^3\text{h}}{25{,}200\,\text{h dm}}}{3.1416}}$$

$$= \sqrt{0.0121\,\text{dm}^2} = 0.1101\,\text{dm} = 11.01\,\text{mm}$$

The required internal diameter should therefore be at least 11 mm. On the basis of Table 3.3 a copper pipe with an internal diameter of 13 mm would be selected (description Cu 15 × 1).

Table 3.3. Dimensions of standard copper pipes

Pipe dimensions (outside dia. × wall thickness)	Internal dia. (mm)	Contents (l/m)
Cu 10 × 1	8	0.05
Cu 12 × 1	10	0.079
Cu 15 × 1 DN 12	13	0.133
Cu 18 × 1 DN 15	16	0.201
Cu 22 × 1 DN 20	20	0.314
Cu 28 × 1.5 DN 25	25	0.491

Note: not all diameters are on the market in all countries.

An examination of temperature-dependent length changes in pipes can be found in Chapter 4, section 4.1.7.

CIRCULATING PUMPS

In general, traditional circulating pumps with three to four control stages are sufficient for small systems (collector surface areas ≤ 20 m²). For detailed design it is necessary to know the overall pressure loss, Δp_{tot}, in the solar system. This is made up of the pressure losses in the collector, in the solar circuit and in the heat exchanger:

$$\Delta p_{tot} = \Delta p_{col} + \Delta p_{solar\ circ} + \Delta p_{heat\ exch}$$

The pressure loss in the collector, Δp_{col}, is dependent on the design, on the selected volumetric flow, and on the type of collector connection. The pressure loss of the individual collectors, which depends on the volumetric flow, can be found in the manufacturer's information (see Figure 3.18). If several collectors are connected in parallel the pressure of the collector field corresponds to that of one collector. If they are connected in series the individual pressure losses are added together.

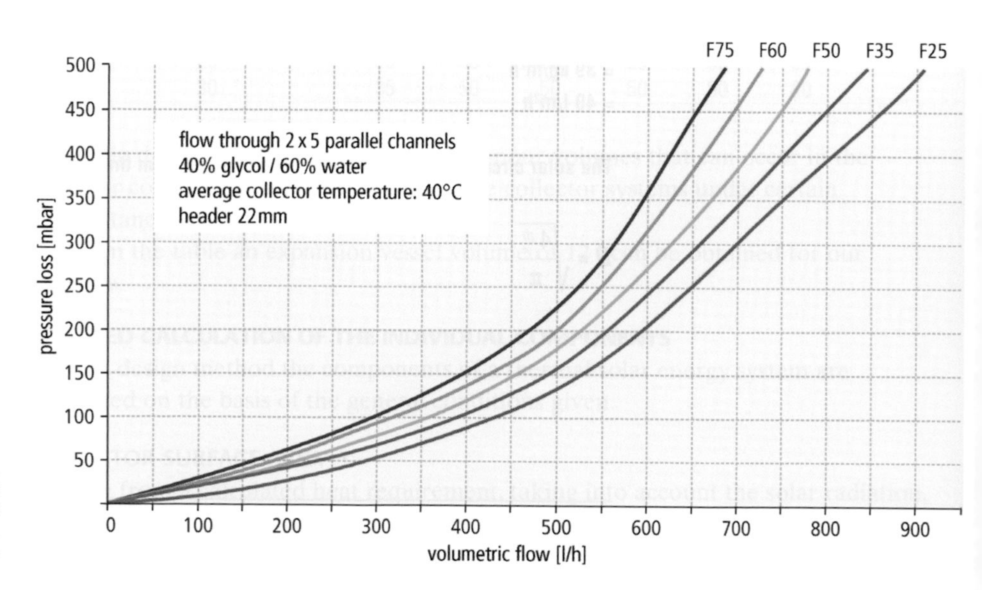

Figure 3.18. Example of pressure loss curves for five glazed flat-plate collectors Source: SOLVIS Energiesysteme, Braunschweig

Example:
On the basis of the calculated collector surface area of 5.5 m^2, the collector with the middle pressure loss curve (a model with an aperture area of 5.08 m^2) is selected. With a volumetric flow of 40 l/m^{2}h we obtain from Figure 3.18:

$$\Delta p_{col} = 50 \text{ mbar}$$

The pressure loss of the solar circuit, $\Delta p_{solar\ circ}$, is made up of the pressure loss in the piping, Δp_{piping}, and the sum of the pressure losses in the fittings, $\Delta p_{fittings}$.

The specific pressure loss per metre of installed pipe length is dependent on the pipe cross-section and the flow speed, and can be established by nomographs. In doing this the pipe material that is used, and the concentration of the water–glycol mixture, must be taken into account. The pressure losses in bends, T-pieces, screwed connections, valves and fittings are taken from the respective tables in the form of pressure loss correction values, or they can be estimated as an overall one third of the piping losses.

Example:
At a volumetric flow of 200 l/h ($\approx$ 5.08 m^2 × 40 l/m^{2}h) the pressure loss for DN 12 (= Cu 15 × 1) from Figure 3.19 is

$$\Delta p_{solar\ circ} = 4.5 \text{ mbar per metre of line length} \times 20 \text{ m} = 90 \text{ mbar}$$

$$\Delta p_{fittings} = 1/3\ \Delta p_{solar\ circ} = 30 \text{ mbar}$$

The pressure loss in the solar circuit heat exchanger, $\Delta p_{heat\ exch}$, can be found in the product documentation according to the type of heat exchanger and the flow speed.

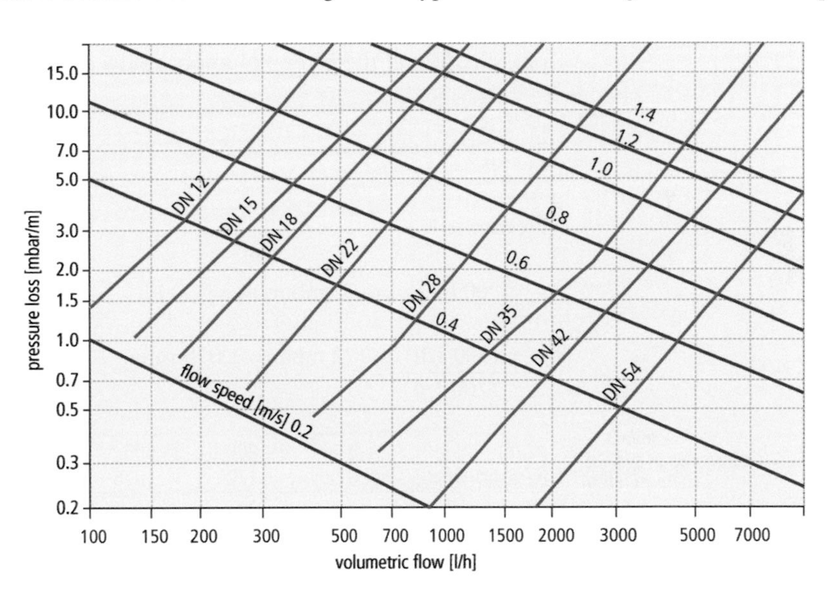

Figure 3.19. Pipe network characteristics for copper, 35% glycol, 65% water, 50°C [22]

Example:
With a volumetric flow of 200 l/h for the plain tube heat exchanger corresponding to the lowest line in the graph (Figure 3.20), with a heat exchanger surface area of 0.2 m^2 per m^2 of collector surface area, and a 5.08 m^2 collector, we obtain a pressure loss of

$$\Delta p_{heat\ exch} = 2 \text{ mbar}$$

We thus obtain for the total pressure loss of the system, at a volumetric flow of 200 l/h,

$$\Delta p_{tot} = 50 \text{ mbar} + 90 \text{ mbar} + 30 \text{ mbar} + 2 \text{ mbar} = 172 \text{ mbar}$$

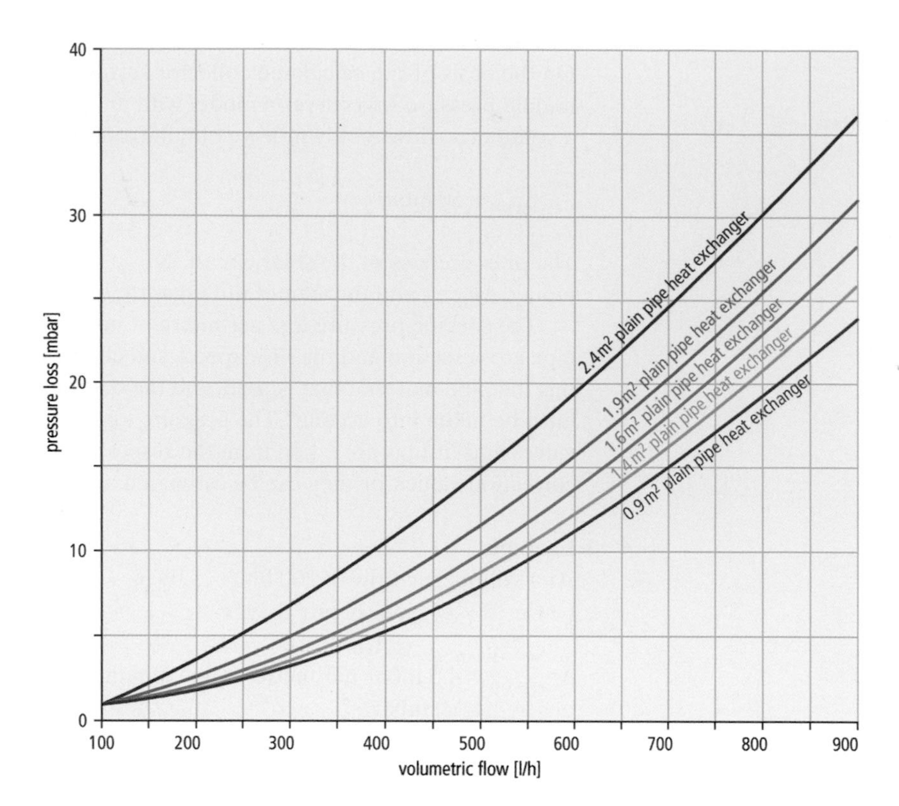

Figure 3.20. Pressure losses of plain tube heat exchangers[23]

For the selection of a suitable pump in the solar circuit the pressure loss characteristics of the system must first be established. For this purpose the total pressure losses for different volumetric flows are calculated from the determined pressure loss at the given volumetric flow according to the equation

$$\frac{v_1^2}{v_2^2} = \frac{p_1}{p_2}$$

Example:

$v_1 = 200$ l/h, $p_1 = 172$ mbar, $v_2 = 500$ l/h

$$p_2 = \frac{(500 \text{ 1/h})^2}{(200 \text{ 1/h})^2} \times 172 \text{ mbar} = 1{,}075 \text{ mbar}$$

Table 3.4. Coordinates of system characteristics

v (l/h)	0	200	500	1000
p (mbar)	0	172	1075	4300

The pressure loss characteristics of the system can then be entered into the manufacturer's pump diagram (Figure 3.21). The intersecting points of the system characteristic lines with the pump characteristic lines at the different power settings give the possible working points. These working points should lie in the middle of the pump characteristic lines if possible, as it is here that the pump operates at its maximum efficiency, and an increase or reduction of the flow is possible by means of the control settings of the pump.

EXPANSION VESSEL AND SAFETY VALVE (SV)

The significance of the membrane expansion vessel for the system safety of filled systems has already been described (see Chapter 2, section 2.4.8.2). (Note that for drainback systems, or systems where the collector fluid is not allowed to boil dry in the collector, an expansion vessel is not needed or may be much smaller.)

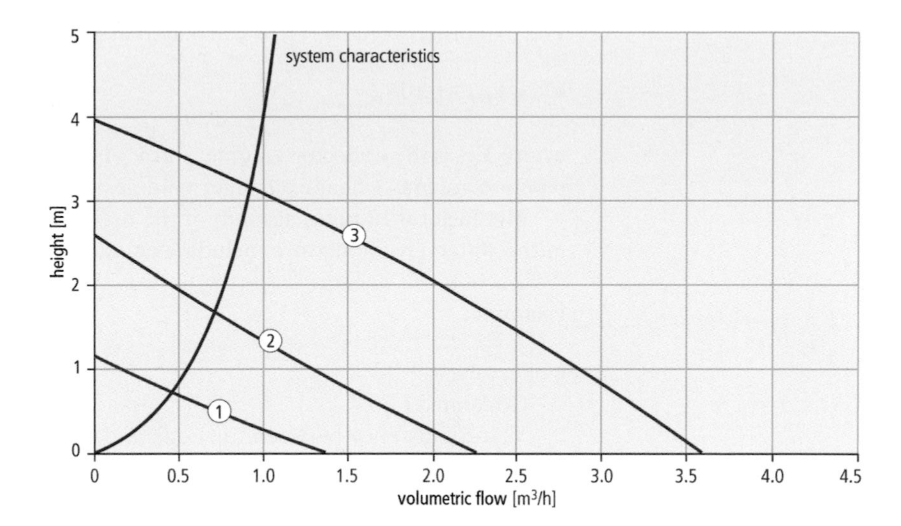

Figure 3.21. Pump characteristics and system pressure loss characteristics. Source: Wilo, Dortmund

For calculation of the size of the vessel, apart from the temperature-related volume changes of the solar liquid, the fluid channels volume of the collector should also be considered to allow the collectors to boil dry in the event of system stagnation. The minimum vessel size is calculated according to the following equation:

$$V_{\mathrm{MEVmin}} = V_{\mathrm{D}} \frac{p_{\mathrm{Bmax}} + 1}{p_{\mathrm{Bmax}} - p_{\mathrm{in}}}$$

where V_{D} is the expansion volume (l), p_{Bmax} is the maximum permissible operating pressure (bar), and p_{in} is the initial (working) pressure with which the system is filled (bar).

The quotient

$$\frac{1}{(p_{\mathrm{Bmax}} + 1)/(p_{\mathrm{Bmax}} - p_{\mathrm{in}})}$$

is called the *pressure factor* or the *utilization* of the vessel. It gives the proportion of the volume of the vessel that is effectively available for the heat transfer medium.

The working pressure should be high enough that under no operating conditions is it possible for air to enter the system because of low pressure. It should be at least 0.5 bar at the highest point in the system. As the vessel is generally below the static pressure, P_{stat} this should also be added to the minimum pressure:

$$p_{\mathrm{in}} = 0.5 \text{ bar} + p_{\mathrm{stat}}$$
$$p_{\mathrm{stat}} = h_{\mathrm{sys}} \text{ (m)} \times 0.1 \text{ bar/m}$$

where h_{sys} is the system height. The calculated maximum operating pressure should be about 0.3 bar lower than the response pressure of the safety valve. The nominal pressure of the safety valve, in relation to the system admission pressure, can be taken from Table 3.5.

Table 3.5. System admission pressure and nominal pressure of safety valve

Preset pressure of vessel (bar)	1	1.5	3.0	6.0
Nominal pressure of safety valve (bar)	2.5	4.0	6.0	10.0

In our example a 4.0 bar safety valve is selected:

$$p_{\mathrm{Bmax}} = p_{\mathrm{BmaxSV}} - 0.3 \text{ bar} = 3.7 \text{ bar}$$

The expansion volume, V_D, is calculated as follows:

$$V_D = V_{col} + 0.10V_A$$

where V_{col} is the collector volume, and V_A is the system volume* = collector volume + pipeline volume + heat exchanger volume.

The factor 0.10 takes account of the maximum temperature-related volume change in the water–glycol mixture including an addition of 10% for safety.

Example:

Collectors:
- Content: 5 l
- Vertical distance between top edge and safety valve: 10 m

Solar circuit:
- Copper tube Cu 15 × 1
- Length: 25 m

Heat exchanger:
- Content: 1.8
- $V_A = 5\ l + 25\ m \times 0.133\ l/m + 1.8\ l = 10.1\ l$
- $V_D = 5\ l + 0.1 \times 10.1\ l = 6.01\ l$
- $p_{in} = 0.5\ bar + 1\ bar = 1.5\ bar$
- $p_{Bmax} = 3.7\ bar$

$$V_{MEV\,min} = 6.01\ l \times \frac{3.7\ bar + 1}{3.7\ bar - 1.5\ bar}$$

$= 12.84\ l$

From the series of standard sizes (10, 12, 18, 25, 35, 50 l) the next largest size is selected: that is, 18 l.

All the components of the solar energy system are calculated in this way.

GRAPHICAL DESIGN WITH NOMOGRAPHS

For an approximation of the dimensions of collectors and stores, nomographs are supplied by several solar system suppliers, which are tailored to company-specific products. Please contact the supplier for more information.

DIMENSIONING WITH A SIMULATION PROGRAM

There are currently several programs available for the dimensioning and simulation of solar thermal systems: see Chapter 10. In many cases, when a system needs to be assembled for a specific situation, it is advisable to use one of these programs together with reliable input data to arrive at the best design of the solar water heater. Some solar water heater manufacturers supply free dimensioning programs on their websites.

3.5.3 Dimensioning of systems for heating domestic water and heating support (central European conditions)

A solar energy system can make a contribution to room heating, especially during spring and autumn. The heating demand and technical requirements for a solar room heating system far much more widely than those for domestic water heating. Partly for this reason, the methods and guidelines for design and performance prediction of solar space heating installations are less developed than those for water heating.

Therefore, and because space heating is mostly an option in temperate and northern climates, we shall limit ourselves to global dimensioning recommendations for central European climates, such as the UK, Germany and Denmark.

For these regions we can differentiate between systems with a *low solar fraction* and those with a *high solar fraction*. Of the first type are solar energy systems for which a maximum of 35% of the full heating requirement of a building is covered by

---

*The volume displaced by vapour in the collector connecting lines is not considered here.

solar energy. These are the main subject of this section. For systems with a high solar fraction, of significantly more than 70% of the full heating requirement, another approach is necessary in which the summer sun's energy is stored in a long-term or seasonal storage system, from which it is taken in the winter.

3.5.3.1 PRECONDITIONS

In order to be able to use a solar thermal system to sensibly support the space heating, several preconditions must be fulfilled with respect to the building and the heating system:

LOWEST POSSIBLE HEATING REQUIREMENT

Existing buildings normally have a very high heating requirement: ~200 kWh/m^2a or more is not infrequent. In this instance it is a fact that a generously dimensioned solar energy system can reduce the energy consumption, but cannot supply a significant contribution to the solar fraction.

In connection with the dimensioning of solar thermal systems with heating support the term *heating load* is often used. This is a performance variable, and is given in W/m^2 with respect to the living area. Just like the heating requirement, the heating load is also dependent on the insulation standard of the building and on the meteorological conditions. It is possible to determine the heating load from the annual heating requirement, the living area to be heated and the number of heating hours in the year (the so-called *full-usage hours*).

Example:

Heating requirement = 200 kWh/m^2a
Living area = 200 m^2
Heating hours = 1800 h/a
Heating load 200 $kWh/m^2a \times 200\ m^2/1800\ h/a$ = 22 kW

With increasing insulation standard of the building, such as in a low-energy house (heating requirement <50 kWh/m^2 of living area per year), the heating requirement is about the same order of magnitude as the energy requirement for hot water heating. Here a combined system can make a significant contribution as heating support, especially in the transitional months during spring and autumn.

LOWEST POSSIBLE FEED AND RETURN TEMPERATURES

Conventional heating systems operate with feed temperatures of 50–70°C. The collectors can attain this high temperature level only in a very few cases during times of low solar irradiance. If, however, large surface-area heat distribution systems such as underfloor or wall surface heating are installed, the low feed and return temperatures (50°C and 30°C) can be generated relatively frequently by the solar energy system. In this way the solar fraction and the amount of oil or gas energy saved can quite easily amount to 20–30%.

FAVOURABLE ORIENTATION OF THE COLLECTOR SURFACES

Whereas for simply heating domestic water the alignment and pitch of a roof surface are favourable over wide ranges (see Chapter 1, section 1.1.3), an angle of incidence of at least 45° and a south-facing surface, if possible, are advantageous for solar heating support because of the low level of the sun and the shorter days during the heating months (see Figure 1.15).

If these preconditions are fulfilled, a solar heating support system can be implemented sensibly by means of an enlarged collector surface area in connection with a combination system.

3.5.3.2 APPROXIMATION FORMULAE

For an initial rough dimensioning of the system the following approximation formulae are given (central European climate):[24]

SYSTEMS WITH LOW SOLAR FRACTION

- Collector surface area: 0.8–1.1 m^2 of glazed flat-plate collector per 10 m^2 of heated living area or 0.5–0.8 m^2 of evacuated tube collector per 10 m^2 of heated living area.
- Store volume: At least 50 l per m^2 of collector surface or 100–200 l per kW of heat load.

SYSTEMS WITH HIGH SOLAR FRACTION

- Collector surface area: 1.5–3 m^2 of glazed flat-plate collector per 10 m^2 of heated living area
- Store volume: 250–1000 l per m^2 of collector surface

3.5.3.3 DIMENSIONING DIAGRAM FOR SYSTEMS WITH LOW SOLAR FRACTION

Manufacturers of solar thermal systems with heating support often already offer product-specific dimensioning diagrams or nomographs. However, these are always tailor-made for the manufacturer's own components and are not generally valid.

3.5.3.4 COMPUTER-AIDED SYSTEM DIMENSIONING

Accurate dimensioning of collector surface areas and store volumes for solar energy systems with heating support is possible only with the help of simulation programs, such as T*SOL, Polysun, or TRNSYS (see Chapter 10). Passive solar gain through windows and from internal heat sources (people, devices, lighting) plays an ever-increasing role in the overall energy balance as building insulation standards increase.

3.6 Costs and yields

3.6.1 Prices and performance

The prices of solar products have decreased steadily over the past decade. For price information on systems in Europe, please refer to the *Sun In Action II* report (on www.estif.org) or the *Soltherm Europe Market Reports* (on www.soltherm.org). Also, the references at the end of Chapter 4 can give price information.

As installation costs still present an unsafe calculation value, particularly for companies new to the solar market, an estimate of the time and personnel costs for the individual installation steps is helpful.

The following example provides help for estimating installation, times and costs (on-roof installation) for a standard system for a one-family home, assuming normal conditions without any particular difficulties. Difficulties that are recognised on site that exceed the normal situation must be incorporated immediately at the calculation stage. Experience plays an important role in recognising and estimating these additional costs.

Please note that this example is for a pumped system with indoor tank, and is meant primarily to provide a checklist. The installation of a thermosyphon system delivered as one compact kit will require activities that are different from those mentioned here.

3.6.1.1 EXAMPLE CALCULATION FOR INSTALLATION

An example calculation is shown in Table 3.6.

3.6.1.2 EXAMPLE OF A LIST OF SPECIFICATIONS

In new buildings the installation cost is reduced in comparison with retrofitting because there are fewer safety measures and no replacing floor coverings etc., and the collector installation can take place during other roofing operations. Within the scope of a financial analysis during the installation of a new heating boiler, a domestic hot water (DHW) store 'credit' is frequently accounted for the solar energy system, as this is already included in the price, and so only the additional costs are taken into account. An example of a quotation is shown in Table 3.7.

3.6.1.3 ENERGY BALANCE AND YIELDS FOR A THERMAL SOLAR SYSTEM

The performance (or solar yield) of thermal solar systems, assuming that the dimensioning has been matched to the hot water requirements, is established by means of the losses on the way from the collector to the tap.

Table 3.6.
Example calculation for fitting a solar energy system into a central European one-family home (on-roof installation, older building, central hot water heating system present, unvented store).[25] All prices are examples.

Description of necessary work	Work in minutes			
	Fitter	Time	Apprentice	Time
Collector. Fix collector surface to roof, remove necessary tiles, screw roof hooks onto rafters, replace roof tiles, transport collector to house wall, lift collector with ropes or belts, install collector in retainers, align and screw in place, position and install air tiles for feed and return lines, install pipeline connectors to collector, install bleeding equipment, insulate piping, pass pipelines through air tiles	2	360	1	180
Solar store. Transport store to cellar, position and install	2	240	1	120
Solar circuit. Install pipelines from roof to heating cellar, fit installation unit for solar circuit (safety valve, expansion vessel, circulating pump, pressure gauge, emptying and filling valve, shut-off valve, non-return valve, feed and return line thermometers), connect to store, check for leaks, flush through, piping insulation, make good any penetrations	1	480	1	480
Auxiliary heating. Install and connect pipes from auxiliary heating to heat exchanger of solar store to heating boiler, install circulating pump, fit non-return valve, insulate pipes	1	60	1	60
Domestic water connection. Fit installation unit for connection to water network (non-return valve, safety valve with connection to discharge pipe, shut-off and drain valve), install and connect cold water line to store, install and connect hot water line, install thermostatic mixing valve to limit domestic water temperature	1	120	1	120
Control system. Install main switch for solar energy system, fit temperature-difference controller, fit temperature sensor to collector and store and connect to controller with sensor cable, connect solar circulating pump to controller, fit temperature sensor for top-up heating and with sensor cable connect to heating controller, connect circulating pump for auxiliary heating to heating controller, connect controller to electrical network.	1	120	1	120
Earthing. Connect feed and return lines of solar circuit equipotential bonding strip	1	30	1	30
Commissioning. Fill solar circuit with solar liquid, bleed, set controller for solar circuit and top-up heating, start up system (technical description of the individual components, general sketch, electrical circuit diagram, operating instructions, maintenance plan), hand over solar energy system to owner including instructions concerning function of the system	1	60	1	60
Time in minutes		1470		1170
Fitter (€0.5/$0.5/£0.35 per min)				
Apprentice (€0.25/$0.25/£0.18 per min)		1470		585
Installation costs	€1051/$1051/£736			

Figure 3.22 shows the balance of a standard solar system with glazed flat-plate collectors.

The average system efficiency for a well-designed thermal solar system with glazed flat-plate collectors is about 35–45%. With global solar irradiance of 1000 kWh/m^2 per annum, each square metre of glazed flat-plate collector surface generates 350 kWh per annum of thermal energy. See Chapter 1 for determination of the global solar irradiance at different locations of the world. If evacuated tube collectors are used, the efficiency is increased to about 45–50%, because of the lower heat losses at the collector.

As already mentioned, the hot water consumption and the size of the system together determine both the system's efficiency and the solar fraction. If the hot water consumption is lower than expected, the solar fraction will be higher than designed, but the efficiency will be lower than designed. On the other hand, if there is a higher hot water consumption than was assumed, a higher solar efficiency is obtained, but the design solar fraction is not achieved.

For quantification of these effects, computer model calculations are necessary. See Chapter 10 for an overview of solar water heating calculation models.

Table 3.7. Quotation text as example[25]

Item	Qty	Description	Amount
1.		Collector	
1.1	2	Collector: Manufacturer, type, surface area 1.5 m^2 Collector characteristics: Give here according to manufacturer`s information	€1250/$1250/£875
1.2	2	Roof installation	
2.		Store	
2.1	1	Solar store, corrosion protection by means of sacrificial anode: Manufacturer, type, volume 300 l	€1100/$1100/£770
2.2	1	Solar heat exchanger: e.g. plain tube, copper, e.g. 1.6 m^2	
2.3	1	Additional heat exchanger: e.g. plain tube, copper, e.g. 1.1 m^2	
2.4	1	Insulation, PU (polyurethane) flexible foam, 100 mm	
3.		Solar circuit	
3.1	2	Connection pipes to collectors: flexible stainless steel corrugated pipe with weather and UV-resistant heat insulation	€500/$500/£350
3.2	2	Bleeding equipment	
3.3	metres	Feed and return lines, copper pipe, fittings, 22 mm, heat insulation, 300 mm, fixing materials	
3.4	1	Installation unit for solar circuit (consisting of: 4 bar safety valve, 18 l expansion vessel, 0–6 bar pressure gauge, feed and return line thermometers, emptying and filling valve, shut-off valve, non-return valve)	
3.5	1	Circulating pump: manufacturer, type, capacity	
4.		Auxiliary heating	
4.1	1	Installation unit for connecting solar store to heating boiler (shut-off valves, non-return valve)	€175/$175/£123
4.2	1	Circulating pumps: manufacturer, type, capacity.	
4.3	metres	Piping with thermal insulation	
5.		Domestic water connection	
5.1	1	Installation unit for connecting to water mains (consisting of: non-return valve, 6 bar safety valve, with connection to drain pipe, shut-off and discharge valve)	€200/$200/£140
5.2	metres	Cold and hot water pipelines to store (pipe materials according to existing pipe network)	
5.3	1	Thermostatically controlled mixing valve for limiting domestic water temperature to 60°C	
6.		Controller	
6.1	1	Temperature-difference controller: manufacturer, type	€250/$250/£175
6.2	3	Temperature sensor (collector, store) and sensor cable	
7.		Excess voltage protection	
7.1	metres	Cable	€75/$75/£53
7.2	1	Fixing set	
8.		Commissioning	
8.1	10 l	Solar liquid: propylene-glycol-based frost protection down to –25°C	€50/$50/£35
8.2		Documentation for system (technical description, general sketch, electrical circuit diagram, operating instructions, maintenance plan)	
9.		Installation	
9.1		Specialist installation of materials given above	€1051/$1051/£736
9.2		Travel and set-up times	
9.3		Commissioning (flushing, filling, settings, and handover of completed system	
9.4		Training and familiarization	
		Total amount	€4706/$4706/£3294
		16% VAT	€753/$753/£527
		Final sum (including 16% VAT)	€5460/$5460/£3822

Note: this table is only to demonstrate the structure of a quotation. Amounts and VAT percentage may vary.

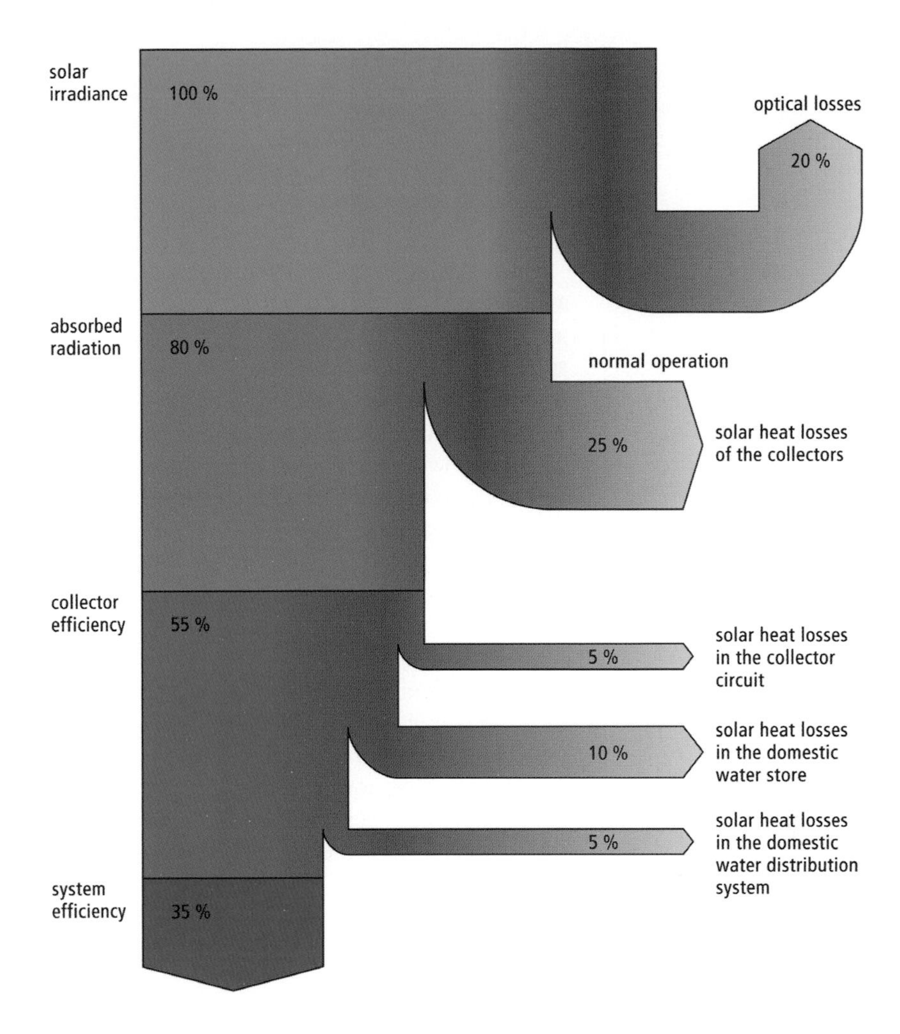

Figure 3.22. Annual solar balance for a thermal solar system with glazed flat-plate collectors

3.6.2 Normalised solar heat costs

The price–performance ratio of a solar system can be described with the help of the *normalised solar heat costs* (that is, the cost per kilowatt-hour). In this way the investment costs (taking account of any possible grants) and operating costs (power for the circulating pump and maintenance costs) are calculated with the energy yield during the service life.

A suitable procedure here is the *annuity method*. In this the annual charge (repayments and interest) is calculated by means of an *annuity factor*, which can then be set against the annual yield. The following example shows in a simplified procedure that the annuity is related only to the capital costs (see also Chapter 5, section 5.6). The annuity factor is given by:

$$a = \frac{(1 + p)^n p}{(1 + p)^n - 1}$$

where p is the effective interest rate (as a decimal), and n is the utilization (years).

Table 3.8. Normalised solar heat costs for a thermal solar energy system (including VAT)

System type	Glazed flat plate 5.0 m²	Evacuated tube 4.0 m²
Complete price: solar system minus grant and store credit	€2250/$2250/£1575	€3450/$3450/£2415
Annuity factor	6.4%	6.4%
Interest and repayments	€180/$180/£126	€226/$226/£158
Maintenance and operating costs	€50/$50/£35	€60/$60/£42
Annual costs	€231/$231/£162	€287/$287/£201
Annual solar energy yield (kWh)	1750	1800
Normalised solar heat costs per kWh	€0.13/$0.13/£0.09	€0.16/$0.16/£0.11

Note: the figures in this table and the text serve as an example only and are not generally applicable.

Table 3.8 shows the results that can be obtained from the annuity method under the following boundary conditions (note: the prices are an example only and are not generally applicable):

- installation of a glazed flat-plate or evacuated tube collector system in London for 60% solar fraction of the hot water requirement
- system costs: €1000/$1000/£700 per m² (glazed flat plate) or €1500/$1500/£1050 per m² (evacuated tube), including VAT
- utilization: 25 years
- effective interest rate: 4%
- maintenance and operating costs per annum: 1% of investment sum
- credit for the store: €750/$750/£525
- grant of 30% of the investment costs. N.B. In England and Wales in 2004, the subsidy level is £400 per system in private households. In Scotland it is up to 30% of system costs. Community schemes attract larger subsidies but these are subject to change.

A comparison with the costs for fossil heat generation shows that solar hot water generation can compete extremely well with electrical hot water heating, even if it still requires a subsidy.

A significant reduction of the system costs is achieved for new buildings (saving on tiled areas, lower installation costs etc.) In comparison with a retrofitted installation the costs arising are up to 20% less.

The resulting normalised solar heat costs resulting from this are reduced to €0.10/$0.10/£0.07 per kWh for the glazed flat-plate collector system and €0.13/$0.13/£0.09 per kWh for systems with evacuated tube collectors.

4 Installation, commissioning, maintenance and servicing

The installation of thermal solar energy systems involves three trade skills:

- roofing
- heat engineering (that is, plumbing and heating appliances)
- electrical.

In other words, specialist knowledge of all three areas is required if an installation is to be successful. The technical documentation of the manufacturer or the specialist literature for an individual component (particularly the installation instructions) are often insufficient.

The area in which the plumber, or the heating fitter, is often in totally new territory is usually the roofing skills. Whenever a collector assembly (field) is installed on a roof and the connecting pipes are laid into the house, work on the roof structure is also involved. It is therefore important to know and to follow the respective standards and regulations. This should help to clarify the question of which tasks the installer of solar systems can carry out, and which require a specialist roofer. For example, intervention by the heating fitter into flat roofs with plastic sheeting is definitely *not* recommended.

As it is not possible to give details of the standards and regulations for all countries in the world, this chapter gives general aspects and details of some common constructions and situations. For more information on a specific country, the reader is encouraged to make use of the references given in section 4.5.

Work tasks that the installer is likely to be familiar with already are summarised briefly, and the special solar-technical requirements are detailed. There is some reference to earlier chapters in which the individual solar components are described. Any existing guarantee obligations with respect to earlier work by skilled tradesmen must be taken into account, particularly during work on the roof or the heating system.

4.1 A brief study of roofing and materials

4.1.1 The purpose of the roof

The principal purpose of a roof is:

- to provide spatial boundaries
- to carry the loads from wind, rain and snow by means of the roof skin
- to keep the weather influences out of the building
- to enhance appearance (shape, colour, material, surface structure).

See Figure 4.1. Because of ever-decreasing fossil energy resources and the evident climate changes, the roof will in future increasingly act as the main structure for supporting energy conversion elements: that is, solar thermal or photovoltaic systems. This means that the roof skin (and the façade) will be subject to significant changes in materials and appearance.

4.1.2 Types of roof

Roofs may be classified as follows, according to the inclination:

- flat roofs: up to 5° pitch
- pitched roofs: $> 5° \leq 45°$ pitch
- steep roofs: >45° pitch.

Numerous sub-types occur, for example with two slopes.

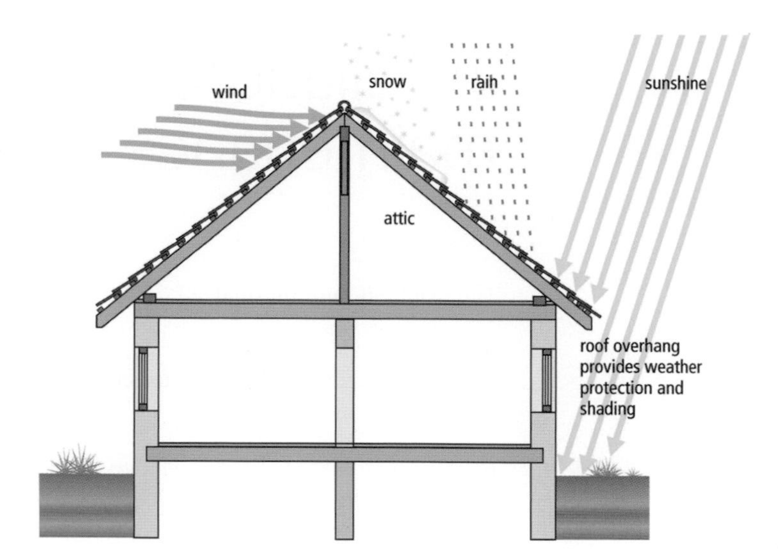

Figure 4.1. The purpose of the roof[26]

4.1.3 The components of the roof

4.1.3.1 ROOF STRUCTURES

This section describes some roof structures that are common in Europe. The list is not intended to be exhaustive.

- *Purlin and rafter.* In the case of roof truss purlins, the rafters rest on the purlins as inclined beams. The ridge purlin takes up the vertical loads and distributes them via the posts and struts; the base purlins correspondingly distribute them to the outer walls, and there may also be intermediary purlins. An extra purlin is sometimes required to attach the collector fixings. See Figure 4.2.
- *Rafter and collar beam.* In the case of the rafter roof, the rafters and the ceiling below them form a rigid triangle. There are no internal constructional elements. In the case of collar beam roofs, each pair of rafters is linked and stiffened with a so-called collar beam. See Figure 4.3.
- *Truss.* These are self-supporting structures (glued trusses, nail plate trusses etc.) and are increasingly found in new build. No changes may be made to trusses without the approval of the structural engineer. For cost reasons, truss roof construction is normally found only up to spans of about 17 m. See Figure 4.4.

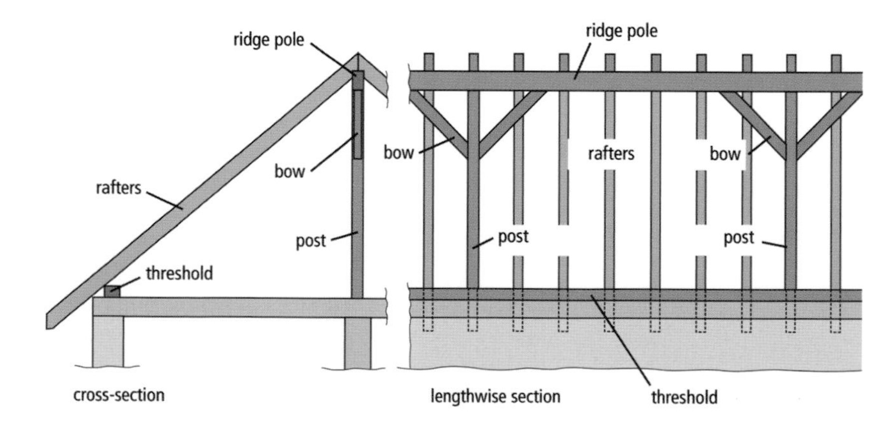

Figure 4.2. Roof truss purlins[26]

4.1.3.2 ROOF SKIN

A differentiation is made here between roof covering and roof sealing.

- *Roof covering.* This consists of individual elements such as tiles, concrete roofing bricks, fibrous cement slabs, bitumen shingles, slates and corrugated bitumen slabs. They must have a prescribed minimum roof inclination according to the type of overlap.
- *Roof sealing* (that is, flat roofs). This consists of, for example, bitumen roof sheeting, plastic roof sheeting, or plastic material applied in liquid form and then hardened.

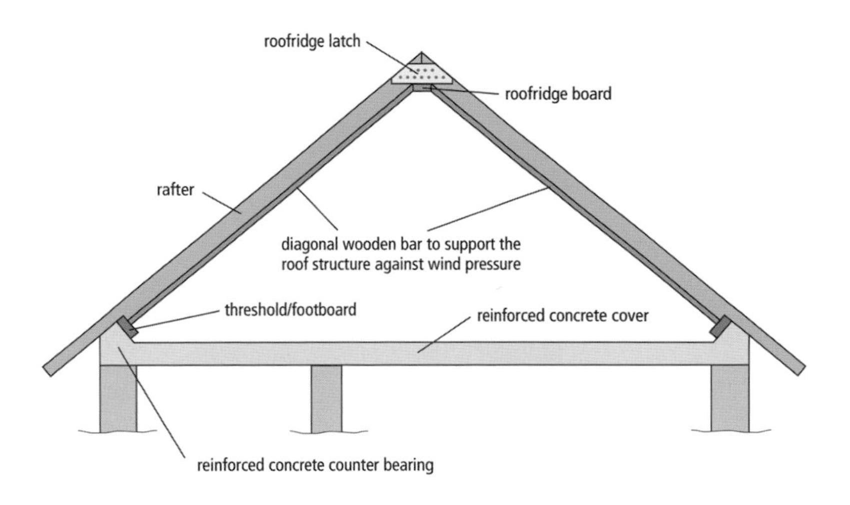

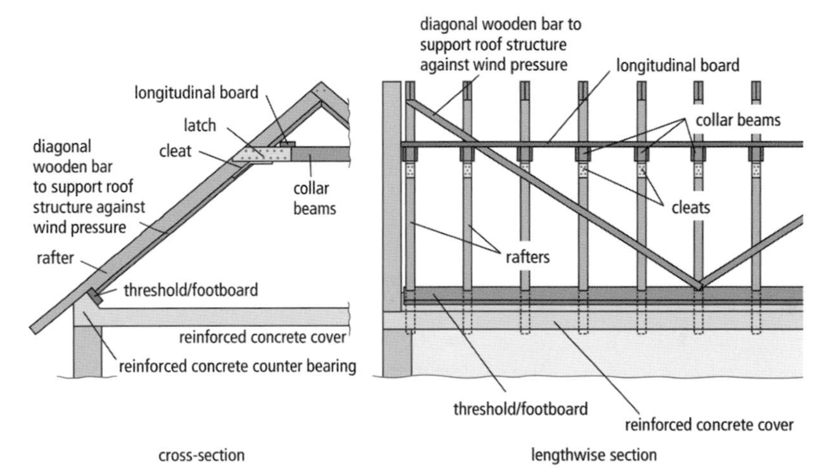

Figure 4.3. Top: rafter roof truss. Bottom: collar beam roof truss[26]

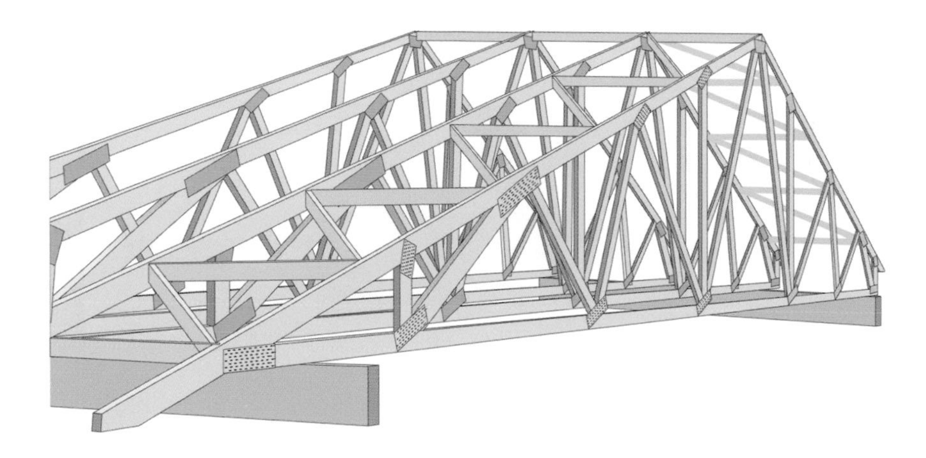

Figure 4.4. Nail plate trusses[26]

Examples of roof coverings and seals are shown in Figures 4.5 to 4.8 and described in Table 4.1. Examples of moulded tiles, specials, roof hooks, snowguards, and walk-on gratings are shown in Figures 4.9 to 4.12.

Figure 4.5. Left: pantiles. Right: wood shingles[27]

Figure 4.6.
Left: natural tiles. Right: plain tiles[27]

Figure 4.7.
Left: welded sheet installation on flat roof[28]. Right: plastic roof sheeting[29]

Figure 4.8.
Metal roofs[30]

Table 4.1. Types of roof covering and sealing

Type	Examples	Material	Type of fixing, processing	Notes
Tiles	Plain tile	Clay	Dry-laid, part-clamped, sarking felt or membrane, cemented	Good back ventilation, otherwise frost damage, medium to high breaking strength, efflorescence, mould formation
	Pantiles (single Roman), double Roman, interlocking	Concrete	Dry-laid, part-clamped, sarking felt or membrane, cemented	Surface weathering, high breaking strength, efflorescence
Slates	Slate I Slate II Circular shingling	Natural slates, stone slabs Fibrous cement[b] Bitumen	Nailed Nailed Nailed, clamped	Discoloration, weathering, low breaking strength RR[a] Shrinkage (when dry), swelling (when damp): drill holes must be larger, efflorescence (smears) RR[a],draws in moisture
Specials	Vents	Concrete, clay	Dry-laid	Can be offset
	Verge/eaves vents	Concrete, clay	Screwed	Impede installation of collectors
	Ridge tiles, bricks	Concrete, clay	Clamped (new build) Cemented (old build)	Impede installation of collectors
	Tiles with air pipe	Plastic, clay	Dry-laid	Impede installation of collectors
	Tiles with perforated base	Plastic, metal, clay	Screwed	Impede installation of collectors
	Special tiles with tread holders	Plastic, metal	Screwed	Impede installation of collectors
Flat sheet metal	Box section profile roof cladding Folded covers Ridge covers, chimney frame	Galvanized steel Zinc, titanium-zinc, copper, aluminium Lead, zinc	Screwed, riveted Fixed with staples, folded Soldered, folded	Zinc-plated box section plate, partially with added plastic coating, screws always in upper chord non-clamped folds, corrosion problems, good back ventilation, high thermal expansion, brittleness
Profiled metal	Plastic-coated corrugation	Fibrous cement[b]	Screwed onto purlins	Not folded, water penetration through wind pressure
	Light corrugated sheets	Polyester	Screwed onto purlins	Not folded, water penetration through wind pressure
	Bitumen corrugated sheets	Bitumen	Nailed onto purlins	Not folded, water penetration through wind pressure
Liquid plastic	Roof impregnation, roof terrace sealing	Polyurethane, acrylate, resin	Cast, painted	Weather-dependent processing, carefully prepared background
Bitumen sheeting	Plastomer welded sheeting	Bitumen	Bonded, welded	Max. life 20 years, may be less than solar system, bitumen corrosion in connection with zinc plates
Plastic sheeting		EPDM	Connected with hot air	Contains plasticizer, brittleness, consider bitumen-compatibility
Organic roof	Thatched roof	Reeds, straw	Placed with wire on lathing	No known collector installation to date
	Grass roof	Film, substrate, special plants	Placed on roof	Only possible with stands

[a]RR = roofer is responsible
[b]Asbestos problem, special waste for these elements

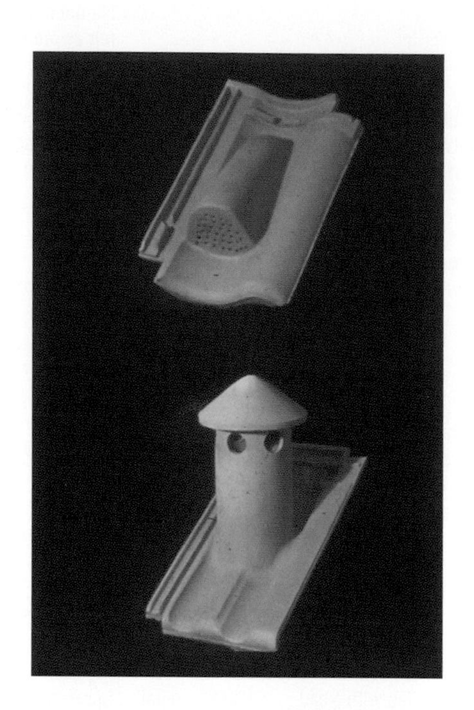

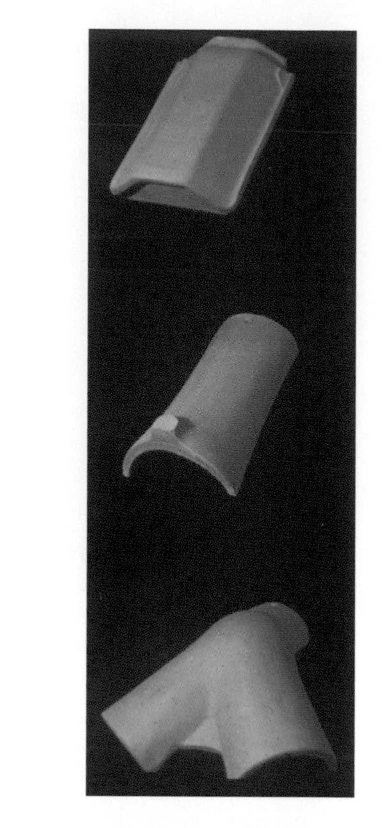

Figure 4.9.
Air tiles, vapour-escape pipes and various ridge tiles[27]

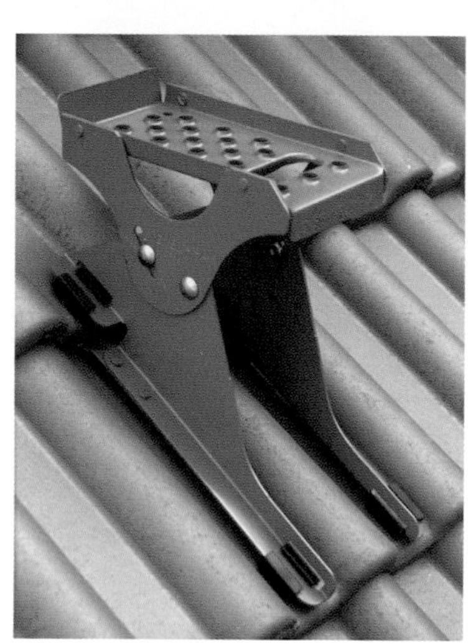

Figure 4.10.
Left: roof hooks. Right: step[30]

Figure 4.11.
Left: roof hooks. Right: walk-on grating[30]

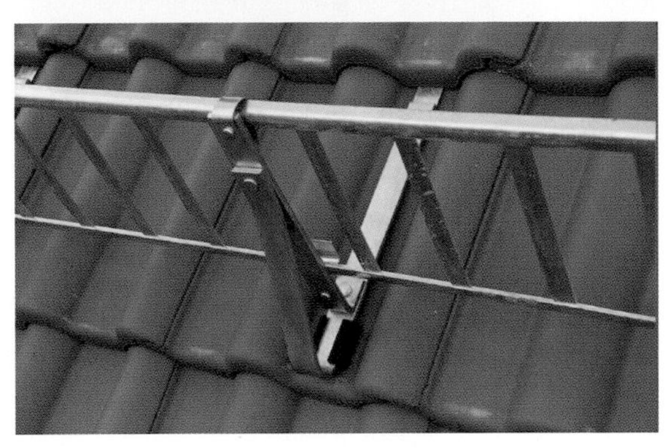

Figure 4.12.
Top: round wood holder.
Below: snowguard[30]

4.1.4 Pitched roofs

A pitched roof is often constructed as follows:

- rafters/thermal insulation
- roofing felt on formwork, otherwise sheeting beneath
- counter-lathing
- roof lathing
- tiles.

Thermal insulation can be installed in one of three ways:

- in the ceiling of the top storey
- between the rafters above head height
- over the rafters. (This is not shown in the figures: there are problems with fixing the collector, as in this case the standard parts cannot be used.)

4.1.5 Roof installations and mountings

During the installation of collectors the available roof surface can be restricted by the following features (Figure 4.13):

- room in roof
- dormer roof
- walk-in skylight
- chimneys, aerials etc.

4.1.6 Flat roofs

On can differentiate between a *cold roof* (thermal insulation under the boarding, ventilated in the intermediate space: see Figure 4.14), and a *warm roof* (thermal insulation above the boarding: see Figure 4.15). The problems of mounting collectors on warm roofs are described in section 4.3.3.2.

4.1.7 Materials

The materials used in the roof location for solar apparatus are subject to severe loads, which they must withstand over a long period, particularly if the system is expected to

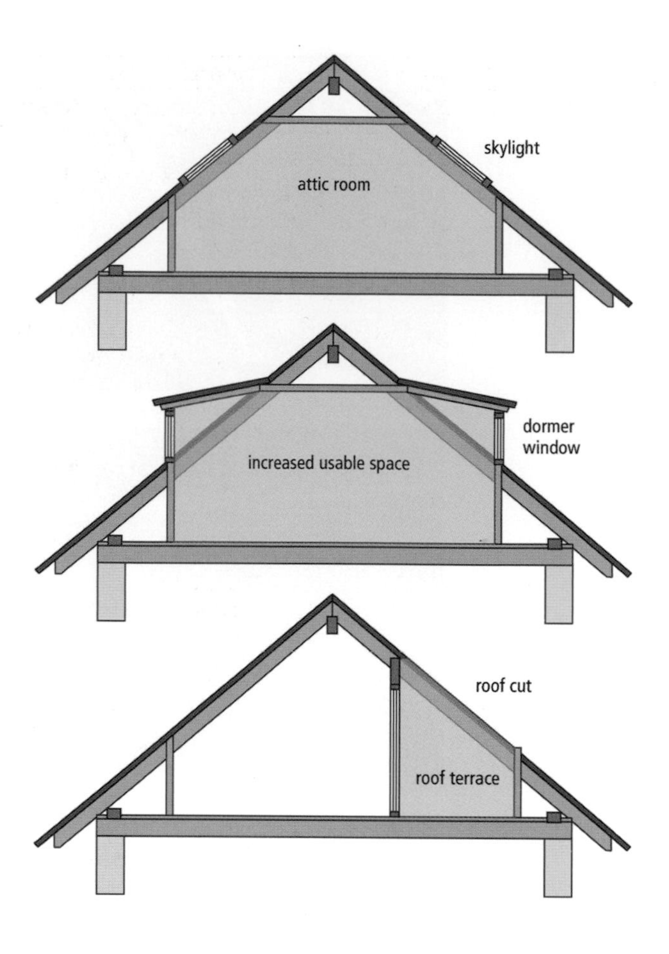

Figure 4.13.
Limitations to tilted roof mounting of collectors[26]

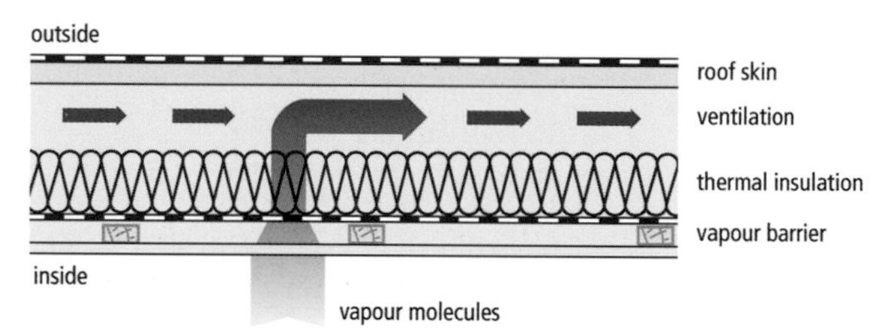

Figure 4.14.
Cold roof[26]

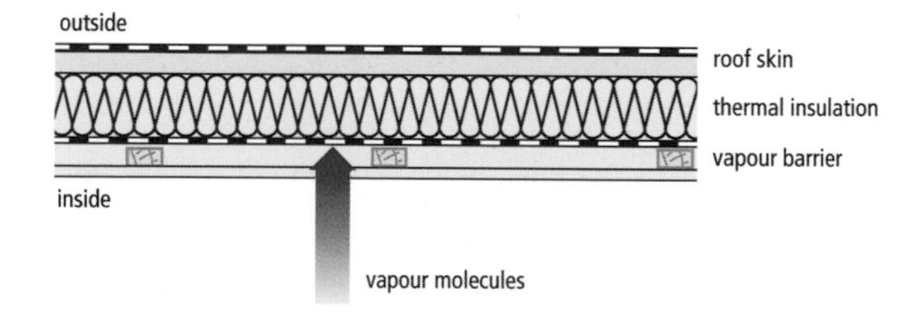

Figure 4.15.
Warm roof[26]

function and provide the predicted yields without faults for more than 20 years. The loads are:

- temperature variations between –15°C and 80°C (and within the absorber up to approximately 300°C)
- UV radiation
- rain, snow, hail
- wind forces
- bird and squirrel damage (in the case of external thermal insulation).

Table 4.2 lists the materials that are normally used:

Table 4.2.
Materials used in roof mounting of solar apparatus

Material	Version	Application
Aluminium	Plain, anodized, powder coated, selectively coated	Collector frames, covering frames, bearers, rails, absorbers
Glass	Float glass, solar glass (prismatic, clear), anti-reflection glass, borosilicate glass	Transparent collector cover, glass cylinder and absorber pipes
Stainless steel	V2A (St 1.4301) chrome, nickel	Roof hooks, rails, screws, nuts, plates, collector frames, absorbers
Stainless steel	V4A (St 1.4571) chrome, nickel, molybdenum, titanium	Screws, nuts, plates
Steel	Galvanized	Roof hooks, rails, screws, nuts, plates, bearers, supports
Copper	Plain, selectively coated	Pipes, absorber strips, full-surface absorbers
Titanium-zinc plate	Plain	Aprons, flashing, soakers, rear wall of collectors
Lead	Rolled, zinc plated	Lead aprons, soakers
Wood	Glued wood Solid wood	Collector frames (in-roof mounting) Additional battens (in-roof mounting), boards for under-packing
Plastics	EPDM, PE, PP, PU rigid/flexible foam, mineral wood, silicon	Thermal insulation of pipes, collector, collector frame, sealing materials, adhesive.

THERMAL EXPANSION

Materials expand during heating and contract during cooling. (The exception to this is water, which has its greatest density at 4°C; with further cooling it expands again.) It is therefore necessary to take suitable precautions in order to absorb these length changes, otherwise damage can occur. The extent of the length changes depends upon:

- **the material**
- **the temperature difference.**

$\Delta l = l_0 \times \alpha \times \Delta\theta$

where Δl is the change in length (mm), l_0 is the original length (m), α is the expansion coefficient (mm/mK), and $\Delta\theta$ is the temperature difference (K).

EXPANSION COEFFICIENTS (mm/mK)

- **polyethylene: 0.2**
- **lead: 0.029**
- **aluminium; 0.024**
- **zinc: 0.02**
- **copper: 0.017**
- **stainless steel: 0.016**
- **steel: 0.012**
- **quartz glass: 0.001.**

EXAMPLE

Four collectors 2.1 m high and 1.2 m wide are installed vertically beside one another. The covering plate projects by 0.25 m on each side.

$\theta_{summer} = 70°C$, $\theta_{winter} = -14°C$

$\Delta l_{zinc} = 5.3 \text{ m} \times 0.02 \text{ mm/mK} \times 84 \text{ K} = 8.9 \text{ mm}$

$\Delta l_{al} = 5.3 \text{ m} \times 0.024 \text{ mm/mK} \times 84 \text{ K} = 10.7 \text{ mm}$

Hence the plate expands in total by about 1 cm. It must not be impeded (for example by solders), otherwise stresses and cracks could occur, and this would result in leaks. Therefore the plates are attached to the roof lathing with retainers or, in the case of zinc, with expansion pieces. For pipes, compensators or expansion bends are used.

CORROSION

We differentiate between corrosion, corrosion signs and corrosion damage.

- *Corrosion (reaction).* This is the reaction of a metallic material with its surroundings, which leads to a measurable change in the material, and can lead to impairment of a metallic component or of a complete system.[31]
- *Corrosion signs (result).* These are the measurable changes in a metallic material through corrosion.[31]
- *Corrosion damage (possible consequence).* This is impairment of the function of a metallic component or a complete system through corrosion. Corrosion can lead to damage, but does not have to.[31]

EXAMPLE

Railway tracks can rust unimpeded in the free atmosphere. However, this does not impair their function; corrosion protection is not required. In contrast, steel radiators in an open heating system will rust because oxygen is continuously drawn in. If a hole occurs in a heater because of corrosion and water runs out, the function of this component – or even of the whole system – will be affected.

CORROSION PROTECTION

If two metals with different electrochemical potentials are in metallic contact, then, in the presence of water, the more base of the two materials will decompose. Therefore different metals should not come into contact with one another in a liquid circuit if possible. If this cannot be avoided in an individual situation, electrolytic separation by means of an intermediate electrical non-conducting material is recommended. Corrosion should particularly be expected if the more base of the two metals is downstream of the more precious metal. Small particles of the more precious metal can then be precipitated on the more base material and can lead to localised corrosion (or pitting). The flow rule must therefore be: copper – in the flow direction – after steel.

EXAMPLES

Roof area

In the flow direction of the rainwater, Cu ions from the copper chimney flashing dissolve a lower-lying aluminium covering frame, which leads to pitting. In contrast, an aluminium covering frame does not have any corrosive effect on a copper gutter beneath it.

Solar absorbers made of aluminium (for example Rollbond) can become unusable in a short time if the inhibitor is not sufficiently effective, or if they are connected to a copper pipeline.

Solar circuit

Steel pipes or fittings made of cast iron should never be mounted downstream of a copper pipeline. However, metallic connections of copper with copper–tin–zinc alloys (red bronze or brass) are unproblematic, because these alloys have a similar electrochemical potential to that of copper.[32]

The corrosiveness of glycol is suppressed by suitable inhibitors. These should be checked from time to time (see Maintenance, section 4.4.5).

Storage tank

Plain ferrous materials are subject to corrosion in water, which always contains oxygen, with the formation of non-protecting rust layers. Hence corrosion protection is necessary: for example, for the domestic hot water tank an enamelling process in connection with external current or magnesium anodes or a plastic coating.

CAPILLARY EFFECT

According to the law of communicating pipes, liquids in pipes that are connected to one another will always be at the same level, independently of the shape of the pipes. The exception to this is that, in a narrow pipe (a *capillary*), the liquid is either at a higher level (for a wetting liquid, such as soldering tin) or a lower level (for a non-wetting liquid, such as mercury).

EXAMPLE

In soldering technology, use is made of this effect in the capillary soldering fittings. The liquid tin is drawn into the narrow gap even against gravity (solder gap width from 0.02 mm to a maximum of 0.3 mm, according to the pipe diameter).

However, there can be problems due to the capillary effect on the covering frames of in-roof-mounted collectors. Here, in the presence of unsoldered side panels, which overlap in the flow direction, water can penetrate.

Table 4.3 gives an overview of the techniques.

Table 4.3
Methods of bonding roof plates

Measure	Unsuitable	Suitable
Bonding	Silicon	Butyl strip can be used with no restriction
Increasing the gap by beading the plate edges	For roof pitches < 30°	For roof pitches ≥ 30° overlapping of at least 100 mm is necessary
Soldering	For aluminium	For zinc (take account of thermal expansion)

4.2 Installation methods and safety

During the installation process, safety regulations must be followed to provide a safe situation for the workers, especially during roof work. There are large differences between individual countries regarding the safety regulations to be followed during the installation of solar water heaters. It is not possible to give up-to-date lists of all countries, and therefore regulations are given for the UK as well as general basic guidelines for central European countries to provide examples. For other countries, sources of information on safety regulations can be found in section 4.5.

4.2.1 UK safety regulations (see also Appendix B)

Advice on the requirements is found in the following Health and Safety Executive publications:

- HSG33: *Health and Safety in Roof Work*
- HSG150: *Health and Safety in Construction.*

When considering roof work of any kind, a risk assessment is mandatory. This entails:

- looking for hazards
- assessing who is at risk, and in what ways
- evaluating and taking action to eliminate or reduce the risk
- recording or communicating the findings
- reviewing the findings.

The Management of Health and Safety at Work Regulations 1999 require employers to carry out risk assessments, make arrangements to implement necessary measures, appoint competent people, and arrange for appropriate information and training.

All roof work is hazardous. Without a risk assessment roof work is also dangerous. In UK, there is a fatal fall from height every three days.

4.2.2 UK personal access and working at heights

Solar collector roof installations typically take less than a day, and access considerations should allow for this. Nevertheless, a temporary platform at or near eaves level is an essential safety item when working at height. On a flat roof, the features of the platform are created on the roof itself, although fall prevention still needs to be added.

Working platform features are as follows:

- minimum width to cover the working area, plus a safety margin while on the roof
- sufficient strength to support loads, including personnel, tools and materials
- sufficient integrity to prevent tools and debris from falling below
- main guardrail a minimum of 910 mm above the fall edge
- toe-board a minimum of 150 mm high
- no unprotected gap between these exceeding 470 mm (that is, use an intermediate rail).

The platform width should not normally be less than that of the whole width of the roof pitch to which the solar collector is to be fixed. A small collector fixed to a very wide roof may suggest an exception to this, but consideration should be given to potential slippage across the roof from the momentum of using tools, for example. The working platform is typically supported by one the following methods:

GENERAL STEEL SCAFFOLDING

- also known as system scaffolding or tube and fitting
- requires trained, technically knowledgeable or experienced installer to provide mandatory inspection report
- installation by trained personnel
- usually already in place during new build
- can support hoist or gin wheel for lifting collector
- can be customized to building shape
- can support external ladder
- should meet NASC (National Access and Scaffolding Confederation) guidelines
- heavy structure with long components.

MOBILE ALUMINIUM ACCESS TOWER (MAT)

- requires trained, technically knowledgeable or experienced installer and user
- inspection report not mandatory where sited for less than seven days
- modifications possible only with manufacturer's permission
- not normally capable of supporting hoist or gin wheel
- requires internal ladder access
- should meet PASMA (Prefabricated Aluminium Scaffolding Manufacturers Association) guidelines
- light structure with compact components.

4.2.2.1 LADDERS

The existing solar trade frequently uses ladders, sometimes without other access equipment; however, the only recommended use of a ladder is for personal access and short-term work (that is, within minutes). In this context the ladder should extend at least 1 m above the landing place and should be tied in to the building to prevent sideslip at top and bottom. The forthcoming EU Directive 89/655/EEC is likely to require ladder work at height only where other safer work equipment is not justified in view of the short duration of use and the low level of risk.

4.2.2.2 WORKING ON THE PITCH

A roof ladder should be used to supplement eaves edge protection on pitched roofs and to provide additional support for work. It also reduces damage to the roof coverings. The ladder should be secured against accidental movement, and the anchorage should not rely on the ridge capping or typical gutters. The use of a ridge iron on the opposing pitch and tying onto the eaves platform support is typical. On long, steep pitches where work is near the ridge, an extra platform or safety harness may be required to reduce injury upon hitting the eaves platform in the event of a fall. All weight-loading calculations should consider a minimum of two workers plus tools and the weight of the solar collector. Furthermore, the forces invoked during drilling operations should be considered, especially where the drill and bit snatch. Working on the fragile roofs sometimes found on commercial buildings is not covered here, but suffice to say extreme caution is required.

4.2.2.3 SAFETY FENCE

A safety fence is mounted on the roof, and the side protection is firmly anchored to the rafters with ropes and special retainers. The protection fence must project beyond the working area by 2 m at each side.

4.2.2.4 PERSONAL SAFETY EQUIPMENT

This is used in addition to work platforms. Such equipment includes:

- safety harness with safety ropes and falling damper
- as above but with height safety device (the safety rope is kept taut the whole time).

4.2.3 Working equipment

The equipment required for working on the roof includes:

- roof hooks, roof-supported ladders

- roofing stool
- ladders.

These are described below.

4.2.3.1 ROOF HOOKS, ROOF-SUPPORTED LADDERS

These are light wooden ladders with cambered rungs. In Germany they are hung from the roof hooks (for safety reasons in the second rung) and placed on the roof surface. In the UK they are hooked over the ridge. They permit movement around pitched roofs with inclinations of up to a maximum of 75°. They may only carry loads of up to 1.5 kN (approximately 150 kg).

4.2.3.2 ROOFING STOOL

In some countries, roofing stools are used for safe working. See Figure 4.16.

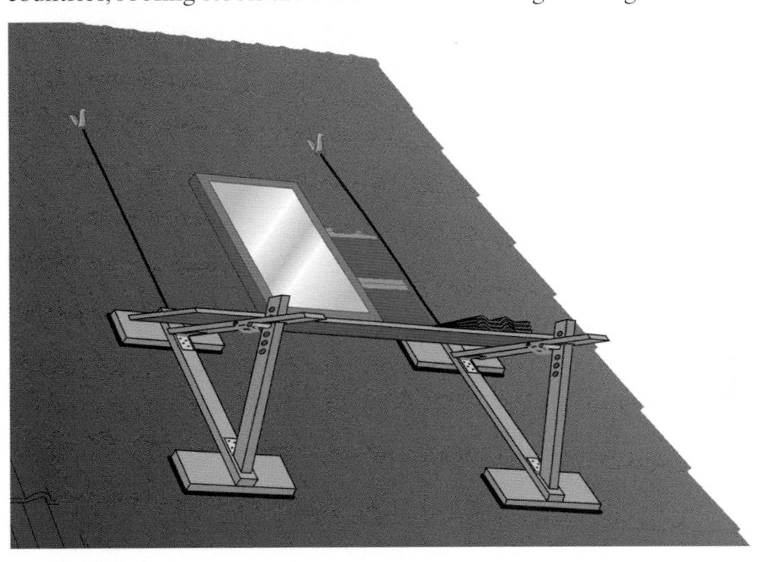

Figure 4.16. Roofing stool[33]

4.2.3.3 LADDERS (ALUMINIUM, WOOD)

Standard ladders are leaned against a building. The correct erecting angle is between 65° and 75°. The ladder must extend 1 m beyond the contact point. It must be secured against slipping, falling and sinking into the ground: that is, use wide feet, rubber mountings on feet, top suspension/fixings, or support by a safety person.

4.2.4 Transport techniques

For transporting the collectors to the roof or over the roof surface to the installation location, there are various options available according to the site conditions. Some common options are as follows.

4.2.4.1 TRANSPORT TO THE ROOF

- *Inclined surface*. This is the most frequently used method of transporting such items to the roof. To create the corresponding inclined surface, two ladders are placed beside one another at the appropriate distance apart. (Protect rainwater gutters from loads.) The collectors are then pulled up with ropes or an electric hoist attached to the scaffold, or carried up with carrying handles.
- *Inclined hoist*. Inclined hoists with electric motors can be used for transporting the collector to the roof either within the scope of roofing work (by agreement with the roofer) or they can be hired.
- *Crane*. For the transport of large-surface-area collector modules, or to reach roof areas that are difficult to access, a crane may well be essential. As crane hire is relatively expensive, accurate time planning and good preparation are necessary.
- *Lift*. Vertical lifts can be used for working on façades.

4.2.4.2 TRANSPORT ON THE ROOF (PITCHED ROOFS)

Relatively safe working on the roof is made possible with:

- roof ladder
- revealing the roof battens.

4.2.5 Installation techniques

4.2.5.1 PIPE INSTALLATION

Pipe installation techniques for solar energy systems include cutting, connecting, fixing and sealing copper and steel pipes as well as the processing of stainless steel corrugated pipe and fast-assembly pipes (twin-tube). Copper pipe is mostly used.

CUTTING

For cutting copper pipe, a pipe cutter is used. It is important to remove the internal burr after cutting.

CONNECTING

- *Soldering (capillary soldering fittings)*. Copper fittings are mainly used for soldered connections. The junctions with mountings and threaded connections are made of brass or red bronze. We differentiate between two soldering processes, soft and hard soldering (brazing). Correspondingly, soft and hard solders are used. See Table 4.4.

Table 4.4. Different types of solder

Type	Solder	Flux	Material connection	Notes
Soft solder	L-SnCu3 L-SnAg5	F-SW22 F-SW25	Cu-Cu Cu-Cu	Up to 110°C permanent temperature load
Hard solder (Braze)	L-Ag2P L-CuP6	No flux needed F-SH1	Cu-Cu Cu-red bronze	Above 110°C

- *Press-fit (crimped)*. Cold, non-releasing connections can be made with the help of compression fittings with O-rings. The advantages of this connecting technique are: no risk of fire; no environmentally damaging soldering materials; and the time saving over soldering.
- *Compression*. Cold, releasing connections can be made with the help of compression fittings with olives. The advantages of this connecting technique are: no risk of fire; no environmentally damaging soldering materials; and the time saving over soldering.
- *Screwed (clamping ring) connections*. Screwed connections can be released easily, they generate no heat during installation, and they significantly reduce the installation time, as for the compression joint. They are used mainly for the collector connections.

FIXING

In fixing the pipelines, the first thing to do is to account for temperature-related length changes (see Thermal Expansion in section 4.1.7). This is done by allowing expansion between the fixed points (clamps with rubber inserts) by installing bends or compensators. For roof openings the line must be installed to be movable (sliding guide).

SEALING

In order to seal some threaded pipes, the outside thread should be wrapped with hemp before screwing together. Teflon tape is not appropriate as, unlike hemp, it does not possess swelling properties. Therefore the high creeping capability of the solar liquid can lead to leaks in the solar circuit.

4.2.5.2 SHEET METAL WORK

The collector flashing system (zinc/aluminium sheeting) is normally prefabricated by the manufacturers. In the best case they simply need to be slotted in place.

The following sheet metal work may be necessary:

- cutting – with sheet metal shears
- connecting – folding, soldering (zinc sheet only), gluing (butyl tape), screwing (plumbers' screws)
- fixing – with adhesive, suspending
- shaping – edging (edging bench).

4.3 Installation

4.3.1 Delivery of material

Traditionally the material for the solar energy system was (and in some countries still is) delivered by the manufacturer, and the remaining installation material has to be procured from wholesalers. Increasingly, wholesalers are also selling solar systems, thereby simplifying the procurement.

When the materials are delivered the following must be done:

CHECK FOR TRANSPORT DAMAGE

Here it is particularly important to check for an intact glazing. For collectors without a metal back panel, damage can easily occur to the thermal insulation plates. There is also a risk of the covering sheets bending.

CHECK THE COMPLETENESS AND CORRECTNESS OF THE DELIVERY

To simplify and support this task, there are:

- ready-to-use packs – that is, complete kits for the installer. It is then only the installation materials (copper pipe, thermal insulation, fixing materials etc.) that still have to be replenished
- detailed advice from the supplier
- installation training courses by the manufacturer or other institutions.

4.3.2 Setting up the site, preparatory work

The final establishment of the collector position, the pipe routes through the house and the location of the store must then be agreed with the customer. The transport route for the collectors is established, sensitive components are secured (for example a glass installation beneath the transport route, or using planking to protect the mounting position on the roof from falling objects), and paths are blocked off as necessary. The materials and tools required for installation are transported to the garage or cellar (if available) and stored there.

4.3.3 Collector installation

In principle, collectors can be:

- integrated into a pitched roof
- mounted on a sloping roof
- set up on stands on a flat roof or a free surface
- mounted in a façade.

Each of these solutions has its advantages and disadvantages. The selection of the one to be used depends on the local conditions, the type of collector, and the customer's requirements. Whereas for pitched roofs and façades the inclination and alignment are more or less preset, the flat roof or free installation (installation on the ground is very rare) allows the possibility of an exact south orientation and a favourable angle of inclination. The limitation to this is that flat-mounted evacuated tube collectors can be varied only within a certain angular range (0–25°), otherwise the individual absorber strips will shade each other when the sun is low.

The roof work during installation should be carried out by specialists, taking account of the safety regulations. The manufacturer's instructions should be followed accurately.

The collector manufacturer generally supplies the required fixing kits. Because of the large number of different roof coverings and collector constructions, the collector installation steps will differ from one another.

4.3.3.1 INSTALLATION ON A PITCHED ROOF

Solar collectors can be mounted on the roof, or between the roofing tiles.

ON-ROOF INSTALLATION

In on-roof installation the collectors are mounted about 5–10 cm above the roof layer. The mounting points are often formed by roof hooks or rafter anchors, which are screwed onto the rafters, or onto a corrugated tile, a standing seam or the like. Roof hooks are formed so that they can be led between two rows of roof tiles or shingles.

In some countries there are roof tiles developed especially for the on-roof mounting of collectors and PV modules. These consist of plastic in various shapes and colours with the bearing element integrated, onto which the mounting rails can be directly screwed. The collectors are fixed onto these. Venting tiles are used for pipe culverts. The supporting structure of the roof (the battens) must be able to take the additional load of the weight of the collectors and the fixing system (approximately 25 kg/m^2). Both glazed flat-plate collectors and evacuated tube collectors can be installed as on-roof systems.

The type of roof cover (shape and colour) should be noted on a checklist during the initial site visit (see Chapter 3, section 3.5.1). The correct roof hooks (and venting tiles) can then be ordered. They are therefore available at the start of the installation, and there will thus be no building delays or interruptions.

The individual work steps for on-roof installation with interlocking tiles are as follows:

1. Set up the transport path for the collectors from the ground to the installation location: create an inclined surface with ladders, push up individual roof tiles to ensure safe movement on the roof or suspend roof ladders from roof hooks, and follow accident prevention regulations (see section 4.2).
2. Mark out the collector field on the roof.

3a. Install rafter anchors.
Uncover the fixing points (rafters): that is, remove the roof covering at these points.
Screw rafter anchors/roof hooks onto the rafters, and if necessary support with pieces of wood so that the roof tiles over which the roof hooks are led are not under load.
Close any holes in the roof covering.

3b. Alternatively, remove the roof covering at the respective points, install special roof tiles and screw firmly.

4. Screw the mounting rails onto the roof hooks or the special roof tiles.
5. Transport the collectors to the roof, place on the mounting rails and fix with screws.
6. Connect together with (prefabricated) thermally insulated pipes. In the case of an extended attic, mount a vent at the highest point in the external area.
7. Position venting tiles, connect the feed and return lines to the collector, and lead them through the roof skin into the house. For this purpose, penetrate the sheeting and the thermal insulation and close them up again properly, for example by gluing or with a sheeting element.
8. Install sensors.
9. Connect the sensor cable to the lightning protection socket (excess voltage protection).
10. Connect to lightning conductor if required.

This description does not replace the installation instructions from the manufacturer.

The advantages of on-roof installation are as follows:

- fast and simple installation, therefore cheaper
- roof skin remains closed
- greater flexibility (it is possible to mount closer to ridge tiles, sheet metal surrounds etc.).

The disadvantages of on-roof installation are as follows:

- additional roof load (approximately 20–25 kg/m^2 of collector surface for flat collectors and 15–20 kg/m^2 for evacuated tube collectors)
- visually not so attractive as in-roof installation
- piping partly installed above roof (weather influence, bird damage).

Practical points to consider:

- When the collector is transported to the roof there is a risk of damage to the roof gutters (metal gutters bending, plastic gutters breaking). The ladders should be set up and secured so that there are no loads on the gutters.
- Older, brittle tiles can be broken when walked on.
- On the roof, soldering bottles of only 0.5 kg should be used; these should be hooked onto the belt.
- Collectors must be suitable for on-roof installation.
- The external insulation must be UV resistant, weather resistant, high-temperature resistant and protected from bird damage. Use heat-resistant flexible foam or mineral wool in the sheet metal jacket.
- Point loads on individual tiles through the roof hooks should be avoided.

Figure 4.17 shows an example of an on-roof installation.

Figure 4.17. Example of on-roof installation. Source: Schüco, Bielefeld

IN-ROOF INSTALLATION

For in-roof installation, the tiles are removed at the corresponding position and the collectors are mounted directly on the roof battens. The seals at the transitions to the roof skin are achieved by means of an overlapping construction. The collectors are thereby mainly integrated into the roof cover by means of special covering frame systems of aluminium or zinc and lead (similar to skylights). Integration into a pitched roof is usually the most elegant solution from the architectural point of view. In principle only flat-plate collectors can be integrated into the roof. An example of an in-roof installation is shown in Figure 4.18.

The individual working steps for in-roof installation with interlocking tiles are as follows:

1. Set up transport route for the collectors from the ground to the installation location: arrange inclined surface with the aid of ladders; push individual tiles upwards to create safe walkways on the roof, or suspend roof ladders from the ridge. Observe the accident prevention regulations (see section 4.2).
2. Mark out the collector field on the roof.

Figure 4.18. Example of in-roof installation. Source: Schüco, Bielefeld

3. Uncover the field (somewhat larger than the collector surface area) and remove the greater part of the tiles.
4. Mount the fixing brackets and stop rails on the rafters.
5. Transport the collectors to the roof, place them on the roof lathing, slide them into the fixings, align laterally and screw into place.
6. If there are several collectors, insert plug connectors with O-ring seals or mount heat-insulated pipe connectors.
7. Connect the feed and return pipes to the collector and lead them into the house through the roof skin. To do this, penetrate the membrane and the thermal insulation and seal properly again, for example using adhesive or by means of extra felt/membrane.
8. Mounting the sensors: insert glazed flat-plate sensors into sensor sleeves. For vacuum tube sensors open the main line, screw sensor to feed (under thermal insulation as close as possible to exit from glass tubes), close main line again.
9. Cover up lower edges, fix lead aprons (usually suspended from within the collector frame).
10. Mount side roofing plates and then the top ones; the side plates must project over the lead aprons.
11. If necessary, insert metal or sealing strips between the collectors.
12. Cover the roof tiles at the sides. If necessary use half roof tiles; if absolutely necessary cut roof tiles and then cover them from above, ensuring sufficient overlap (at least 8 cm).
13. Shape lead aprons around the roof tiles.
14. Store left-over tiles as reserves or dispose of them.

This description does not replace the installation instructions by the manufacturer

The advantages of in-roof installation are as follows:

- No additional roof loads are applied.
- It is visually more appealing (roof covering frames can be obtained in different colours from some manufacturers).
- Pipes are laid beneath the roof cover.
- Saving of roof tiles (new build), reserve tiles (old build).

The disadvantages of in-roof installation are as follows:

- More expensive materials and mounting work.
- The roof skin is 'broken through', creating possible weak points.
- There is a possible need to transport excess tiles away (costs).
- It is less flexible: because of the covering frames there must be greater distance from ridge tiles, window and chimney surrounds.

4.3.3.2 MOUNTING ON A FLAT ROOF

In principle, collectors on flat roofs should be set up at an appropriate angle (see Chapter 1). For this purpose, flat roof stands are available made of galvanized steel or aluminium with the corresponding setting angles. (The exception to this is direct flow-through evacuated tube collectors: see below.) Because of the surfaces that are exposed to the wind, these collectors must be secured from lifting up and falling, or from sliding.

There are three options:

- *Counterweights* (concrete thresholds, gravel troughs, box section sheet with gravel filling). Approximately 100–250 kg per m^2 of collector surface for flat collectors and about 70–180 kg/m^2 for heat-pipe collectors (maximum 8 m mounting height above ground level according to building height); beyond this larger loads are necessary.
- *Securing with thin guy ropes*. The precondition for this is the availability of suitable fixing points.
- *Anchoring to the flat roof*. Here a suitable number of supports are screwed to the roof and sealed. Bearers are fitted to these supports, on which the flat roof stands that carry the collectors are mounted.

(See Figures 4.20 and 4.21 for examples.) In every case the bearing strength of the roof must be checked first.

The working steps for flat roof mounting in detail (for example flat collectors on box section sheet with gravel filling) are as follows:

1. Set up transport route for collectors from the ground to the installation location: create sloping surface with ladders, follow accident prevention regulations (see section 4.2).
2. Mark out collector field on roof.
3. Lay out the building protection mats and the box section sheet.
4. Set up the flat roof stands, connect to box section sheets.
5. Transport collectors to roof.
6. Install collectors on the flat roof stands.
7. Make piping connections between collectors, mount vent at highest point.
8. Connect the feed and return pipes
9. Mount the collector sensor.
10. Install the gravel filling on the trapezoidal sheet (10–15 cm) in order to increase the weight.

This description does not replace the installation instructions by the manufacturer.

Direct flow-through evacuated tube collectors can be mounted flat; see Figure 4.19.

Figure 4.19. Direct flow-through evacuated tube collector on a flat roof. Source: Viessmann, Allendorf

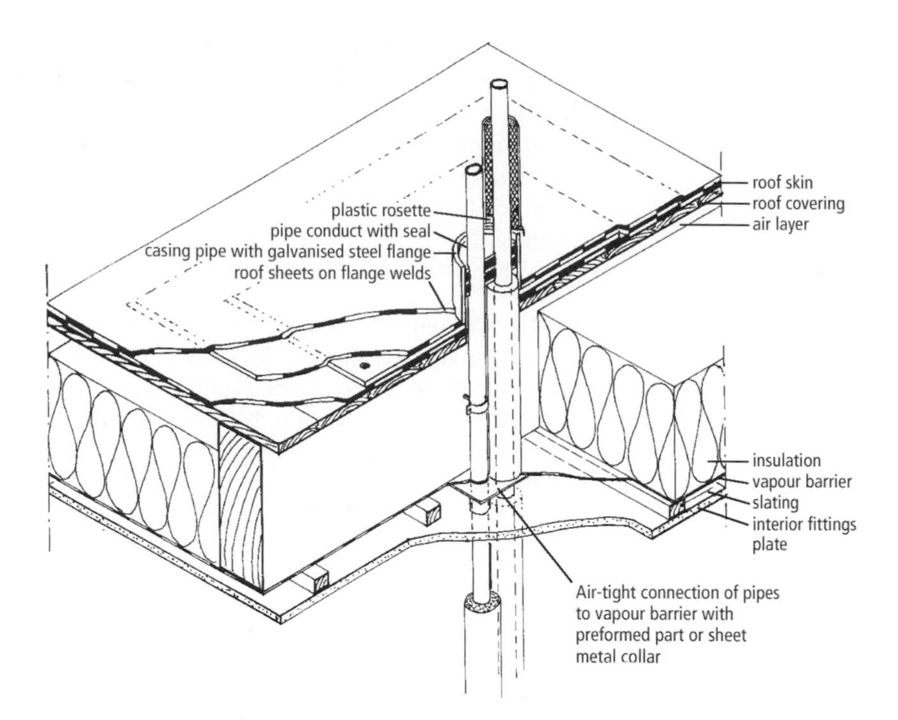

Figure 4.20.
Roof opening.
Source: Sol.id.ar, Berlin

Figure 4.21.
Government Press Office, Berlin.
Source: Viessmann, Allendorf

The individual installation steps are:

1. Establish collector positions.
2. Lay out the building protection mats and cast concrete blocks (paving slabs).
3. Assemble collector and distributor boxes.
4. Assemble bottom rails.
5. Install evacuated tubes.

For the horizontal installation of vacuum tube collectors there are two options for aligning the collectors:

- tubes longitudinal facing the equator, absorber horizontal
- tubes transverse facing the equator, absorbers adjusted by about 20–30°.

Anchoring or fixing with guy ropes is not necessary; only counterweights are required on building protection mats (approx 40 kg/m^2 depending on height).

The advantages are:

- fast and simple installation (low installation costs)
- no stand costs
- no penetration of roof skin at the fixing points

- low roof loading because of lighter counterweights, as there are lower wind loads
- in the case of listed buildings the collector field can be installed so that it cannot be seen.

The disadvantages are:

- higher costs for evacuated tube collectors
- lower yield when the sun is at low level.

The collectors are connected together hydraulically according to the circuit diagram. The fixing of the pipes and the collectors can take place on cast concrete blocks. The header pipes must be led into the building. There are two options to do so: the first is pipe installation in an unused chimney flue. The advantages of this are:

- no sealing problems on the roof
- no roof breakthroughs necessary
- pipe installation possible right into the cellar, and frequently directly into the heating room.

Preconditions are:

- free chimney flue
- the chimney must not be offset
- information has to be provided to the chimney sweep.

The second option is pipe installation through a roof opening. Practical points to consider in flat roof installation are as follows:

- Special care is required when working on flat roofs, as the roof skin can be easily damaged, and this can lead to leaks and consequential damage. For example, sharp-edged objects can easily break through (roofing felt pins, bent remnants of zinc plate, etc.). In the summer, bitumen becomes soft and deep indentations can arise from collector corners and edges.
- If the supporting structure is made on site it is necessary to ensure matching accuracy and good corrosion protection. Galvanized materials should not be processed afterwards – that is, drilled or cut to length – as the zinc plating that would have to be applied afterwards to the ungalvanized places is not as durable as hot-dip galvanizing.
- An advantage in flat roof installation is that the collectors can usually be set up with the optimum orientation and inclination.
- In areas that have a lot of snow, a gap must be left between the roof surface and the bottom of the collector, which is set according to the winter snow level.
- For flat roof installation the maximum load and the distance to the edge of the roof must be set with care.

INSTALLATION ON WARM ROOFS (THERMAL INSULATION ABOVE THE BOARDING)

For warm roofs the fitting of the heavy counterweights for flat collectors is not possible. For this reason a roof anchoring system is necessary.

- *Old buildings*. Cut out roof skin and insulation material, screw stand feet to the concrete roof/ceiling joists, replace insulation material, close roof skin, bond in stub pipe to provide a good seal.
- *New buildings*. In this case the stub pipe is fitted first; then the thermal insulation and the roof skin are fitted, and the stub pipe is bonded in to provide a good seal.

In both cases cold bridges are created, with the risk of condensation formation. It is therefore better to use stub pipes with thermal separation.[34]

In the case of direct flow-through evacuated tube collectors, lower counterweights are required. If these are distributed over a large surface area they can also be installed above the roof skin on warm roofs, for example sheet metal with gravel filling.

MUTUAL SHADING

If the collectors are set out in several rows behind one another, the distance between the rows must be great enough so that no or as little shade as possible occurs.

If the collectors are too close to each other, the amount of indirect radiation is reduced, and there may be direct shading. For direct shading, see Chapter 1. To estimate the reduction of irradiation caused by shielding of indirect radiation, the following formula, which is valid for most types of climate, may be used:

$$\Delta I = \beta \times \frac{f_{\text{dif}}}{180 - \alpha}$$

and

$$Y = \frac{1}{\tan(\beta)}$$

where ΔI is the reduction in irradiation (dimensionless), β is the shielding angle (see Figure 4.22) (degrees), f_{dif} is the fraction of diffuse radiation for a specific location (for instance, in central Europe this is about 0.6, in southern Europe it is about 0.2), and α is the tilt angle of the solar collectors (degrees).

For instance, assuming a project in central Europe (60% diffuse radiation), with a collector tilt angle, α, of 36°, in order to limit the reduction of irradiation, ΔI, by mutual shading to 10%, we would obtain a maximum shielding angle, β, of 24°.

This corresponds to a distance, Y, from the bottom of one collector row to the top of the next row that must be at least 1/tan (24°) = 2.2 times the collector height, h (see Figure 4.22). Thus, if the height, h, is 1 m, Y should be at least 2.2 m.

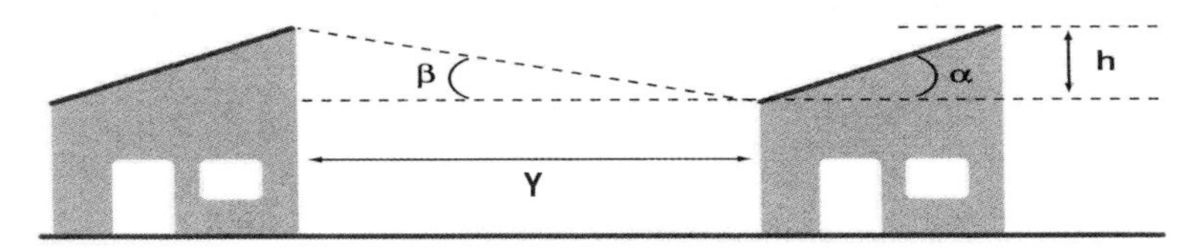

Figure 4.22. Recommended distance between collector rows for flat roof installation

4.3.3.3 INSTALLATION ON THE FAÇADE

In principle, flat collectors and evacuated tube collectors can also be mounted on façades, especially at higher latitudes, where the tilt angles are higher. See Figures 4.23 and 4.24. Façade installation, however, currently plays only a minor role. However, with the background of a possibly higher solar fraction in the winter and in particular as an architectural design element (for example coloured absorbers) it is becoming increasingly popular in some countries.

The glazed flat-plate collectors used should be mounted in the same way as on the flat roof. They are frequently screwed to the wall with the same stands that are used on the flat roof.

Evacuated tube collectors are attached to the wall with their collectors and their base rails, sometimes as transverse tubes with the absorber and sometimes with vertical tubes.

For façade installation the following points should be observed:

- the bearing strength of the wall
- shading
- pipe installation
- wall ducts
- appearance.

A collector that is installed on the façade receives a lower annual global solar irradiance than a roof installation. However, it has a more uniform yield profile per year, and is subject to lower thermal loads: that is, there are fewer stagnation periods. As can be seen in Figure 4.25, a façade system with a collector surface area of more than 12 m² achieves the same yield as a comparable roof system with lower material stresses. In addition, positive effects can be achieved in the architectural façade configuration.

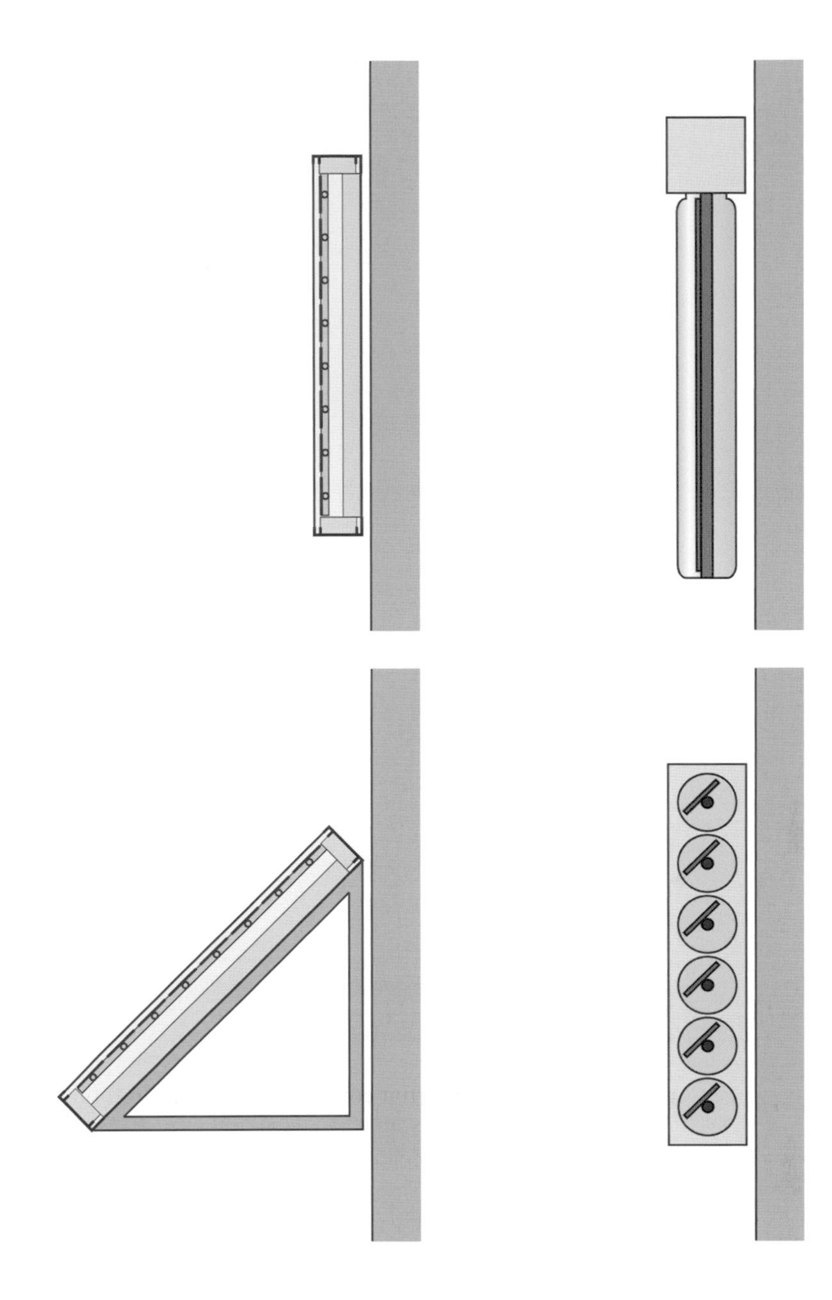

Figure 4.23.
Façade installation. Left: flat plate arranged vertically/inclined. Right: evacuated tube vertical (vertical pipes); horizontal (absorber set at 45°)

Figure 4.24.
Façade installation of a large surface collector in Bielefeld. Source: Wagner & Co, Cölbe

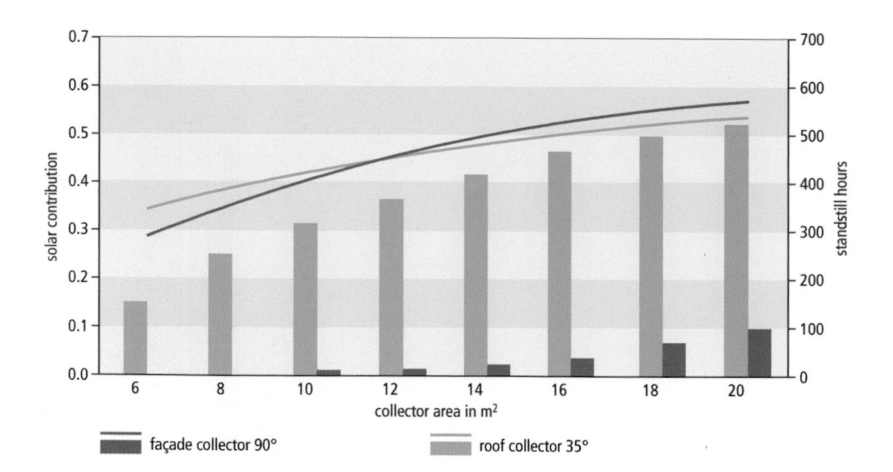

Figure 4.25.
Comparison of façade and roof-mounted systems for domestic water heating and heating support, location Würzburg[35]

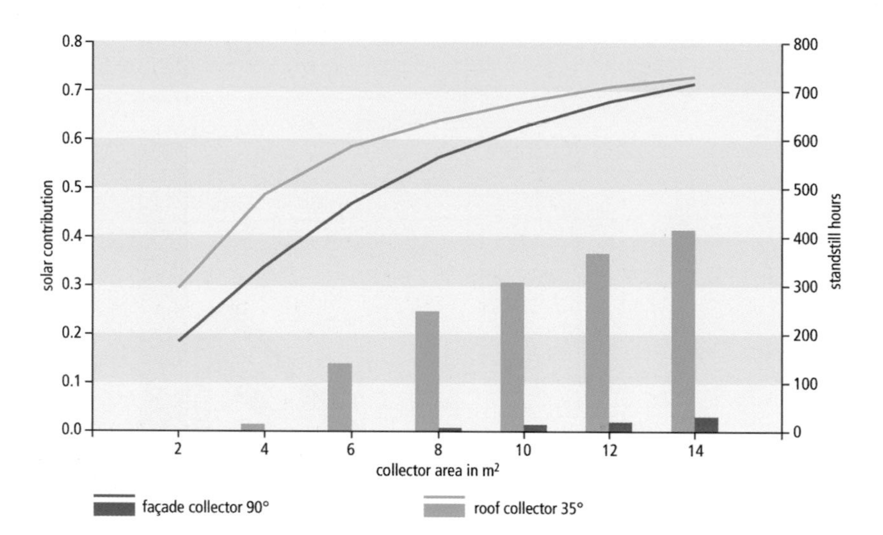

Figure 4.26.
Systems for domestic water heating, location Würzburg[35]

4.3.3.4 CONSTRUCTION OF THE COLLECTOR FIELD

If several collectors are connected together as a complete surface the question of the connecting arrangement arises. The objective is to achieve a uniform flow rate throughout the whole field. Non-uniform flow rates cause different temperatures in the collector field, with negative effects on the control behaviour and the collector yields. In principle we can differentiate between parallel and series connection. The two connecting arrangements can also be combined.

PARALLEL CONNECTION

In this case all the collectors are installed between two header pipes – one distributing and one collecting header pipe. To achieve uniform flow rates the header pipes should have a lower flow resistance (that is, a larger tube cross-section) than the collectors. All flow paths should have the same length, this is called Z-connection or Tichelmann connection.

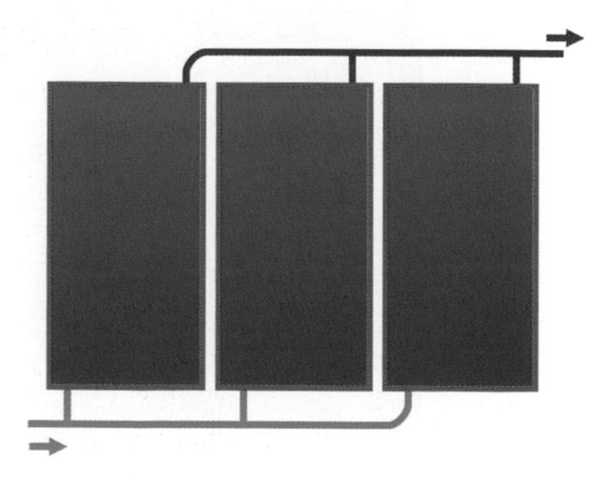

Figure 4.27.
Parallel connection

SERIES OR ROW CONNECTION

As the name indicates, with this type the collectors are connected successively. The flow rate that is necessary in order to deliver the heat arising increases in proportion to the number of collectors, but the flow resistance increases exponentially. Because of the significantly increased pumping power required, the number of rows in connected collectors is limited. This circuit arrangement ensures uniform flow through the collectors.

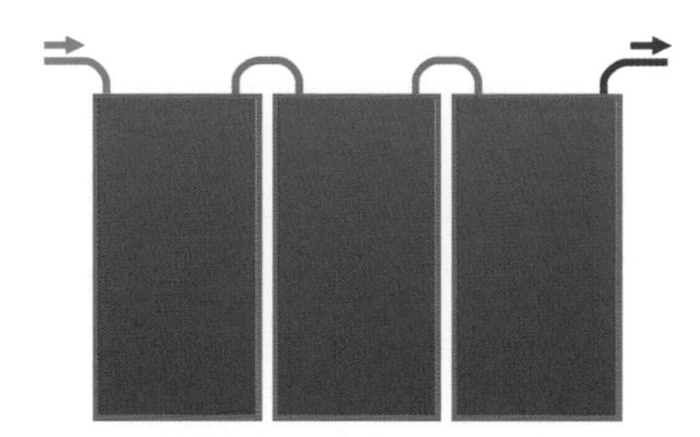

Figure 4.28. Series connection

COMBINATION OF SERIES AND PARALLEL CONNECTIONS

Such combinations make it possible to combine the advantages of both connection principles. By successively connecting collectors, the flow resistance in the individual rows is significantly increased over that of the header pipes. It is then possible to dispense with a Tichelmann connection and the related long pipelines. The objective of combination connections is a uniform flow through the collector fields with a low flow resistance (low pumping power).

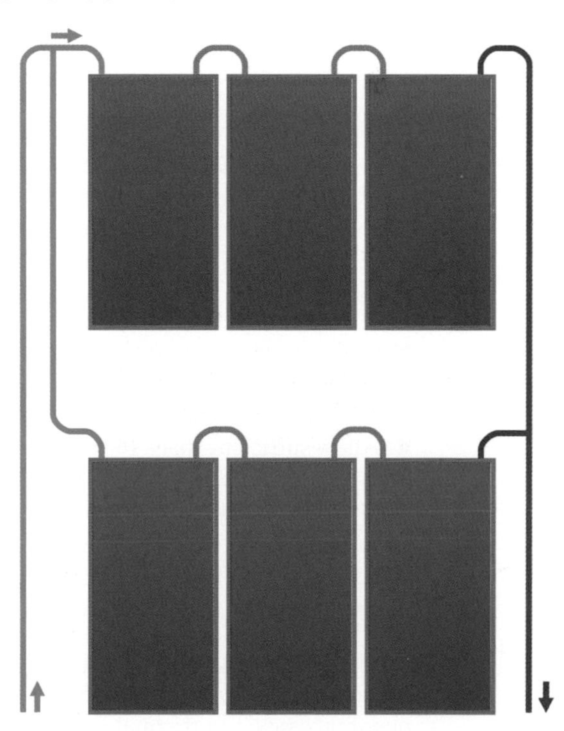

Figure 4.29. Combination series and parallel connection

4.3.4 Installation of the solar circuit

The well-proven materials and connecting techniques used for classical heating and sanitary fittings can be used for installing the solar circuit, as long as they meet the following requirements:

- Temperatures of over 100°C may arise.
- The solar medium is a water–glycol mixture in the ratio 60:40.
- Fittings are sometimes mounted externally.

Other points to note:

- With such high temperatures, plastic tubes cannot be used because of their poor temperature resistance.
- Glycol in connection with zinc leads to the formation of slime.
- The use of steel tubes is possible in principle, but they are expensive to process (welding, bending, cutting, cutting threads, applying hemp). They are used for larger solar energy systems.
- Stainless steel corrugated pipe is rarely used. It is mainly used for self-build installation, as in this case it is possible to dispense with soldering. In any case it is more expensive than copper pipe.
- Copper pipes have become popular for small systems. Common types of connection are hard and soft soldering. Various solders and fluxes are available for this (see section 4.2.5).

Soft soldering is permitted up to a permanent temperature load of 110°C. As higher temperatures are present in the solar circuit, hard soldering is frequently demanded. In practice, however, it has been found that soft soldering is mainly used for solar energy systems for one- and two-family homes with success. That is not surprising, as temperatures of more than 110°C outside the collectors seldom occur (exception: pipes near the collectors). On the other hand large solar energy systems, especially those with evacuated tube collectors, should always use hard solders.

The advantages of soft soldering are as follows:

- The material load is lower than with hard soldering.
- Soft soldering is quicker.

Further connecting techniques are:

- *Press fittings*. Using a pressing clamp, the press fittings (made of copper with sealing elements of EPDM) cannot be released, and are linked with the copper pipe. This technique is also used for stainless steel pipes.
- *A clamping ring connection*. This is a releasable screw connection, which is reliable, and temperature- and glycol-resistant.

What needs to be considered when installing the pipes?

- Select the shortest possible paths.
- Lay the lowest possible length of pipe in the external area (high heat losses, more expensive thermal insulation).
- Allow sufficient space for retrofitting the thermal insulation.
- Provide ventilation options (a sufficient number with good accessibility).
- Make sure that the system can be completely emptied.
- In the case of long straight pipe runs (from approximately 15 m) install expansion bends.
- Provide sound insulation.

4.3.4.1 THERMAL INSULATION OF PIPES

The heat gained in the collectors has to be delivered to the store with the least possible losses. The thermal insulation of the pipes is therefore very important (see Heat Losses in Piping and Stores in section 3.5.2.5).

Essential factors are:

- sufficient insulation thickness (up to DN20: 20 mm, from DN22 to DN35: 30 mm, for greater diameters the insulation thickness equals the nominal width of the pipe, with thermal conductivity, λ = 0.04 W/mK)
- no gaps in the insulation (also insulate fittings, tank connections etc.)
- correct selection of material (temperature resistance, UV and weather resistance, low heat capacity values).

Table 4.5 lists heat-insulating materials with high temperature resistance.

Table 4.5. Heat-insulating materials with high temperature resistance

Make	Max. temp.	λ value (W/mK)	Remarks
Aeroflex	150°C	0.040	Not UV-resistant
Armafllex HT	170°C	0.045	Not UV-resistant
Mineral wool, insulating shells	>250°C	0.035	Sensitive to moisture

When pipes are installed in closed ducts, as in a chimney for example, the heat insulation should be installed before the pipes are inserted. When installing unslit over the pipe (to avoid gluing) insulating 'hoses' are pulled over the pipes before soldering. They should be pushed back and retained, for example with a gripper, in order to protect them from damage by heat. After soldering and cooling, the gripper can then be removed, the insulating hose pulled back to its original length and any gaps closed.

In the external area they are best protected by a metal jacket (galvanised steel plate, aluminium) from moisture, UV radiation and bird damage.

Pipe installation options are as follows:

- installation in a redundant chimney shaft/supply shaft
- installation through the individual floors (surface or flush mounted); ceiling openings are required here, and special care is required in the case of underfloor heating
- installation below the roof skin, down the façade (for example in a downspout), through the outer wall or into the cellar.

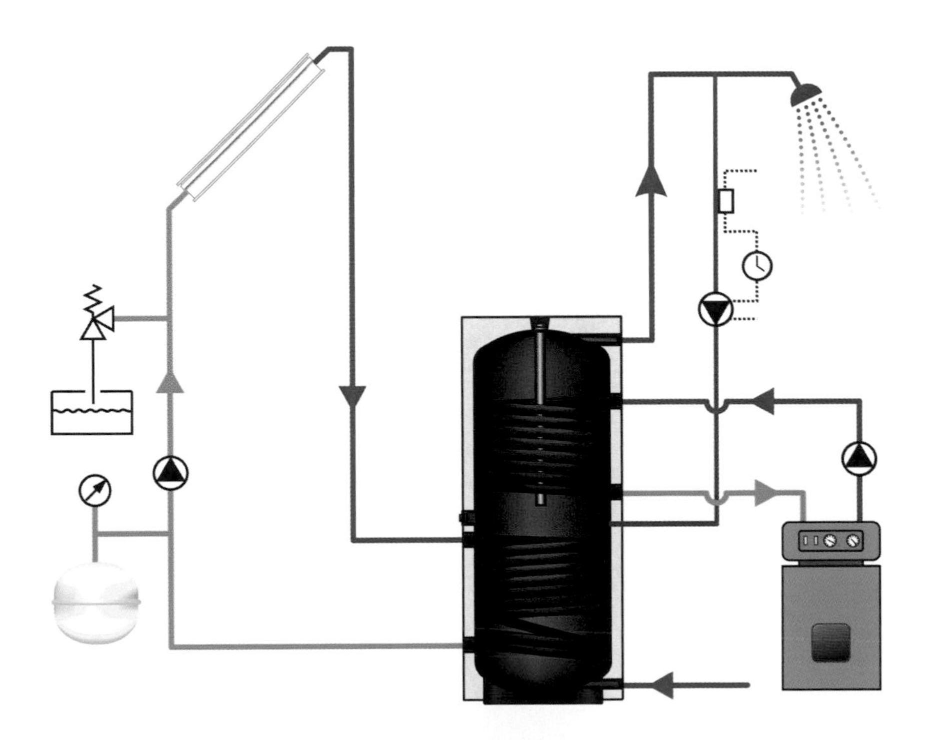

Figure 4.30. Twin-coil system with domestic water storage tank

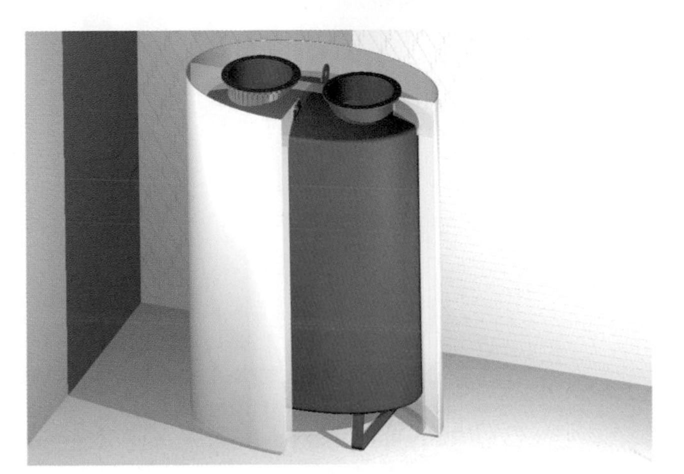

Figure 4.31. The Hybrid XXL Power oval thermal store for narrow doors. Manufacturer: Sailer Solar Systems, Blaubeuren

For low-flow systems the twin tube is frequently used. This is a thermally insulated double pipe made of soft copper from the roll with integrated sensor cable (see Chapter 2, section 2.4.1).

4.3.5 Store installation

The loading capacity of the floor has to be considered (store weight including the water volume). If the store is mounted on the ceiling, the ceiling may have to be strengthened or the load be distributed if necessary.

The diameter of the tank is restricted by the smallest door on the way to the installation location. Removable thermal insulation is an advantage, as the store is narrower and can therefore be transported more easily. Enamelled tanks are sensitive to impact. The height of the tank is determined by the free height at the installation location, because waste water or heating pipes can reduce the height. The dimension of the store when tilted must also be considered.

4.3.5.1 CONNECTION OF SOLAR CIRCUIT

The solar circuit is always connected to the lowest heat exchanger (when more than one heat exchanger is used), as the whole store needs to be available during high insolation. In this position it is also in the coldest area of the store, which increases the efficiency of the solar system.

TYPES OF HEAT EXCHANGER

- *Permanently installed internal plain tube heat exchanger*. The solar feed pipe is connected to the upper tapping, the solar return pipe to the lower one: that is, the solar medium flows through the tank in parallel with the temperature layers in a counter-flow to the convection flow in the tank. Sometimes they are also mounted the other way round, which would lead to uniform heating of the standby area but lower efficiency.
- *Permanently installed internal copper-finned tube heat exchanger*. The fins increase contact with the storage water.
- *External plate exchanger*. This requires a separate pump and control.

Figure 4.32 gives examples.

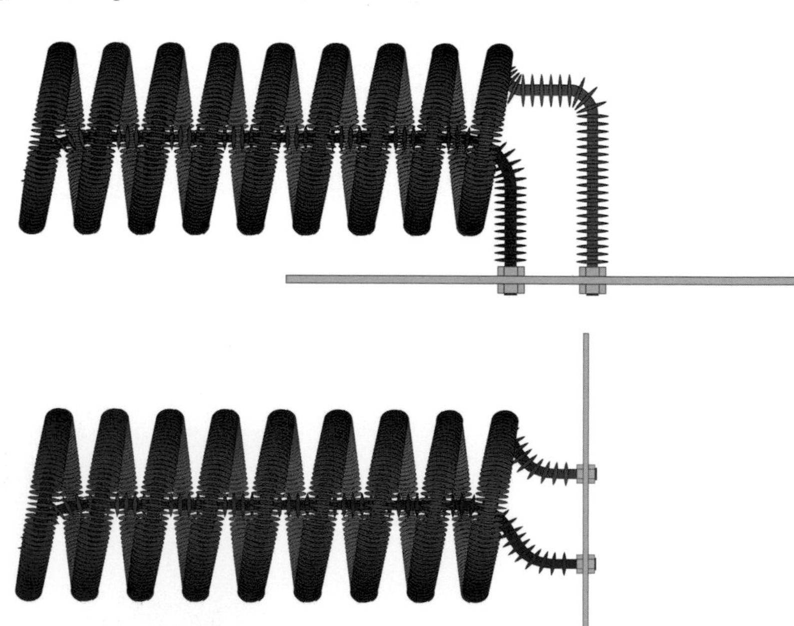

*Figure 4.32.
Top: Finned tube heat exchanger, cranked. Bottom: Finned tube heat exchanger, horizontal*

4.3.5.2 CONNECTIONS FOR ADDITIONAL HEATING SYSTEM

The additional heating system(s) is always connected to the upper heat exchanger(s) to ensure that peak demands can be met without solar gain.

4.3.5.3 COLD WATER CONNECTION

The cold water connection is installed at the bottom of the store. For unvented stores, the inlet combination, consisting of shut-off valve, non-return valve and safety valve, should be installed in the cold water pipe. If necessary an additional pressure reducer should be installed (if needed to reduce the mains pressure). The safety valve must be installed so that it cannot be shut off.

To reduce water losses through the safety valve during the charging process it is also mandatory to install either a membrane expansion vessel or internal bubble top which is suitable for domestic water.

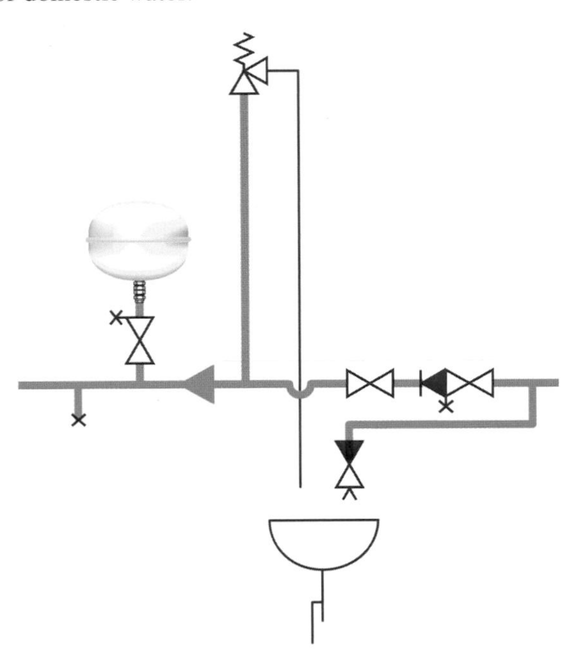

Figure 4.33.
Cold water connection with fittings

4.3.5.4 HOT WATER CONNECTION AND INSTALLATION OF SECONDARY CIRCULATION

The hot water connection is usually at the top of the store. A hot water mixer should normally be installed to provide protection from scalding.

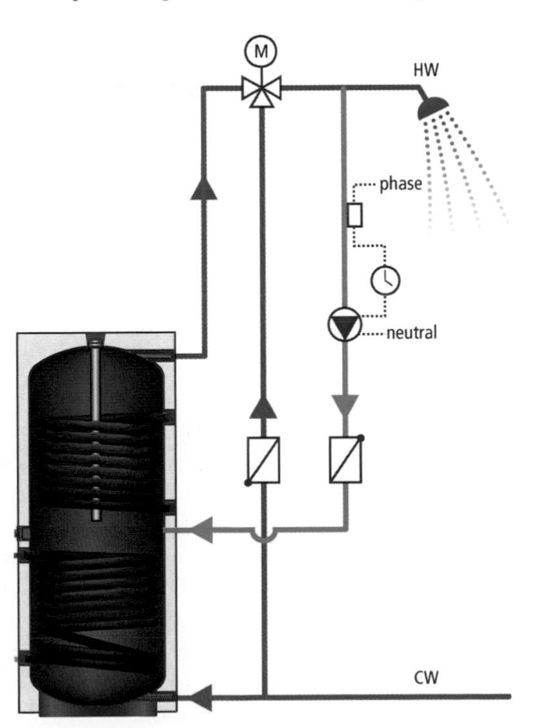

Figure 4.34.
Hot water connection with all fittings and secondary circulation

Secondary hot water circulation always impairs the necessary temperature layering effect in the store. It should therefore be led back at the half-height level of the tank, and must never be connected to the cold water feed line. This would lead to complete mixing of the whole tank. (The exception to this are special circuits that heat up the whole installation periodically to prevent Legionella contamination.)

To reduce secondary circulation heat losses, there are a number of options in addition to 100% thermal insulation:

- installation of a time clock
- installation of a thermostat
- installation of a bypass pipe with temperature-controlled three-way valve between the circulation and the hot water pipe
- installation of a sensor
- installation of a pressure switch/flow monitor.

4.3.5.5 INSULATION OF THE STORE

The quality of the thermal insulation strongly influences the performance of the whole system. Hence some manufacturers supply excellent insulating materials. Apart from PU rigid foam shells, which cannot fit closely enough to the tank to provide a good heat seal, the materials used vary from CFC-free PU flexible foam jackets to even better EPP or Melamine resin flexible foam with a *k*-value of 0.3 W/m²K (see Chapter 2, section 2.3.2.5 and Figure 4.35).

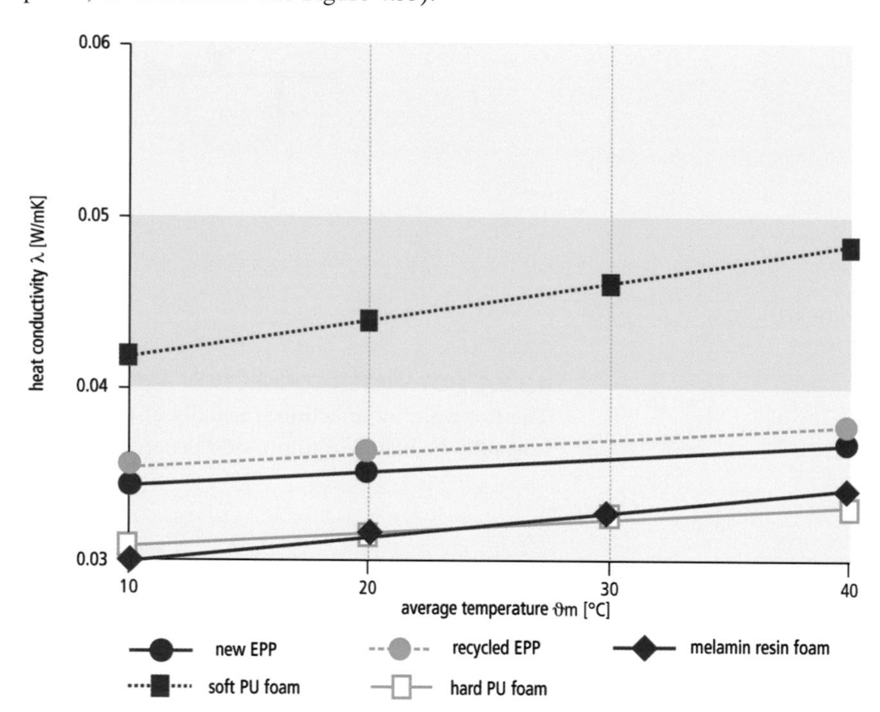

Figure 4.35. Thermal insulating values (EPP: Expanded polypropylene)[36]

Apart from using a high-quality insulating material it is also important to ensure good installation:

- Use thermal insulation for the base area of the tank.
- Use a closely fitting insulating jacket.
- Reinforce the upper cover thickness to a minimum of 15 cm.
- Insulate without gaps the pipe connections, flanges and stoppers.

Tanks of 300–400 l capacity are frequently supplied ready insulated (with hard PU foam) and jacketed. PU foam has good insulating properties. It must, however, be fitted tightly, otherwise the tank will be cooled to by air circulation.

The side connections on the tank are fitted with either a convection brake (Figure 4.36) or a syphon arrangement to reduce heat losses through cooling in the pipe and self-induced circulation.

4.3.5.6 INSTALLATION OF OTHER STORE TYPES

Combined stores, stratified, direct and single-coil stores with a downstream-compatible direct heater are sometimes installed. If there is any doubt about the best option, the

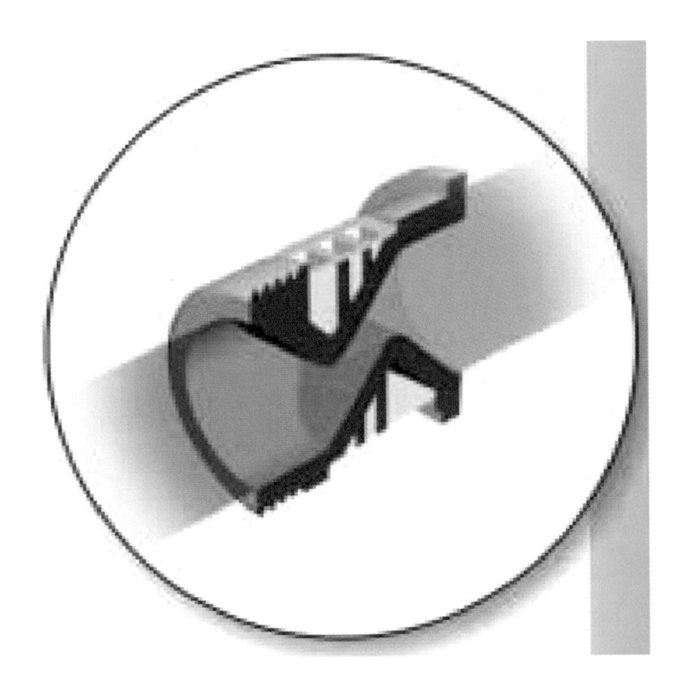

Figure 4.36.
Convection brake. Manufacturer: Wagner & Co., Cölbe

manufacturer's technical advice service should be contacted. This is always better than having to modify the system later when it is found not to function correctly.

4.3.6 Installation of fittings

4.3.6.1 FITTINGS IN THE SOLAR CIRCUIT

COMBINED FILLING AND EMPTYING TAPS

Draining valves should be installed for filling and emptying, as well as for flushing the system at the lowest point of the feed and return pipes. Each one should be equipped with a hose connection.

CIRCULATING PUMP

The pump is installed in the return flow line so that the thermal shock remains as low as possible. It should be fitted between two shut-off valves (pump gate valves, ball valves) so that it can easily be removed. It should be arranged so that the heat transfer medium can be transported vertically from bottom to top: hence included air can flow upwards and escape through the air vent. The pump must be installed without any mechanical stresses.

RETURN-FLOW PREVENTER (CHECK VALVE)

To prevent convective liquid circulation from taking place in the solar circuit when the circulating pump is switched off, which would withdraw heat from the store and transfer it to the surroundings via the collector field, the solar circuit is fitted with a return-flow preventer. If possible this should be mounted immediately downstream of the circulating pump. Since with larger tube cross-sections so-called single-pipe circulation can take place, many manufacturers recommend the installation of a second return-flow preventer in the feed line. The return-flow preventer must be capable of being opened for flushing and emptying the solar circuit.

EXPANSION VESSEL (MEV)

The expansion vessel is fitted to absorb the volume increase in the solar liquid when heated, and to return it again to the system when it cools down. It should be selected to be large enough so that it can additionally take up the whole liquid volume in the collectors and in the case of evacuated tube collectors also a portion of the content of the feed and return lines. This ensures that during stagnation the solar liquid is not released (inherent safety). The expansion vessel should be suspended in a non-insulated branch of the solar circuit return line in order to reduce the thermal load of the membrane. The lower side with the valve must be accessible for checking and for setting the admission pressure. Expansion vessels with volumes greater than 12 l are fixed with their own retainers. Before installation the admission pressure, p_{adm}, should be set;

$$p_{adm} = h_{sys} \times 0.1 \text{ bar/m} + 0.5 \text{ bar}$$

Example:
System height, h_{sys} = 7m (distance between MEV and top edge of collector)
Hence $p_{adm} = 7 \text{ m} \times 0.1 \text{ bar/m} + 0.5 \text{ bar} = 1.2 \text{ bar}$

An admission pressure of 3 bar is usually set in the factory, so that the required value can be set by releasing a corresponding amount of gas with a pressure tester. A pressure pump containing nitrogen is used to increase the admission pressure.

SAFETY VALVE

In order to prevent a non-permitted pressure increase in the solar circuit, the installation of a safety valve is specified. When the response pressure is reached the valve opens and releases solar liquid. This response pressure must not exceed the maximum operating pressure permitted for the collector and the expansion vessel. The safety valve has to be mounted in such a way that it cannot be shut off with respect to the collectors: that is, the solar circuit leg between the safety valve and the collectors must be free of shut-off elements. The safety valve should be installed in a vertical position, and a pipe should be led down from its blow-off opening into a container, in which the escaping solar liquid can be collected.

VENTS

Frequently, after filling the system, entrapped air is found at various points in the solar circuit. In addition, air in dissolved form is found in the solar liquid, which is released again upon heating. The entrapped air gathers at the highest points of the piping system. If no countermeasures are taken this would impair the liquid circuit. It is therefore necessary to install vents. If automatic vents are used it must be ensured that the selected version is suitable for the maximum temperatures occurring at the installation site and suitable for the antifreeze. Vents for temperatures up to 150°C can be obtained for this purpose. Vents installed close to the collectors must be fitted with a shut-off element (ball valve) because of possible vapour escapes. Vents are mainly installed at the highest point of the system (see Chapter 2, section 2.4.6). Good accessibility must be ensured. If air locks can form at particular points of the solar circuit additional vents should be installed.

AIR SEPARATORS

Air separators are built into the solar liquid circuit. They increase the flow cross-section, thereby reducing the flow speed permitting the air bubbles to rise. These can then escape through an installed vent (see Chapter 2, section 2.4.6). Versions are available for horizontal and vertical mounting.

DISPLAY INSTRUMENTS (THERMOMETERS, PRESSURE GAUGES, FLOWMETERS)

To monitor the operation of the system, the display of the solar feed temperature and the operating pressure (with identification of the maximum permitted pressure) is specified. The pressure gauge should be installed so that it can be read when the system is filled. Furthermore, display instruments are recommended to show the solar return temperature and the volumetric flow rate in the solar circuit.

SHUT-OFF FITTINGS

Various versions of shut-off fittings exist, including angle and straight seat valves, ball valves, slide valves, and butterfly valves. When installing the valves it is important to note the direction of flow. Ball valves possess the lowest flow resistance.

DIRT FILTERS

The use of a dirt filter can be dispensed with for small systems (soft soldered) and suitably flushed solar circuits. Their installation is recommended for hard-soldered systems as, over time, loosening scale can impair the function of particular components (pumps, safety valves, gravity brakes, mixing valves). The dirt filter should not be 'hidden' beneath the thermal insulation but should be easy to find and accessible for maintenance work.

4.3.6.2 FITTINGS FOR THE DOMESTIC HOT WATER PIPING

DRAIN TAP

For emptying the store during maintenance work and repairs a drain tap should be installed at the lowest point of the cold water feed pipe.

THERMOSTATIC MIXING VALVE

Because of the high storage tank temperatures (>60°C) that arise during favourable weather conditions, it is necessary to limit the maximum water temperature to protect the system user from scalding. In some countries higher temperatures are allowed, but for solar water heaters it is always recommended to install a thermostatic mixing valve between the cold and hot water pipes to limit the output temperature. A characteristic of a solar water heater is that sometimes the delivered water may be much hotter than at other times. Users might become accustomed to a certain temperature and burn themselves when suddenly the temperature out of the tap is much higher, for example after a period of high solar irradiation and low consumption.

The connections of the mixing valve are marked, and must not be exchanged. The maximum temperature can be set on the mixing valve. This should be the same as the set temperature on the additional heating. As already said, in some countries the supplementary heater may be set at 45°C, whereas in other countries the water temperature must be at least 60°C because of Legionella regulations. If the storage tank water is warmer, cold water is added. It is also recommended that a thermometer is installed after the mixing valve.

DIRT FILTER

The function of the thermostatic mixing valve and the pressure reducer can be impaired by small particles. It is therefore recommended that a dirt filter is installed in the cold water pipe upstream of these fittings in the flow direction.

4.3.7 Installation of sensors and controllers

All work on electrical equipment must be carried out according to the appropriate regulations.

4.3.7.1 INSTALLATION AND CONNECTION OF SENSORS

The correct installation of sensors is an important precondition for the trouble-free functioning of a thermal solar energy system. The correct location and a good thermal contact are equally important (firm seating, thermal conducting paste):

- The collector sensor is either fixed directly onto the absorber (usually it is premounted in the factory), or installed in an immersion sleeve at the hottest point in the collector feed. When laying cables, care should be taken to ensure that the cables do not come into contact with the hot pipes.
- The store sensor should be fitted at the height of the solar circuit heat exchanger. This is either done with the help of a clamping bar, which is attached to the outside of the store wall and is covered by the tank's thermal insulation, or it is fitted in an immersion sleeve made of brass or stainless steel.

If the cable lengths of the temperature sensor are insufficient they must be extended with a cable with a minimum cross-section of 0.75 mm^2. The cross-section depends on the length, and can be found in the manufacturer's documentation.

Sensor cables should not be laid together with mains grid cables in a pipe or cable duct, as the electromagnetic fields will influence the measured values. In addition, a sensor-connecting socket should be installed with excess voltage protection (lightning protection socket).

4.3.7.2 CONTROL UNIT INSTALLATION

The housing should first be attached to a wall in the vicinity of the solar station. According to the connection assignment (circuit diagram) the temperature sensors (collector and tank sensors) as well as the solar circuit pump should all be connected to the respective terminals on the terminal strip of the solar control unit. After that, the mains connection is made and the housing is closed; the controller can then be started up.

Before opening the control unit housing it is important to ensure that it is isolated from the mains voltage.

4.4 Starting up, maintenance and servicing

The necessary steps to start up a thermal solar energy system are:

1. Flush out the solar circuit (Figure 4.37).
2. Check for leaks.
3. Fill with solar liquid.
4. Set pumps and controller.

4.4.1 Flushing out the solar circuit

A thorough flushing process removes dirt and residual flux from the solar circuit. Flushing should not be carried out in full sunshine or during frost, as there is a risk of evaporation or freezing. The flushing process initially takes place via valves 1 and 2 (see Figure 4.37). Valve 1 is connected to the cold water line by a hose; a further hose on valve 2 is laid to the drain. All fittings in the solar circuit should be set to through-flow (gravity brake, shut-off taps). Finally, in order to flush out the heat exchanger valve 2 is closed, after attaching a hose to it valve 4 is opened, and valve 3 is closed. The flushing process should last for about 10 minutes.

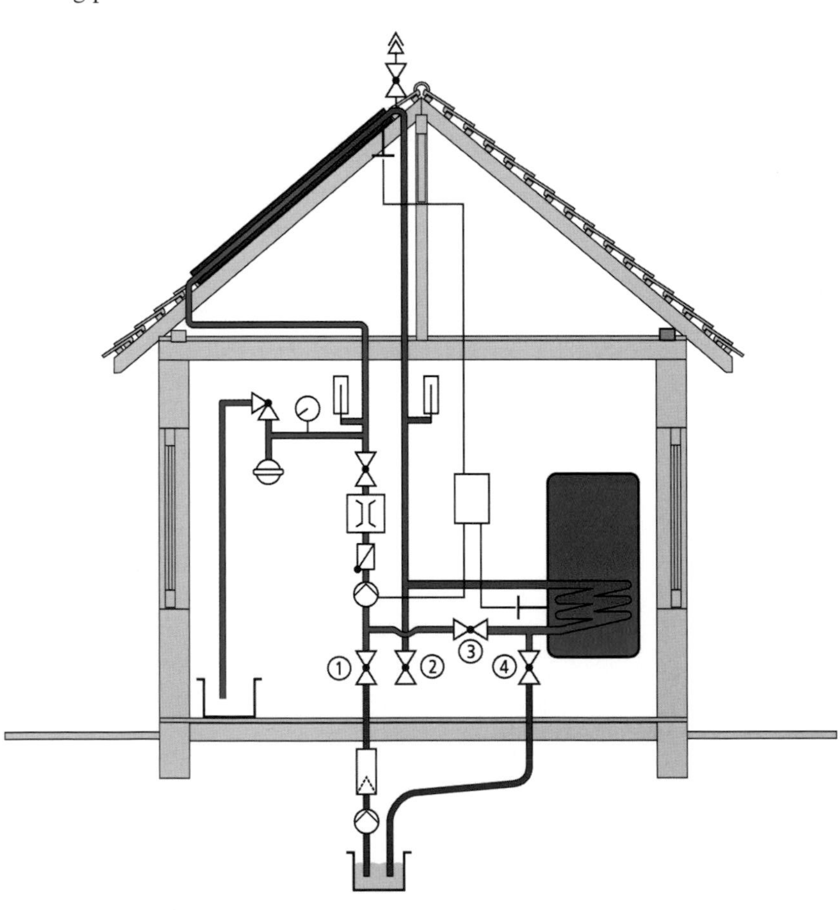

Figure 4.37. Solar circuit with fittings for flushing and filling–flushing process.[37]

4.4.2 Leak testing

The pressure test takes place after flushing. For this purpose valve 4 is closed, and the system is filled with water through valve 1. The system pressure is then raised to a value just below the response pressure of the safety valve – maximum 6 bar. Then valve 1 is closed, the pump is manually started, and the solar circuit is vented via the vents or the pump (vent screw). If the pressure falls significantly as a result of bleeding, it must be increased again by additional filling. The system is now ready to be tested for leaks (visually and by hand). A leak test using the pressure gauge is not possible because of irradiation-caused pressure variations over the course of a day. At the end of the leak test the function of the safety valve can be tested by increasing the pressure further. The solar circuit should finally be fully emptied again by opening taps 1 and 2. By measuring the amount of water that runs out it is possible to establish the amount of antifreeze required to make up a water–antifreeze mixture of for example 60:40. As some water always stays in the solar circuit (for example in the collectors or the heat exchanger) the measured amount of water should be correspondingly increased.

4.4.3 Filling with solar liquid

After mixing the antifreeze concentrate with water to achieve the desired level of frost protection (or using pre-mixed antifreeze) the solar liquid is pumped into the solar circuit through valve 1. As the solar liquid – compared with water – is much more likely to creep, it is necessary to recheck the system for leaks. A general procedure to release the air from the solar liquid is as follows – but always check the installation instructions for the specific product, as differences may occur.

1. When pumping the solar liquid in the system and the mixing container, a large part of the air is already removed. To be effective the ends of the hoses must be completely submerged in the liquid. When no more air bubbles come out, valve 4 can be closed.
2. Reduce the pressure to system pressure (= static pressure + 0.5 bar) plus an allowance for pressure loss through further bleeding.
3. Switch on the circulating pump. Switch it on and off several times at 10 minute intervals.
4. To bleed the circulating pump, unscrew the brass screw on the face.

An alternative method of bleeding the system can be used where there does not need to be a top vent. The high rate of flow takes the air bubbles down with it again. The system is bled in this case via a vent bottle, which is integrated into the feed line of the solar station. In this bottle the relatively high flow rate of 60 l/m^2h is severely reduced owing to the increase in cross-section. The air gathering in the upper area of the vent bottle can be discharged through a manual vent.

If the pressure falls below the system pressure as a result of bleeding, solar liquid should be added accordingly. After several days the shut-off tap under the vents is then closed.

4.4.4 Setting the pump and controller

The volumetric flow in small systems is usually about 40 l/m^2h (high-flow operation); in systems with stratified stores it is 15 l/m^2h (low-flow operation). The pump should be capable of generating the pressure required in its medium performance range. With full irradiation, this leads to a temperature difference between the feed and return lines of about 10–15 K in high-flow operation and 30–50 K in low-flow operation. The actual volumetric flow can be controlled with the help of a taco-setter or a flowmeter.

The switch-on temperature difference of 5–10 K and the switch-off difference (hysteresis) of about 2 K should be set on the controller. In this way, on the one hand the heat generated in the collector is transferred to the store at a useful temperature level, and on the other hand no unnecessary pump energy is used.

4.4.5 Maintenance

Solar thermal systems require very little maintenance; however, a regular check is recommended.

During the course of maintenance the users should be asked about their level of satisfaction. The maintenance work is described in a maintenance report, and should contain the following in detail.

4.4.5.1 VISUAL INSPECTION

The visual inspection involves checking the collectors and the solar circuit for visual changes:

- *collectors*: contamination, fixings, connections, leaks, broken glass, tarnishing on panes/tubes
- *solar circuit and storage tank*: tightness of thermal insulation, leaks, check/clean any dirt traps, pressure, filling level.

4.4.5.2 CHECKING THE FROST PROTECTION

The frost protection of antifreeze fluids is checked with a hydrometer or refractometer. For this purpose a given amount of solar liquid is removed. Either the temperature to which the system is protected is shown directly, or a specific density is read off. This allows the actual content of antifreeze mixture to be established from a density–concentration diagram and thereby the freezing point.

For drainback systems, the level of the collector fluid needs to be checked: if necessary fill the circuit to its proper level.

4.4.5.3 CHECKING THE CORROSION PROTECTION

SOLAR CIRCUIT

Checking the corrosion protection of the solar liquid (if it contains antifreeze liquid) is done indirectly by establishing the pH value. Test strips are suitable for this with which the pH value can be read from a colour scale. If the pH value falls below the original value (as should be mentioned in the commissioning report of the installation) to under 7, the frost protection mixture should be exchanged.

STORE TANK (ONLY FOR TANKS WITH ANODES)

The magnesium sacrificial anode can be tested by measuring the current between the detached cable and the anode using an ammeter. If the current is over 0.5 amps, there is no need to renew the anode. In case of a powered anode, only the LED indicating proper functioning needs to be checked.

4.4.5.4 MONITORING THE SYSTEM PARAMETERS

The pressure and temperature, and the controller settings, must be checked. During operation the system pressure varies, depending on the temperature. After complete bleeding it must not vary from the set value by more than 0.3 bar. It must never fall below the admission pressure of the MEV vessel. The temperature difference between feed and return lines should in full irradiance conditions not exceed 20 K (otherwise there may be air in the solar circuit or blockages due to contamination – however, for low-flow systems the temperature difference may be higher) and not drop below 5 K (otherwise limescale will form in the heat exchanger) in high-flow systems. The controller settings and functions must be tested. If provided, within the scope of maintenance the data for system function and yield monitoring can also be recorded. Among these are:

- operating hours of the solar system pump
- quantity yield.

The solar circuit pump should have an approximate annual running time according to the sunshine hours at the respective location (example: London 1700 hours).

4.4.6 Detecting and rectifying faults

4.4.6.1 TYPICAL ERROR SOURCES

COLLECTORS

- Minimum overlap of roof tiles/covering sheet not ensured (8 cm): leaks in roof.
- Use of different materials on the roof (copper above aluminium cover plates): corrosion risk.
- Collector shaded: low feed temperatures.
- Collector sensor shaded by a chimney, for example, or not correctly inserted into the collector (especially for evacuated tube collectors): controller responds too late.

- Top vent forgotten: incomplete bleeding.
- External thermal insulation not UV- or weather-resistant or protected from bird damage: heat losses.
- Feed and return lines confused: system short circuit.
- Over-dimensioning of collector surface area: stagnation temperature reached too frequently (unnecessarily high temperature load on collector materials and solar liquid).
- No follow-up inspection access to roof and collectors: maintenance, cleaning and repair not possible.

SOLAR CIRCUIT

- Poor bleeding: no heat transport possible because of blocking by air bubbles.
- Volumetric flow rate too low: low heat yield.
- Feed and return lines confused: pump damaged by overheating.
- Proportion of antifreeze in the water–glycol mixture too high (>50%): impaired heat transport. (For drainback systems) fluid level too high: damage from frost or overheating; fluid level too low: no collector yield.
- (If installed in winter) residual water from pressure testing overnight in collector: damage to collector from frost.
- Filling in full sunshine: vapour formation, complete filling not possible.
- Insufficient thermal insulation, with gaps: heat losses.
- System pressure too low: poor heat transport and possibly even interruption of solar circulation.
- Expansion vessel dimensions too small: blowing of the solar liquid out of the safety valve.

STORAGE

- Temperature sensor fitted too high/too low or poorly attached to the tank: poor system control.
- Feed and return lines confused at the solar circuit heat exchanger: poor heat transfer to the storage tank water, low collector efficiency because of high return temperatures.
- Over-dimensioning of storage tank: frequent need for additional heating.
- Store too small: the solar energy system switches off frequently as the maximum tank temperature has been reached.

CONTROLLER

- False setting of the controller parameters: too early/late switching on/off of the solar circuit pump.

ORGANIZATION

- Lack of discussion with the customer with respect to other trades (electrical, roofing, plumbing): interruptions to installation sequence.
- Communication problems between customer and installer: system ends up different from what customer expects.

4.4.6.2 EXAMPLES OF FAULT DETECTION AND RECTIFICATION

SYSTEM PRESSURE DROPS

See Table 4.6.

- The better the system is vented when filled (by pumping the solar liquid through the system and through a tank filled with the same liquid until no more bubbles come out), the lower will be the number of faults caused by air in the system – and possibly no faults at all.
- Pressure variations of 1–2 bar due to temperature variations during system operation are normal. Pressure comparisons are only meaningful and useful for testing purposes at identical system temperatures.
- The system pressure should not decrease below the value of the static pressure + 0.5 bar, as otherwise a partial vacuum could be created in the upper area of the solar system and air may be drawn in.

Table 4.6. Fault detection – system pressure drops

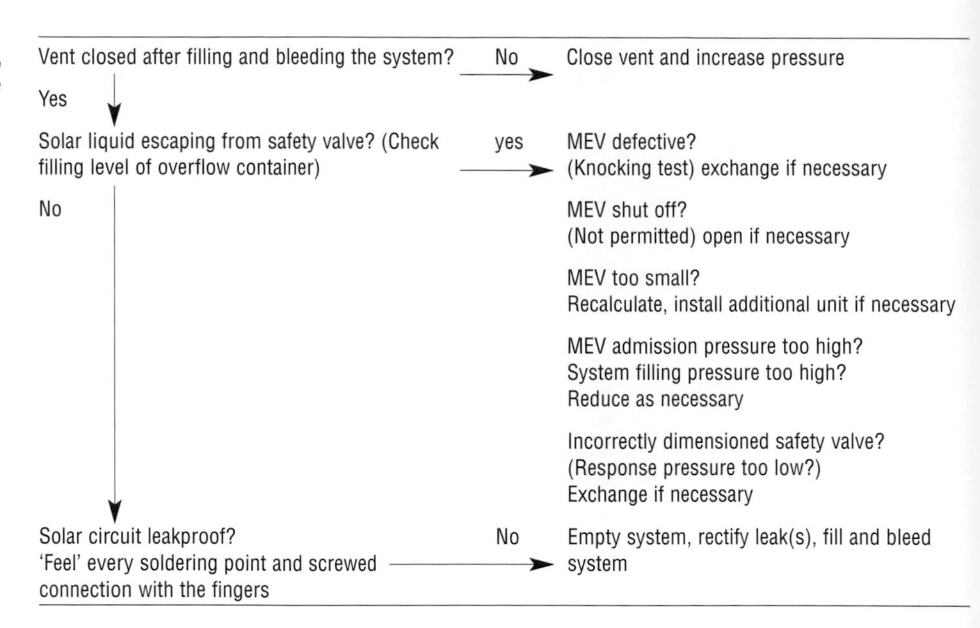
Vent closed after filling and bleeding the system?
No
Close vent and increase pressure
Yes
Solar liquid escaping from safety valve? (Check filling level of overflow container)
yes
MEV defective?
(Knocking test) exchange if necessary
No
MEV shut off?
(Not permitted) open if necessary
MEV too small?
Recalculate, install additional unit if necessary
MEV admission pressure too high?
System filling pressure too high?
Reduce as necessary
Incorrectly dimensioned safety valve?
(Response pressure too low?)
Exchange if necessary
Solar circuit leakproof?
'Feel' every soldering point and screwed connection with the fingers
No
Empty system, rectify leak(s), fill and bleed system

THE SUN IS SHINING BUT THE PUMP IS NOT RUNNING

See Table 4.7.

Other reasons:

- The store is hot, but below the maximum temperature. The sun's radiation is not sufficient to reach the switch-on temperature difference at high level. The behaviour of the system in this case is satisfactory.
- Twin storage tank system: the controller switches over to the hotter store, and the switch-on temperature difference has not yet been reached. Wait.

Table 4.7. Fault detection – the sun is shining but the pump is not running

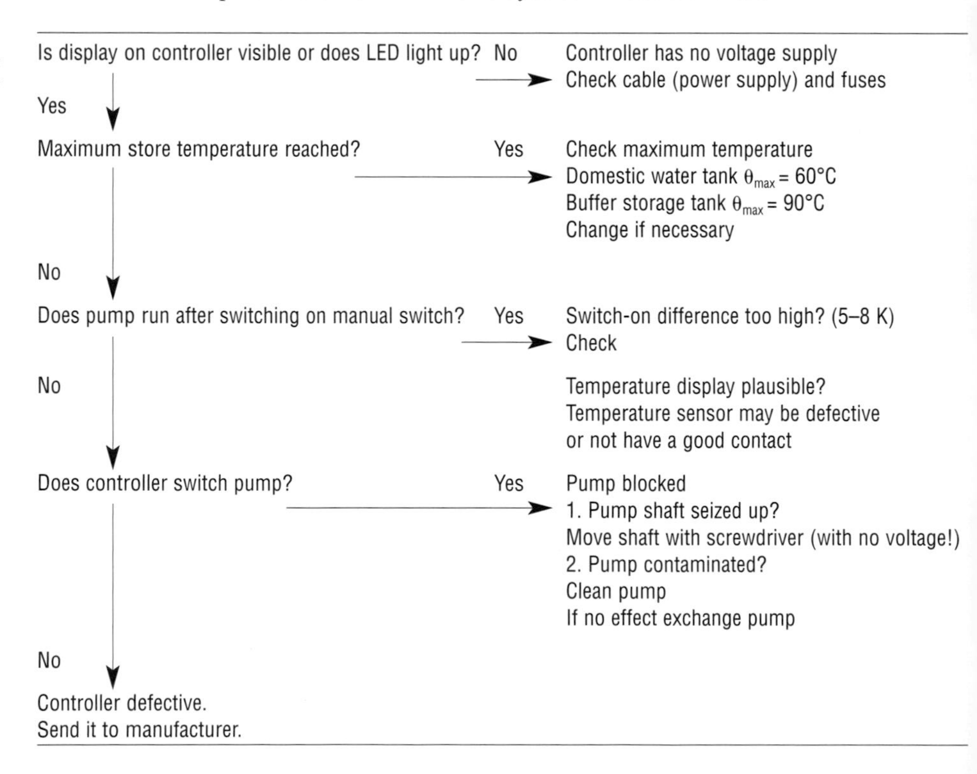
Is display on controller visible or does LED light up?
No
Controller has no voltage supply
Check cable (power supply) and fuses
Yes
Maximum store temperature reached?
Yes
Check maximum temperature
Domestic water tank θ_{max} = 60°C
Buffer storage tank θ_{max} = 90°C
Change if necessary
No
Does pump run after switching on manual switch?
Yes
Switch-on difference too high? (5–8 K)
Check
No
Temperature display plausible?
Temperature sensor may be defective
or not have a good contact
Does controller switch pump?
Yes
Pump blocked
1. Pump shaft seized up?
Move shaft with screwdriver (with no voltage!)
2. Pump contaminated?
Clean pump
If no effect exchange pump
No
Controller defective.
Send it to manufacturer.

TEMPERATURE DIFFERENCE BETWEEN COLLECTOR AND STORAGE TANK VERY HIGH

The difference between the feed and return temperatures should be about 10 K (high-flow) and 30 K (low-flow) when the pump is running. See Table 4.8.

Table 4.8. Fault detection – very high temperature difference between collector and storage tank

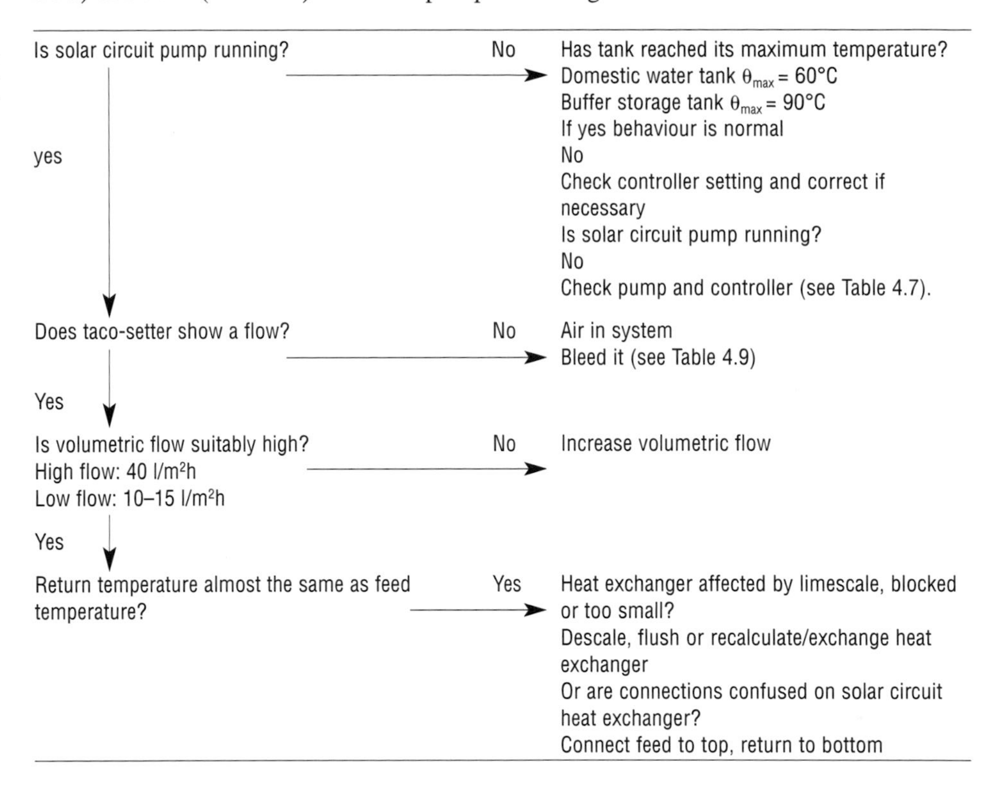

Check	Answer	Action
Is solar circuit pump running?	No	Has tank reached its maximum temperature? Domestic water tank θ_{max} = 60°C Buffer storage tank θ_{max} = 90°C If yes behaviour is normal No Check controller setting and correct if necessary Is solar circuit pump running? No Check pump and controller (see Table 4.7).
yes		
Does taco-setter show a flow?	No	Air in system Bleed it (see Table 4.9)
Yes		
Is volumetric flow suitably high? High flow: 40 l/m²h Low flow: 10–15 l/m²h	No	Increase volumetric flow
Yes		
Return temperature almost the same as feed temperature?	Yes	Heat exchanger affected by limescale, blocked or too small? Descale, flush or recalculate/exchange heat exchanger Or are connections confused on solar circuit heat exchanger? Connect feed to top, return to bottom

PUMP RUNNING BUT TACO-SETTER OR FLOW GAUGE SHOWS NO FLOW RATE

See Table 4.9.

Table 4.9. Fault detection – pump running but taco-setter or flow gauge shows no flow rate

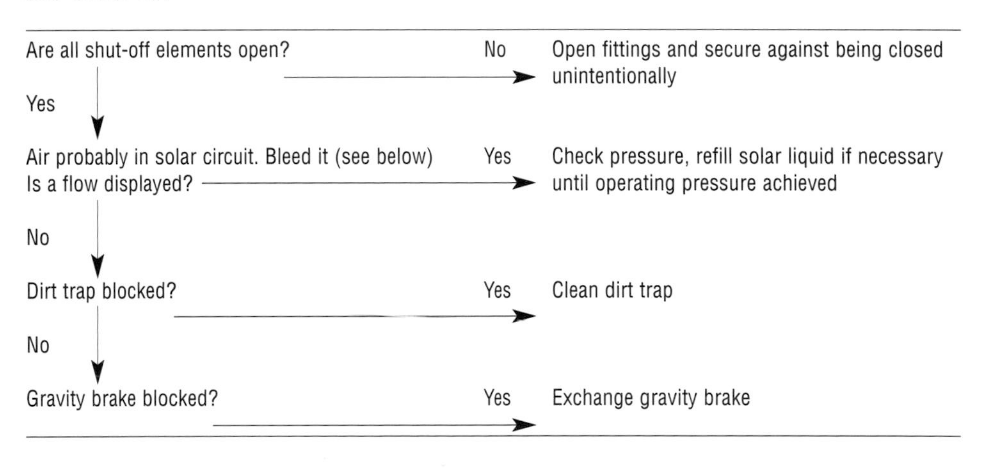

Check	Answer	Action
Are all shut-off elements open?	No	Open fittings and secure against being closed unintentionally
Yes		
Air probably in solar circuit. Bleed it (see below) Is a flow displayed?	Yes	Check pressure, refill solar liquid if necessary until operating pressure achieved
No		
Dirt trap blocked?	Yes	Clean dirt trap
No		
Gravity brake blocked?	Yes	Exchange gravity brake

The procedure for bleeding the solar circuit is as follows (because of the risk of vapour formation in times of high solar radiation this should be done if possible in the early morning or late afternoon):

1. Open all vents.
2. Increase system pressure with the help of a filling pump.
3. Operate solar circuit pump with brief pulsating movements at maximum capacity.
4. If necessary undo screw on recirculating pump, let air out and close again.
5. Close all vents after successful bleeding.

STORAGE TANK COOLS TOO MUCH OVERNIGHT WITHOUT HOT WATER CONSUMPTION

See Table 4.10.

Table 4.10.
Fault detection – storage tank cools too much overnight without hot water consumption

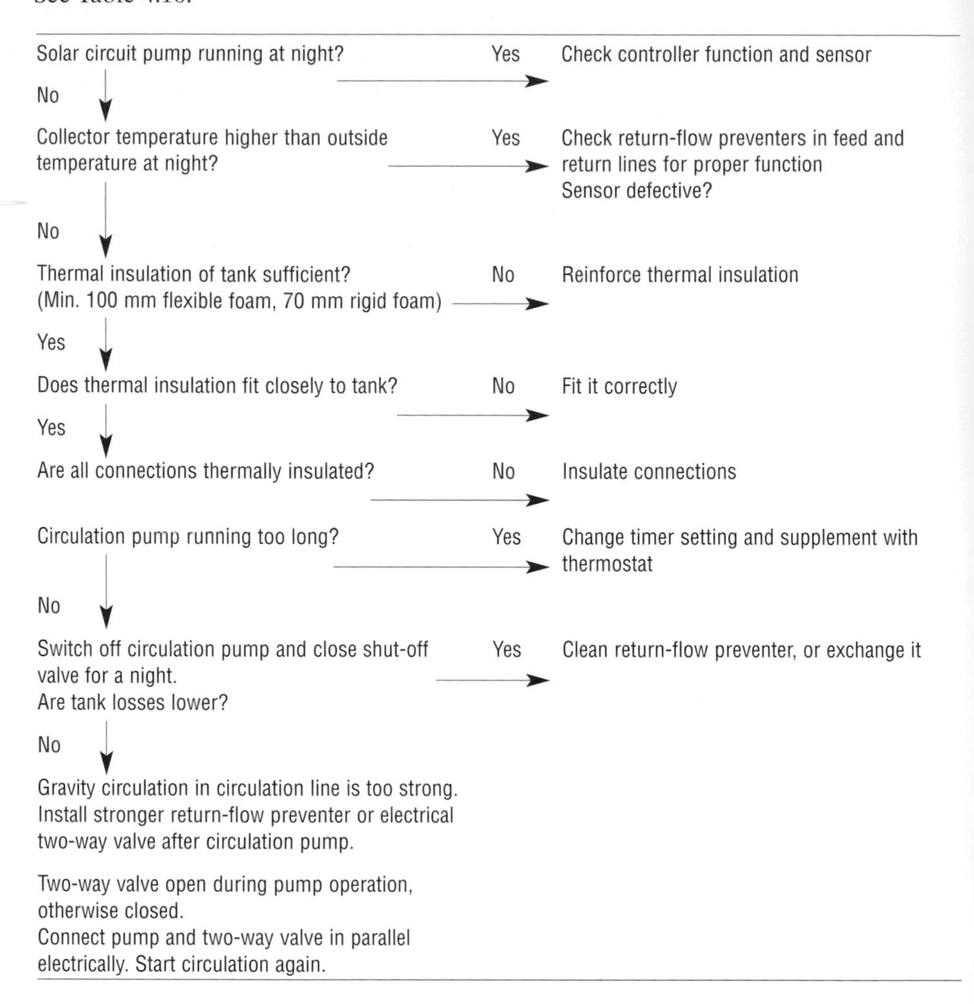

Check	Answer	Action
Solar circuit pump running at night?	Yes	Check controller function and sensor
↓ No		
Collector temperature higher than outside temperature at night?	Yes	Check return-flow preventers in feed and return lines for proper function Sensor defective?
↓ No		
Thermal insulation of tank sufficient? (Min. 100 mm flexible foam, 70 mm rigid foam)	No	Reinforce thermal insulation
↓ Yes		
Does thermal insulation fit closely to tank?	No	Fit it correctly
↓ Yes		
Are all connections thermally insulated?	No	Insulate connections
Circulation pump running too long?	Yes	Change timer setting and supplement with thermostat
↓ No		
Switch off circulation pump and close shut-off valve for a night. Are tank losses lower?	Yes	Clean return-flow preventer, or exchange it
↓ No		
Gravity circulation in circulation line is too strong. Install stronger return-flow preventer or electrical two-way valve after circulation pump. Two-way valve open during pump operation, otherwise closed. Connect pump and two-way valve in parallel electrically. Start circulation again.		

4.5 Information sources for specific countries

4.5.1 USA

- The Florida Solar Energy Center has an on-line tutorial on installation of solar water heaters: http://www.fsec.ucf.edu/solar/install/dhwinstl/dhwinstl.htm
- The US Federal Energy Management Program has an extensive introduction page on the technical details and installation of solar water heaters: http://www.eere.energy.gov/femp/prodtech/sw_water.html
- The Solar Energy Industries Association (SEIA) provides lists of solar equipment manufacturers and dealers on http://www.seia.org
- The American Society of Sanitary Engineering lists and sells relevant publications: http://www.asse-plumbing.org/Standards.htm
- The US Solar Rating and Certification Corporation has a certification programme of solar products, with directory of certified products: http://www.solar-rating.org/

4.5.2 Australia

- Australian Standard AS 3500.4 provides installation details. Available at Standards Australia: http://www.standards.com.au
- The Ithaca College of technical and further education (TAFE) in Queensland has produced a solar energy course for installers (http://www.tafe.net).
- The Sustainable Energy Industry Association SEIA has an accreditation programme for installers of solar water heaters: http://www.seia.com.au/Accreditation/BCSEAccreditation/Appendices/Units/BCSE402.htm

- The Sustainable Energy Authority of Victoria (http://www.seav.vic.gov.au) has a solar rebate programme, and has an installers' accreditation programme in cooperation with the Master Plumbers' and Mechanical Services Association of Australia: the GreenPlumbers (solar hot water): http://www.greenplumbers.com.au
- Many manufacturers in Australia supply instructions on installation of their products. On http://www.seav.vic.gov.au a list can be found of manufacturers in the Victorian rebate programme.

4.5.3 Canada

- The Canadian Solar Industries Association offers various fact sheets on solar energy on http://www.cansia.ca/factsheets.htm

5 Large-scale systems

5.1 The fundamentals of designing the system size

5.1.1 Possible application areas

Large solar energy systems for water heating can be used in a multitude of buildings, which have either a corresponding domestic hot water (DHW) usage, a whole year's heating requirement, or both. For example:

- hospitals
- hairdressers
- old people's homes
- fitness centres
- holiday homes
- breweries
- camping sites
- laundries
- student lodgings
- dairies
- flats
- butchers
- schools
- bakers
- barracks
- large kitchens
- office buildings
- public houses
- sports complexes
- hotels
- indoor swimming pools
- showers in open-air swimming pools
- floor heating in stables
- agriculture (cattle troughs)
- car washing facilities
- lorry washing systems.

5.1.2 Initial data required for planning the solar system

Because of the varying provision of power and energy from the sun, knowledge of the consumption quantities, the consumption profile and the desired hot water temperature is required in the dimensioning of a solar energy system. These data then feed into the dimensioning task.

The values for hot water consumption given in the literature have been found to be too high in practice: their use often leads to over-dimensioning in the planning of solar energy systems. The consumption values are frequently old, and have been calculated with the objective of providing security of supplies. Determination of the required data for the consumption profile should, if possible, be carried out by long-term measurements of the hot water consumption. If this is not possible, or is too time-consuming, then the following method can be used for acquiring the consumption data and using them for calculations in simulation programs.

5.1.2.1 CONSUMPTION PROFILE AND QUANTITY

Data must be collected to differing extents according to the various potential users (see Table 5.1). For example, Figure 5.1 shows a typical questionnaire used to obtain the required data for hot water consumption.

Table 5.1. Examples of the data required for calculating solar thermal systems

	Daily	Weekly	Annual	Occupational profile seasonal	Strongly variable daily requirement for HW	Operating times of secondary HW circulation	Other users in times of low water consumption
Multi-family residences	X	X	X			X	
Administration buildings	X	X				X	X
Hotels, restaurants	X	X	X	X	X	X	X
Schools, sports halls	X	X	X	X	X	X	X
Sports complexes	X	X	X	X	X	X	X
Commerce, industry	X	X		X	X	X	X
Open-air swimming pools, indoor pool showers	X			X	X	X	X
Open-air swimming pools, indoor pool			X				X
Camping sites	X	X	X	X	X	X	X

The consumption profile data thus obtained are used, for example, in simulation programs to calculate the system yield and behaviour. For this purpose the data are inserted as a table into the simulation program. Alternatively the different simulation programs offer consumption profiles for different buildings and user groups.

In order to permit planning where there is no possibility of measuring the water consumption, or in the event of uncertainties, it is possible to access standard published DHW consumption data of buildings. The following rule-of-thumb values have been determined for multi-family residences in many countries:

> 20–30 l (at 60°C) per day per person (which is roughly equivalent to 30–50 l per person per day at 45°C; see Chapter 3, sections 3.5.2.4 and 3.5.2.5).

Moreover, the occupation of the flats (one dwelling unit = average 3½ persons) must be examined in more detail if possible. In one example a consumption of 40 l per day per person and average occupation of 3 people was assumed. The actual consumption however, instead of the estimated 120 l per day per flat, was in fact only 45 l per day per flat when measured. Obviously this led to far too large a solar water heater.

5.1.2.2 REQUIRED HOT WATER TEMPERATURE

Of great significance for the future energy yield of the solar system is the target domestic hot water temperature of the application. The lowest temperature level likely to be required is 25°C for heating the water in cattle troughs. In one/two-family homes (and also some production applications) 45°C is sufficient. However, in many countries drinking water regulations require higher temperatures. Larger buildings usually require a temperature of 60°C at the storage outlet. Wherever it is reasonable and safe within regulations, the domestic water heating should be restricted to the lowest possible temperature level in order to save energy.

The heat yield of the solar collector and the whole system significantly increases with a reduction in the required temperature in the collector circuit. A reduction in the water temperature generated by the sun causes a distinct increase in the energy yield per m^2 of collector per annum in the entire system.

5.1.3 System planning and design

At the beginning of the planning phase, rough figures on the dimensions of the most important system components – that is, the collector field and heat stores – are required for the preliminary planning and cost estimate. For dimensioning purposes there are different target sizes. The systems are designed either for a particular solar fraction using existing structural conditions (for example the size of the roof, or the

1. **Description** (Indicate one)

One family home	☐	Multifamily residence	☐		
Refectory	☐	Student lodging	☐	Old-peoples' home	☐
Hotel	☐	Restaurant	☐	Camping site	☐
Nursery	☐	Sports hall	☐	School	☐

Other (Add description)..

2. **Hot water consumption profile**

State peak consumption*: litres/hour at temperature °C

or persons/hour**

*Assume DHW is mixed down to 38°C at appliance ** valid for showers

% of daily total	Mon	Tue	Wed	Thur	Fri	Sat	Sun	Example
Between.......o'clock to............ .o'clock								20%
Between.......o'clock to............ .o'clock								40%
Between.......o'clock to............ .o'clock								25%
Between.......o'clock to............ .o'clock								15%
Total daily consumption in litres (= 100%)								1000 l

If each month is the same, write 100% = normal consumption. Otherwise indicate % days in month absent i.e. holidays

	Jan	Feb	Mar	Apr	May	Jun	Jul	Aug	Sep	Oct	Nov	Dec
Consumption in %												

Secondary domestic hot water circulation (indicate one): Yes No

Running times: From ...... o'clock to ...... o'clock and from ...... o'clock to ...... o'clock and from ...... o'clock to ...... o'clock and from ...... o'clock to ...... o'clock

Total existing hot water consumption: State if calculated or measured: .... l/annum

3. **Special factors**

Base consumption (e.g. caretaker's flat or other) .... persons

Total no. of showers: ....

Max. water flow per shower: ...... l/h

Type of shower (select as many as apply):

Total no. of shower cubicles:............

Max. water flow per cubicle: .... l/h

Coin-operated:

Timer controlled:

Hand operated:

Hot-fill dishwashers/washing machines: Consumption:. ...... litres/day

Hotel

Overnight rooms:

Single rooms: ...... with showers: .... and washbasins:....... and baths: ....

Double rooms: .... with showers: .... and washbasins:....... and baths: ....

.... bed rooms: .... with showers: .... and washbasins:....... and baths: ....

Restaurant

Capacity: .... persons Breakfasts:.........../week Lunches:......../week

Evening meals: ..../week .... toilets with in total .... hot water washbasins

Figure 5.1. Questionnaire template to establish hot water consumption

size desired by the customer), or for the highest possible energy yield per square metre of collector field.

For example, the owner of a one-family home may prefer to procure a solar system with a high solar fraction so that in the summer he or she can enjoy solar-heated water with the heating boiler turned off. The administrator of a multi-family residence, who would rather keep the operating costs low, asks the designer for a low solar heat price and hence low costs per square metre. Another investor, with a restricted budget, might prefer to achieve high primary energy substitution. As preheating systems achieve higher system utilization in this case, two preheating systems in two of his houses fulfil this requirement better than one solar system with a high solar fraction on one of the houses.

During client discussions, solar systems with high system utilization and the resulting high kWh yields per square metre of collector field are often stated to be the aspired objective. On the other hand, solar energy systems with high solar fractions are unjustifiably described as over-dimensioned. Both design variants are equally good if the prescribed objective (the brief) is achieved by the planning. A deviation from the expected yields that is due to incorrect consumer figures or a later change in the hot water requirement can also not be blamed on the solar system. What is decisive in the assessment of the function or performance of a solar thermal system is comparison of the actual yields with the results of simulation calculations on the basis of the actual hot water consumption.

5.1.3.1 DESIGN OBJECTIVE: HIGH PRIMARY ENERGY SUBSTITUTION

In order to achieve the greatest possible saving of conventional fuels in the project under consideration (that is, high primary energy substitution), the degree of solar coverage in the whole system must be selected to be as high as possible. Such a design is frequently sought in the case of smaller solar systems for one- and two-family homes, or in solar systems for space heating. In temperate climates, for a desired solar fraction for domestic water heating above 60%, the conventional heating can be put out of operation in the summer, which can contribute greatly to the satisfaction of the operator and to avoiding stand-still heat losses in the heating boilers. In larger solar systems, a degree of solar coverage of 50% should be targeted to achieve the design objective.

In tropical climates a solar coverage of 75–100% can be chosen.

5.1.3.2 DESIGN OBJECTIVE: LOW SOLAR HEAT PRICE

Decisive for the efficiency of the solar system is the highest possible system utilization, which leads to a high (specific) kWh yield per square metre per annum. As already described, high system utilizations are achieved with low solar fractions. These systems are also described as *preheating systems*. As this type of system usually achieves only preheating of the domestic water, even in the summer, the conventional heating system remains in operation for the full year. In temperate climates the degree of solar coverage for this design objective is between 10% and 45%; for further considerations it will be assumed to be 25%.

In tropical climates this is a less logical option unless the load is irregularly distributed over the year (for example, the holiday season for a hotel). In such cases, custom calculations need to be made.

5.1.3.3 DESIGN WITH THE HELP OF AN APPROXIMATION FORMULA

Starting point values are given in Table 5.2 for dimensioning the collector field and storage tank sizes. The following factors are used for this:

- a hot water temperature of 60°C
- continuous domestic warm water draw-off

Table 5.2. Approximation formula for dimensioning collector fields and storage tanks for two types of climate

	Temperate climates (1000 kWh/m²a, pronounced seasons)		**Tropical climates (2200 kWh/m²a, evenly spread over the months)**
	Solar fraction 25%	**Solar fraction 50%**	**Solar fraction 80%**
Collector surface area	0.5 m² per 50 l HW consumption (60°C) per day	1.25 m² per 50 l HW consumption (60°C) per day	0.6 m² per 50 l HW consumption (60°C) per day
Storage tank size	30–50 l tank volume per m² of collector surface	50–70 l tank volume per m² of collector surface	40–60 l tank volume per m² of collector surface

- location with average solar radiation (approximately 1000 kWh/m^2)
- south alignment of the collector field and a setting angle of 40°.

The values apply to flat collectors with good performance ($\eta_0 = 0.8$, $k_{eff} < 3.5$ W/m^{2}K). System-related data must then be determined on the basis of building-specific simulations. Further simulation calculations are then required to convert to different temperature levels. The store volumes should be dimensioned fairly generously within the scope of the economic options and the space available.

Detailed building-related design should take place with the aid of recognized simulation programs (see Chapter 10).

5.2 Systems

5.2.1 Systems with domestic water store(s)

The simplest systems for solar energy systems with up to 30 m^2 of collector field are one or two-store systems with domestic water stores and the option of thermal disinfection. The design of the stores corresponds with those in a one or two-family home. As an alternative to immersed heat exchangers providing auxiliary heating, it is possible to use an external store-charging unit (see Figure 5.2). This auxiliary heating method also permits thermal disinfection of the whole domestic water storage system. It has the following advantages and disadvantages compared with systems with buffer storage tanks:

ADVANTAGES

- simple system design
- best utilization of low collector temperatures
- fewer system components
- low costs if materials are sensibly selected
- no need for discharge regulator or heat exchangers.

DISADVANTAGES

- lower heat yields because of thermal disinfection
- under certain conditions domestic water stores may involve higher costs than buffer storage systems.

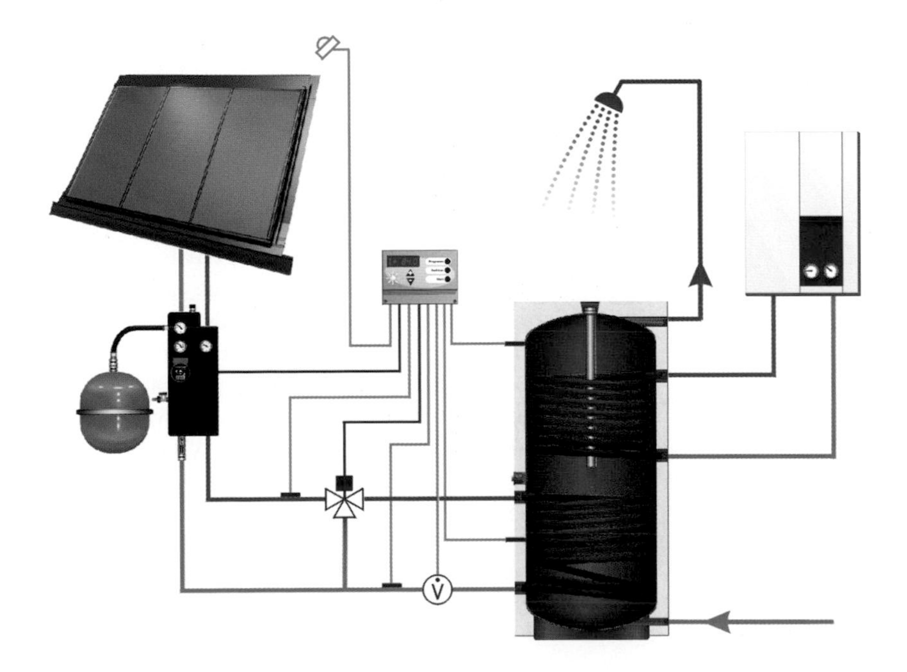

Figure 5.2. Solar energy system with domestic water tank and the possibility of thermal disinfection

Thermal disinfection brings the whole storage system to a high temperature level. In order to minimise the solar yield loss, this should be done in the late afternoon just before a large amount of warm water is drawn off. This ensures the lowest possible storage temperatures in the evening and on the following day, which permits the system to switch on at the earliest possible time next morning. With this system size, both immersed and external designs of heat exchanger can be used.

As the domestic water is heated by the solar system via an internal heat exchanger, a further heat exchanger can be dispensed with, compared with a system with intermediate buffer storage tanks. Thus there are lower temperature losses (temperature difference in the second heat exchanger). In the lower part of the domestic water store, the cold water temperature orientates itself to that of the domestic water supply. In this way the solar energy system can make its contribution even at low temperature levels, and thus achieve high yields.

The need for such thermal disinfection measures is obviously strongly dependent on local drinking water regulations. As thermal disinfection brings about extra energy consumption, it should be avoided when regulations allow.

5.2.2 Systems with thermal buffer stores

An increase in the temperature level in the whole domestic water storage tank automatically leads to higher energy losses in the system considered. With large domestic water stores that serve as solar energy storage systems the losses are greater than in conventional systems for hot water heating with correspondingly smaller stores. Moreover, heating of the energy store leads to a reduction in the collector circuit utilization because of the heat losses associated with the higher temperatures. The energy yield may decrease to an extent of 15% with boundary conditions that are otherwise the same.

Common to the systems described in this section is the fact that the heat gained in the collector circuit is first stored in a buffer thermal store, and is led to the domestic water store only when required. In order to obtain a high system yield, similar to those for systems with more direct energy storage, the system variants described in the following sections are used.

The use of external heat exchangers and – with the exception of stratified stores (and their internal charging) – the type of charging is also common to the systems.

When a minimum solar radiation is reached, the collector circuit pump first starts and heats up the collector circuit. If a useful temperature then exists at the entry to the collector circuit heat exchanger, the buffer circuit charging pump is switched on, and the buffer tanks are charged by means of a switching system, for example a three-way valve or a stratified-charging device in the buffer tank.

5.2.2.1 BUFFER STORAGE SYSTEMS USING THE STORAGE CHARGING PRINCIPLE

If the temperature in the hotter buffer storage zone (the right-hand buffer store in Figure 5.3) reaches a useful temperature level for heating the domestic water in the standby store, then the buffer circuit charging pump and the DHW store-charging pump are engaged. Ideally, the DHW store should be divided into a standby zone for auxiliary heating and a (lower) zone for charging by the solar system.

The upper standby zone is held continuously at the temperature level required for safe supply by the conventional auxiliary heating. Taking account of regulations, the whole domestic water storage area can be thermally disinfected (not shown in the figure). In this way top-up heating of the domestic water and thermal disinfection are carried out exclusively in the DHW store by conventional heating. In comparison with the different concepts for auxiliary heating of the domestic water to the required withdrawal temperatures, this has proved to be the most expedient with respect to the maximum utilization of the solar system.

ADVANTAGES

- Discharging of the buffer tank is possible independently of the current water consumption (see section 5.2.2.2).
- High system utilization is achieved for the whole system with auxiliary heating only in the domestic water tank (the solar zone is not heated via the auxiliary heater, and is always at the lowest possible temperature level).
- The discharge heat exchanger can be kept relatively small and inexpensive.

DISADVANTAGES

- The DHW store requires a solar zone in the lower area, or the temperature of the auxiliary heating must be reduced to allow charging by the solar system even at lower temperature levels.

- The discharge of the buffer storage tank must be controlled so as to keep temperatures as low as possible in the bottom zone of the colder buffer tank.

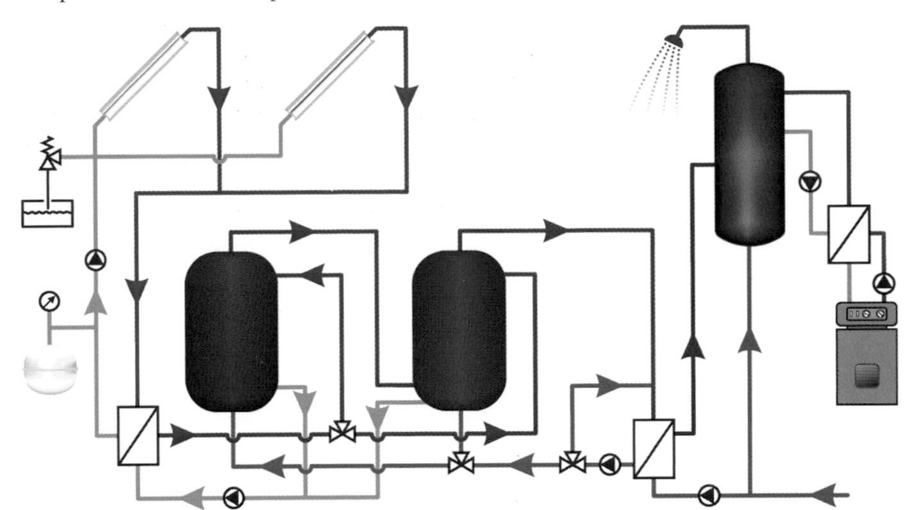

Figure 5.3. Buffer storage system using storage charging principle; top-up heating in the standby tank

5.2.2.2 BUFFER STORAGE SYSTEMS USING THE ONCE-THROUGH-FLOW PRINCIPLE

With this type of system (Figure 5.4) the buffer thermal store is discharged using the once-through-flow principle. When domestic water is withdrawn, and if at the same time a useful temperature level exists in the hotter buffer, this is then discharged via a circuit pump. The solar heat is then transferred to the domestic water. By controlling the volumetric flow in the discharge circuit, the buffer water can thus be cooled to, for example, 5 K above the cold domestic water entry temperature. The precondition for this, however, is very accurate matching of the volumetric flow in the buffer tank discharge circuit to the momentary withdrawal volume, which requires precise and rapid control systems.

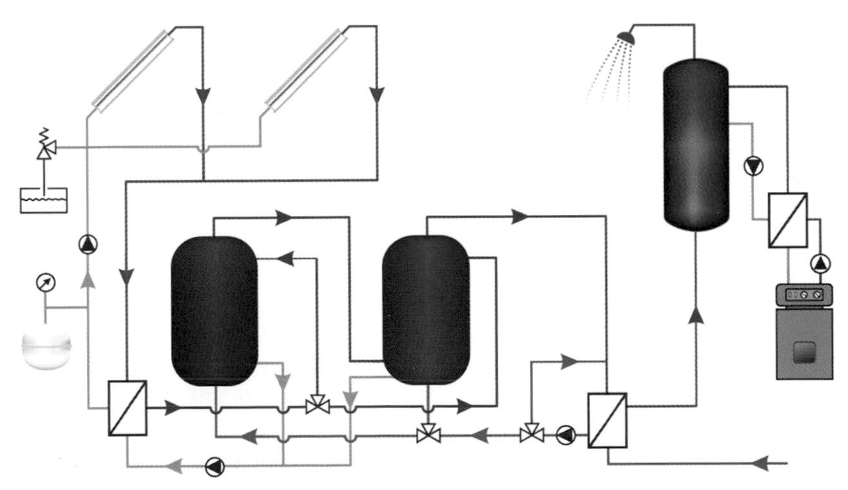

Figure 5.4. Buffer storage system using the once-through-flow principle

Cooling of the buffer tank water is possible for both small and large withdrawal rates, and the colder buffer storage zone thus takes on the lowest possible temperatures. The solar yield is increased because of the heat transfer from the solar circuit at this low temperature level.

ADVANTAGES

- Existing domestic water tanks can continue to be operated unchanged after the installation of the solar system. Even pure once-through-flow systems can be supplemented by a solar system.
- The solar system can be linked very easily into the existing domestic water network through the installation of the discharge circuit heat exchanger.
- A higher solar yield can be achieved through the cold return to the buffer tank if the discharge controller operates perfectly.

DISADVANTAGES:

- Control of the discharge must operate very accurately; in practice this does not usually happen.
- If the withdrawal rates vary considerably and the buffer tank volume is very high, the once-through-flow system becomes sluggish and cannot supply sufficient heat from the buffer.
- The discharge circuit heat exchanger must be designed for medium or maximum withdrawal peaks, and is thus large and expensive.
- Heat exchangers cause pressure losses, which lead to pressure variations in the domestic water supply with changing withdrawal rates, which can cause unpleasant temperature variations with mixer taps.

5.2.3 Integration of circulation systems

See Figure 5.5. In order to use solar heat to cover the heat losses of a circulation line, the following general conditions should be observed:

Again, drinking water regulations greatly influence the temperature regime at which circulation systems are and should be operated. In general, temperatures of >55°C are to be maintained. This temperature level is led back into the DHW store at a given position. Depending on the running time of the secondary circulation, a temperature increase may occur in the colder zone of a store. In addition, this circulation leads to a mixing process that also results in a lower solar yield.

Heating up of the (lower) solar zone of the store by the circulation circuit should be avoided as far as possible. The temperature level, which is significantly higher than the cold water temperature, would otherwise lead to a strong reduction in the collector circuit utilization.

For preheating systems (or systems with low solar fractions) in connection with secondary circulation and a return of >55°C, the linking of a circulation circuit into the solar system makes little sense. In a solar system with a low solar fraction, a temperature level >55°C is achieved on only a few days in the summer, which makes coverage of the circulation losses at this temperature level practically unnecessary.

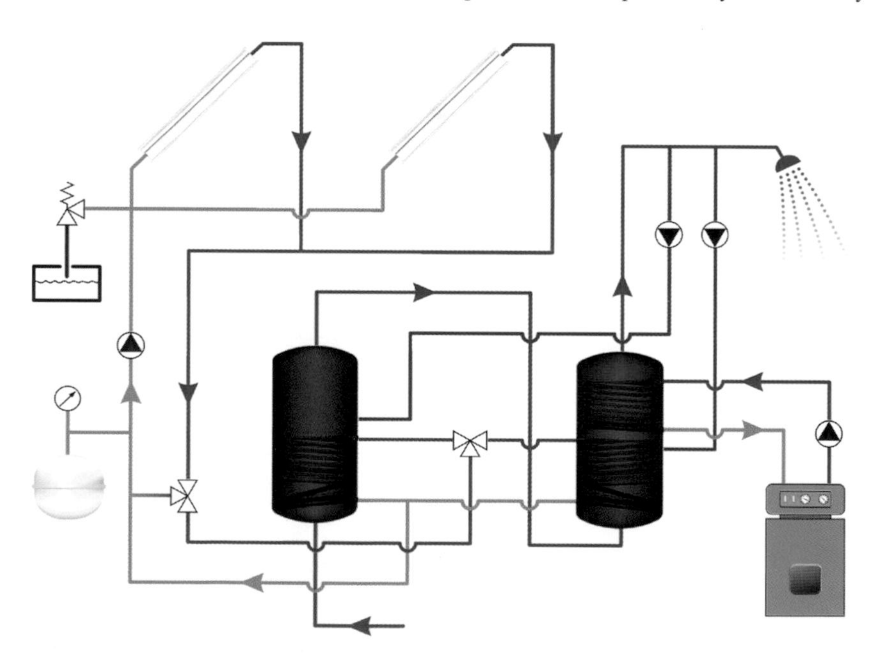

Figure 5.5. Integration of the circulation circuit, according to temperature, into either the middle of the height of the left-hand preheater tank or the right-hand standby tank

5.3 Control of the systems

Basically, in solar systems with buffer stores two systems have to be controlled and regulated:

- the collector circuit/storage charging circuit
- the store discharge circuit.

For simplification purposes, instead of differentiating between regulation and control in the following, the term 'control' will be used.

5.3.1 Collector circuit/storage charging circuit

5.3.1.1 BASIC FUNCTION

See Figure 5.6. If there is a useful temperature difference between the collector field and the solar store, the collector circuit pump and possibly the store-charging pump are switched on, and the store is charged. If store charging is no longer possible owing to a decrease of the temperature difference, the controller switches off the pumps, and unwanted discharging of the solar store via the collector field is thus prevented. Almost all control concepts are based on the principle of temperature difference measurement and on corresponding programs for controlling the different systems.

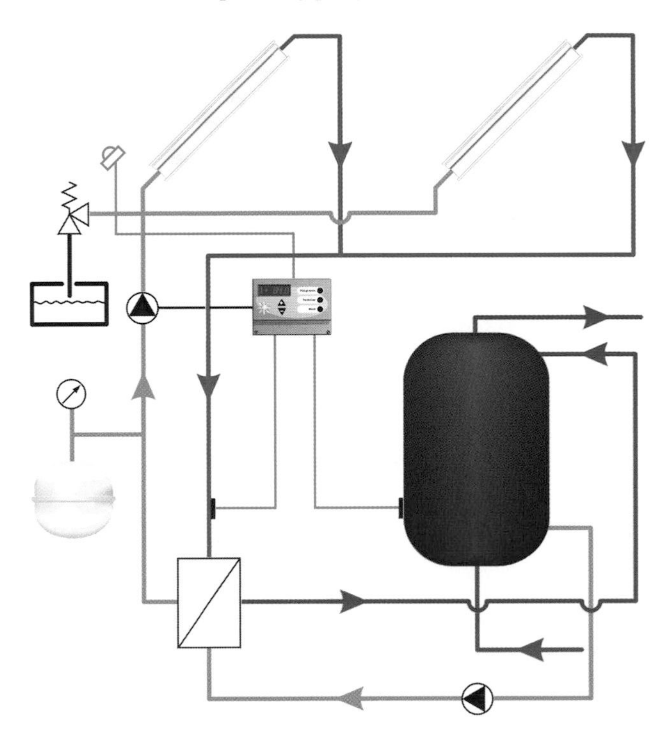

Figure 5.6. Control of storage tank charging

The heat exchangers used in the collector circuit also influence the switching of the pumps. A small heat exchanger in the store with a reducing temperature difference between the heat transfer medium and the store can only supply an ever-smaller heat quantity to the storage water. In spite of higher temperatures in the collector, the controller must switch off the collector circuit pump in this case to avoid inefficient circulation around the collector circuit. It is therefore necessary to adapt the switch-off temperature difference to the heat exchanger performance.

5.3.1.2 CONTROL STRATEGY

Most systems and control concepts found on the market today follow the strategy of charging the solar store when an adjustable temperature difference between the collector field (or feed temperature on external heat exchanger) and the store (temperature in the lower zone) is reached. Thus, in systems with DHW storage, charging from the collector circuit starts for example as early as from a temperature of 13°C in the collector (8°C cold water entry plus 5 K temperature difference). During charging, this temperature level increases until the maximum storage temperature (60–95°C) is reached.

The collectors, circuits and heat transfer medium have a given heat capacity. The longer the lines in the collector circuit and the greater the pipe diameter, then the greater the heat capacity in the collector circuit. The system inertia therefore increases. In systems of the size under consideration it is usual to first allow the collector circuit to 'run up to heat' via a bypass circuit (this avoids cooling of the store). The bypass circuit is not used in small solar systems with short line paths because of the lower heat capacities in the collector circuit.

5.3.1.3 BYPASS CIRCUIT WITH IMMERSED HEAT EXCHANGER AND THREE-WAY VALVE

Apart from the collector circuit pump, the collector circuit controller also manages a three-way valve (Figure 5.7). Until a useful temperature is reached at the line just before the inlet to the store, the collector circuit is switched to bypass operation and is led past the store heat exchanger. This switching system is used only in connection with immersed heat exchangers.

5.3.1.4 BYPASS CIRCUIT WITH EXTERNAL HEAT EXCHANGER AND TWO PUMPS

Apart from the collector circuit pump, the collector circuit controller can also manage the storage circuit charging pump (Figure 5.8). Until a useful temperature is reached at the line shortly before the inlet to the external heat exchanger, only the collector circuit pump is operating. If the useful temperature is reached, the storage circuit charging pump starts to operate.

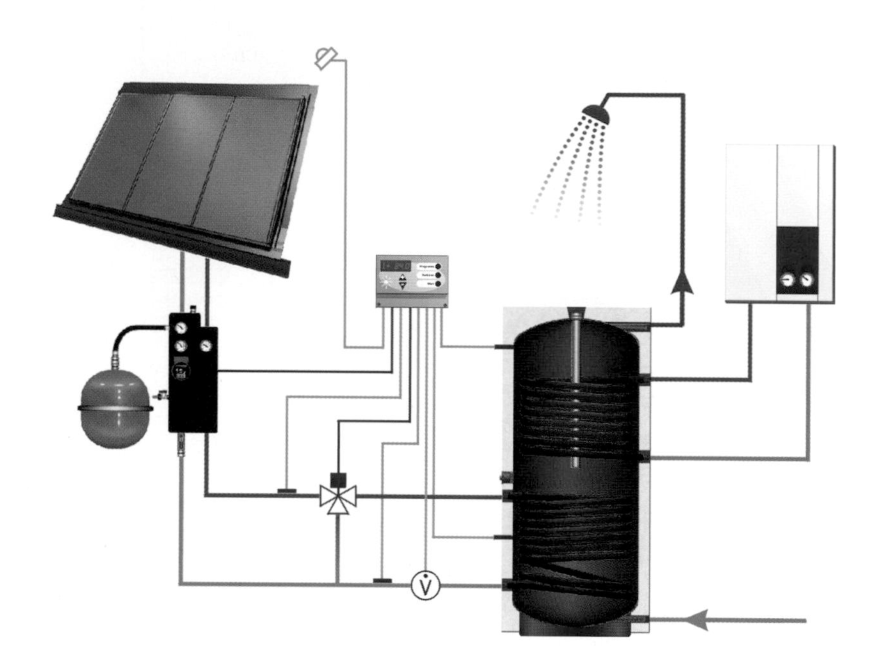

Figure 5.7.
Bypass circuit with three-way valve

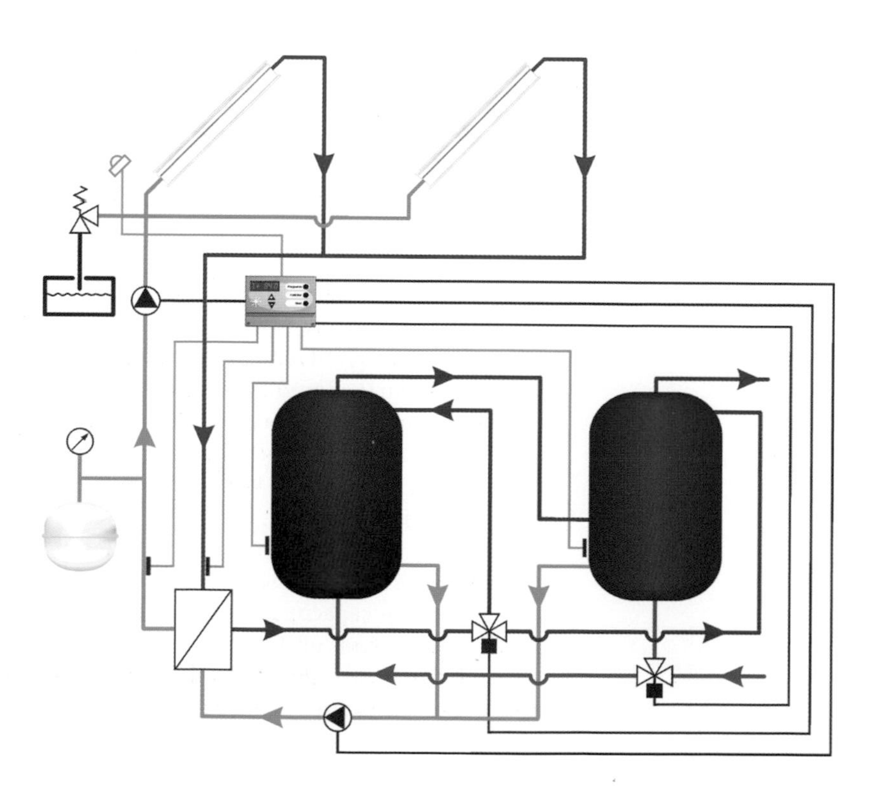

Figure 5.8.
Bypass circuit with separate control of collector and buffer tank circuit pumps

This variant is examined in more detail below, in view of the widespread use of external heat exchangers

CONTROL OF THE COLLECTOR CIRCUIT BY MEANS OF RADIATION AND TEMPERATURE SENSORS

A radiation sensor can be attached to the collector field and connected to the collector circuit controller. With sufficient solar radiation, the collector circuit pump is started up. To start the pump it is advisable to compare the measured radiation value with the store temperature. During system operation, the controller continuously gathers information on the activity of the collector circuit pump with corresponding radiation values as well as the store temperature. For instance, at a store temperature of 50°C, the collector circuit pump does not start up until the solar radiation exceeds 500 W/m^2; at a store temperature of 10°C, the pump switches on as early as at 150 W/m^2; and so on. This avoids having the collector circuit pump start up at a solar radiation level that is too low for further charging of the store. Moreover, the controller is matched to the changed gain and loss variables in the collector circuit through the continuous gathering of the new temperature and radiation values.

If the heat transfer medium in the collector circuit at the temperature sensor in front of the heat exchanger inlet reaches a useful temperature difference of $\Delta\theta$ = 2–5 K above the store temperature, the charging pump is started. If $\Delta\theta$ falls below 1–3 K the buffer circuit charging pump is stopped. Moreover, if the intensity of the sun's radiation measured at the radiation sensor decreases below a threshold, the collector circuit pump is also switched off. The system is not restarted until the intensity of the sun's radiation increases again.

If the collector circuit pump is started at a fixed value, for example 130 W/m^2, this can lead to extremely long operating times of the pump without yields. For example, the solar radiation can be just slightly above this value for the whole day but the collector will still not reach a useful temperature level. Also, when the store is already very hot, a renewed starting of the collector circuit pump in the bypass can be pointless (for example, if the sun comes out after an afternoon shower, the collector temperature will not exceed the store temperature even by late evening).

Shading of the radiation sensor must be avoided. For example, the solar cell can be put out of action by bird droppings: this type of fault should be considered in the case of defects.

CONTROL OF THE COLLECTOR CIRCUIT BY MEANS OF TEMPERATURE SENSORS

The collector circuit pump can be switched on if the temperature at a temperature sensor in the collector is 5–7 K above the temperature at the lower store sensor. If a temperature of 2–5 K above the store temperature is measured at the sensor in front of the heat exchanger inlet, the store-charging pump is also started. If the temperature difference falls below 2–4 K the pumps are switched off one after the other.

5.3.1.5 STORAGE TANK CHARGING AT DIFFERENT TEMPERATURE LEVELS

See Figure 5.9. To achieve useful temperatures in the stores rapidly, they should be divided into different temperature levels. This can be done by series connection on the discharge side, in which the store is charged alternatively into the cold or hot storage area. With the use of valves and pumps to control the different tanks the temperature levels should be limited to two, independently of the system size, as with the increasing use of pumps and valves the susceptibility to faults also increases greatly, without the solar yield being significantly increased.

Charging of the two levels can alternatively take place by means of a three-way valve or a second charging pump. After a useful temperature for charging a storage level has been measured at the temperature sensor in the collector circuit in front of the heat exchanger inlet, the three-way valve is opened at the corresponding charging level. With the use of two storage tank charging pumps instead of the three-way valve the corresponding storage tank charging pump is started to charge the usable level. A priority storage area (charging level) is usually defined in the controllers for charging. This charging level is then loaded as a priority, and only if no further temperature increase is possible is the second charging level used.

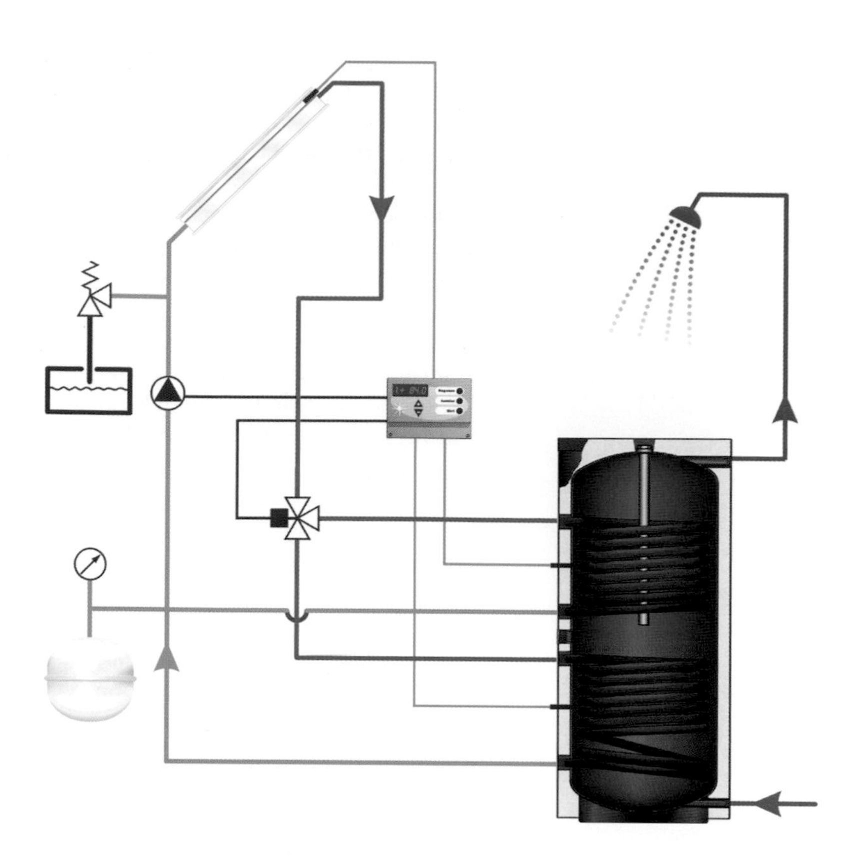

Figure 5.9.
Charging a storage tank with different temperature levels (immersed heat exchanger)

5.3.1.6 STORAGE TANK LOADING IN THE CASE OF TANKS WITH STRATIFIED CHARGING EQUIPMENT

See Figure 5.10. When using layer charging equipment it is also possible to charge several stores in a series-connected system. The heated water always exits through the charging equipment in the warmer store at the charging level corresponding to its temperature. In this way, in the case of high temperature levels the warmer store is charged in the upper area. If the charging temperature level falls, the heated water exits the charging equipment at ever-lower levels. If the temperature level of the loading circuit in the warmer tank can no longer be used, the water is fed into the corresponding temperature level of the second (colder) tank via the alternative route of the lowest layer.

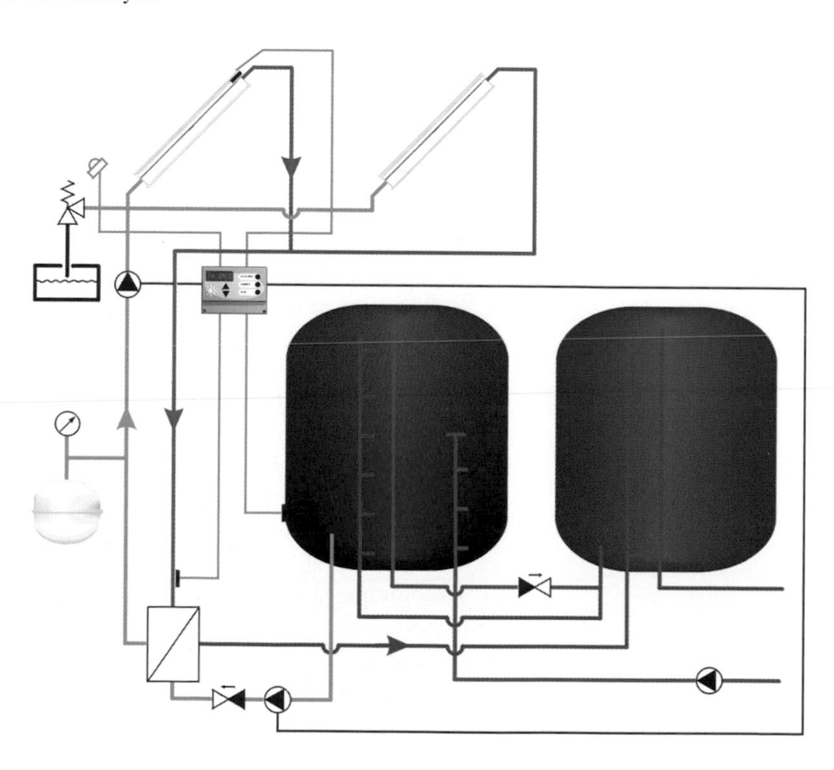

Figure 5.10.
Storage tank loading for tanks using stratified charging equipment

This storage loading system requires a controller that has been especially matched. This only measures the temperature in the coldest tank; charging of the different levels takes place automatically through the layer charging equipment without the use of further valves or pumps (see also Chapter 2, section 2.3.2.1).

5.3.2 Buffer tank circuit discharging

According to the discharging strategy, correspondingly adapted control concepts are used according to either the storage charging principle or the once-through-flow principle.

To avoid possible limescale build-up in the discharge heat exchanger of the buffer storage system and, moreover, to establish a maximum temperature for the design of the heat exchanger, the temperature in the inlet of the discharge heat exchanger is limited for example to 55°C (but taking into account drinking water regulations). This takes place through a mixing process by a thermostatic mixing valve in the inlet or a mixer in the return line from the heat exchanger. The thermostatic mixing valve is set to the desired maximum inlet temperature to the heat exchanger. The three-way valve or the three-way mixer in the return line is controlled by means of a fixed-value controller or, for smaller systems, also by a thermostat head with remote sensor. The accuracy of the controller thereby depends on the running time of the valve/mixer and the speed of the signal processing in the fixed-value controller. For systems with discharge controllers for controlling the once-through-flow principle, which react very quickly, the feed restriction is therefore often dispensed with, as its control speed is too low in comparison with the discharge regulation.

5.3.2.1 ONCE-THROUGH-FLOW PRINCIPLE

The objective of this principle is the best possible cooling of the buffer water during discharging or the maximum temperature increase in the domestic water. To meet both objectives the volumetric flow of the discharge circuit must be matched to the (variable) volumetric flow of the domestic water circuit. This is done by cycling (or speed controlling) the discharge pump.

The discharge controller measures the temperature difference between the top of the (warmest) buffer tank and the cold water. If the buffer is 2–5 K warmer than the cold water entry, the pump for the buffer discharge circuit is started up at, for example, 10% of its nominal flow rate. In this way the inertia of the discharge circuit is first overcome, and a useful temperature is achieved in the discharger heat exchanger for immediate use at the hot water taps. The controller monitors the flow in the domestic water circuit. If a flow is measured here the buffer discharge pump is started and the volumetric flow from it is regulated according to the criterion 'coldest possible return flow'. For this purpose, with the discharge pump running, the temperature difference is measured between the entry temperature of the cold water and the exit temperature of the return flow.

If the buffer circuit discharge pump is cycled by the controller (or is speed-regulated), a selectable temperature difference is set up between the cold water entry and the return flow of the buffer discharge circuit of, for example, 5 K. The buffer circuit can thus be discharged down to the temperature of the cold water entry plus 5 K.

For this control variant it is necessary to use precise and fast-reaction sensors. Moreover, the controller must possess high scanning rates when acquiring measured values. The sensors must be installed according to the manufacturer's instructions using immersion sleeves. To avoid pressure losses or possible error sources the measurement of the flow should take place if possible in the domestic water circuit by means of temperature comparisons instead of flow monitors or the like. Additional manufacturer's instructions concerning the positioning of sensors must be taken into account with respect to the yields.

Note: In practice these systems do not usually operate without faults. They should therefore only be used in discharge units that have been preconfigured by the manufacturer. In any case great care is necessary when installing them.

5.3.2.2 STORAGE TANK CHARGING PRINCIPLE

The discharge controller compares the temperatures in the (warm) buffer and in the bottom area of the domestic water store. If the temperature in the buffer is 2–5 K higher, the pumps are started to discharge the buffer. If the temperature difference falls below 1–2 K both pumps switch off.

During charging, the temperature in the domestic water store increases and hence also in the buffer discharge return. To avoid too severe an increase in the return flow temperatures in the buffer discharge, the temperature in the lower zone of the domestic water store is often limited. To discharge the buffer store, temperature difference controllers should therefore be installed to limit the maximum temperature via the measuring sensor (= lower storage sensor). The maximum temperature can be set for example to 30°C.

If the domestic water storage cools down as a result of more water being drawn off, then charging continues if a useful temperature level exists in the buffer tank.

To avoid undesirable heating up of the colder buffer tank during the charging operation, an additional three-way valve can be installed in the return line of the discharge circuit. If a higher temperature is then measured at the temperature sensor than in the colder buffer, then the three-way valve switches over the flow to the warmer buffer tank.

The three-way valve is not necessary for buffers with stratified charging equipment, as the buffer discharge return line is led into the respectively matched temperature level through the stratified charging equipment.

5.4 Heat exchangers

5.4.1 Design types

5.4.1.1 IMMERSED HEAT EXCHANGERS

Stores with immersed heat exchangers are most often installed in solar systems for one- and two-family homes. These internal heat exchangers are wound with (Cu-) plain, finned tube or plain steel or stainless steel tube. For their design, and for a logarithmic mean temperature difference of 10 K (see also the following pages), the following approximate formulae are valid:

- plain tube heat exchangers: 0.2 m^2 area per m^2 of collector field
- finned tube heat exchangers: 0.3–0.4 m^2 area per m^2 of collector field.

In the case of plain tube heat exchangers the energy is transmitted via the surface of the tube. As the temperatures of the entire surface of the tube and that of the medium inside the tube are very close, the same temperature difference between the medium in the tube and the surrounding medium is available over the whole surface. On the other hand the average temperature difference between the surface of a finned tube heat exchanger and the surrounding medium is lower than that for plain tube heat exchangers because of the lower temperature at the ends of the fins (see Figure 5.11).

One square metre of surface on the plain tube heat exchanger can therefore transmit more energy than one square metre of the surface of a finned tube heat exchanger. The surface area of a finned tube heat exchanger is, however, significantly increased by the fins, so that a finned tube heat exchanger – in spite of a lower transmission performance per square metre – is more compact than a plain tube heat exchanger with the same performance.

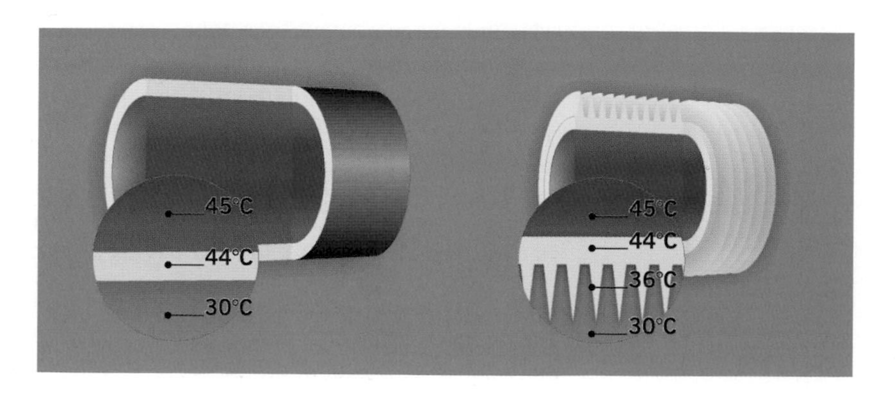

Figure 5.11. Temperature progressions in plain and finned tube heat exchangers

In large solar energy systems, immersed heat exchangers take up a significant volume in order to increase the efficiency of the collector field, especially if the logarithmic mean temperature difference is to be limited to 5 K. If several stores are used, then one heat exchanger with the full transmission performance is required for each store, which leads to higher costs. In larger solar systems, therefore, external heat exchangers are usually installed so that all stores can be charged with just one heat exchanger. Even if a further pump is required for these systems, this variant is preferable.

5.4.1.2 EXTERNAL HEAT EXCHANGERS

For external heat exchangers we similarly differentiate between tubular and flat plate heat exchangers. Tubular heat exchangers made from stainless steel are mainly used in solar energy systems for heating the water in swimming pools.

In the case of flat plate heat exchangers (Figure 5.12) we can differentiate between screwed and soldered models. In the *soldered* version, pressed stainless steel plates are soldered together. In the *screwed* version, the stainless steel plates are fitted with seals and then screwed together with threaded rods. Soldered plate heat exchangers can normally be obtained up to particular performance sizes, and in the smaller performance area are cheaper than the screwed variants. For the solar energy systems considered in this chapter, soldered plate heat exchangers using the counter-current principle are preferred.

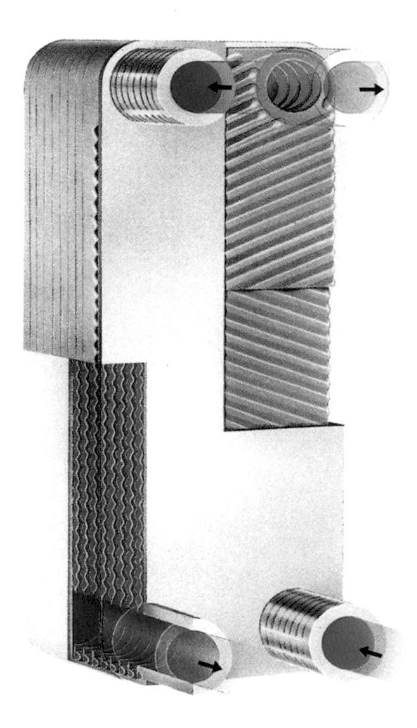

Figure 5.12
Cross-section through a plate heat exchanger (SWEP, Hildesheim)

To avoid corrosion damage, soldered plate heat exchangers should not be used for heating swimming pool water.

Heat exchangers differ from one another on the basis of the plate geometry, the once-through-flow and the construction. A heat exchanger can therefore not be replaced by another model from the same or another manufacturer without a recalculation.

5.4.1.3 COMPARISON OF IMMERSED AND EXTERNAL HEAT EXCHANGERS

Advantages of immersed heat exchangers:

- Simple system construction with few components.
- Without domestic water being drawn off, mixing takes place only by convection in the individual tanks.

Disadvantages of immersed heat exchangers:

- With several stores, a heat exchanger dimensioned for the full performance of the collector field is required for each tank: this leads to high costs.
- Highly stratified charging systems can be used only with special heat exchangers.

Advantages of external heat exchangers:

- With several stores, costs are lower than for immersed heat exchangers.
- Stratified charging systems are simpler to implement.

Disadvantages of external heat exchangers:

- Additional components; complicated installation on site.
- In the event of an unfavourable arrangement of the inflow on the charging side, disturbance to the temperature stratified effect in the tanks.

5.4.2 Collector circuit heat exchangers

See Figure 5.13. Collector circuit heat exchangers are designed for a maximum performance of 600 W/m^2 of collector field.

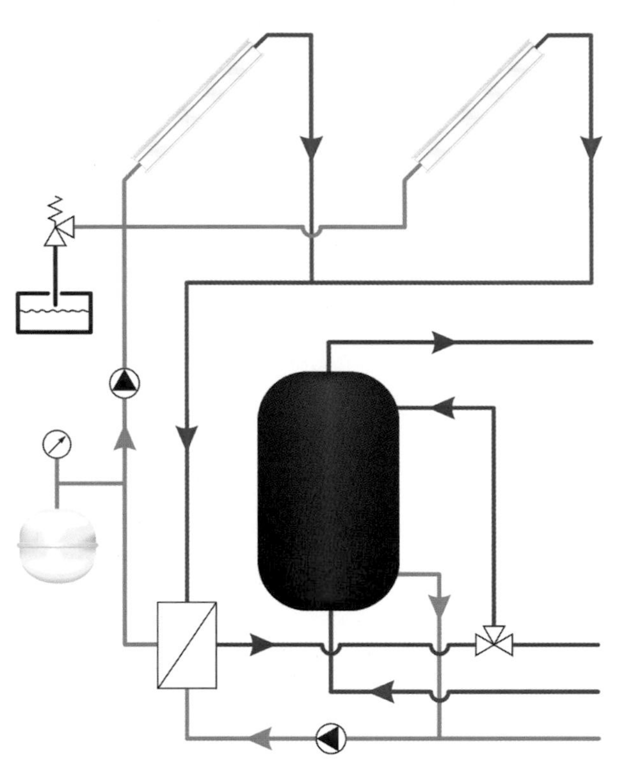

Figure 5.13. External collector circuit heat exchanger

The figure 600 W/m^2 is a guide value that arises from averaged values for the irradiated power of the sun and the efficiency of the collectors. The assumptions for this are an irradiance of 1000 W/m^2 of collector field and an assumed efficiency of 0.6.

Through the dynamics of the irradiated power and the operating conditions in the collector circuit, this value is rarely achieved. On the other hand, the value of 600 W/m^2 of collector field can indeed be exceeded briefly. For these cases the temporary under-dimensioning of the heat exchanger and the resulting increase of the temperature level in the collector circuit are acceptable, in order to keep the heat exchanger costs within a sensible range.

The temperature spread of the collector circuit (inlet/outlet of heat exchanger) is obtained from the following formula:

$$\Delta\theta = \frac{\dot{Q}_{col}}{\dot{m} c_{G,W}}$$

where $\dot{Q}_{col}$ is the collector performance (W), $\dot{m}$ is the mass flow (kg/h), and $c_{G,W}$ is the specific heat capacity of the solar fluid (Wh/kgK).

Collector systems have different temperatures at different times of the day depending upon the solar radiation and the charging condition of the store. For dimensioning the heat exchanger, the temperature at the start of storage charging is used. At the outlet of the heat exchanger (collector circuit return line) a temperature of 5–10 K above the cold water temperature is used for systems with buffer stores; for systems with domestic water stores a temperature of 0–5 K above the cold water temperature is used. These temperature differences occur because a buffer circuit that is connected between the collector circuit and the domestic water system is at a higher temperature than the cold water temperature because of the necessary temperature difference for heat transfer. The collector circuit has a higher temperature than the tank temperature owing to the necessary additional heat transfer.

For central Europe, with an average cold water temperature of 12°C, this would amount to temperatures of 22°C and 17°C respectively. In the example in section 5.4.3, the temperature pairs are used for systems with buffer storage tanks. In warm climates the cold water temperature may be higher – up to 25°C or more.

The inlet temperature into the heat exchanger is derived by adding the outlet temperature in the primary solar circuit and the temperature spread arising (see formula above). On the secondary side of the heat exchanger the temperature spread θ_A–θ_E is equal to the temperature spread θ_A–θ_E on the primary side, where θ_A = exit temperature and θ_E = entry temperature. At the entry to the secondary circuit (buffer charging circuit) the temperature is 17°C. The volumetric flow on the secondary side is obtained from the calculation of the heat exchanger.

Note: Because of the different designs of heat exchangers every manufacturer achieves different values for flat plate heat exchangers with the same exchange surfaces. Therefore the heat exchangers used must be calculated with a program from the respective supplier.

5.4.3 Buffer tank discharge circuit heat exchangers

For systems that, apart from the collector circuit heat exchanger, require an additional heat exchanger to transfer the heat to the domestic water circuit, the following considerations are necessary:

- To avoid limescale build-up, a constant entry temperature should be set before the inlet to the heat exchanger by means of a thermostatically controlled three-way mixing valve (see Figure 5.14). This should be between 55°C and 60°C or, for designs with auxiliary heating in the buffer tank, between 65°C and 70°C.
- In selecting the heat exchanger it is necessary to consider the possibility of subsequent cleaning, bearing in mind particularly the lime content of the water. Considering the better cleaning possibilities and the avoidance of a type of crevice corrosion in the heat exchanger soldering, screw-fitted heat exchangers are recommended here. These can be reactivated by (cost-intensive) dismantling and cleaning if the flushing/cleaning processes are ineffective. However, for lower performance levels, the much higher price compared with that of soldered heat exchangers makes their use undesirable.

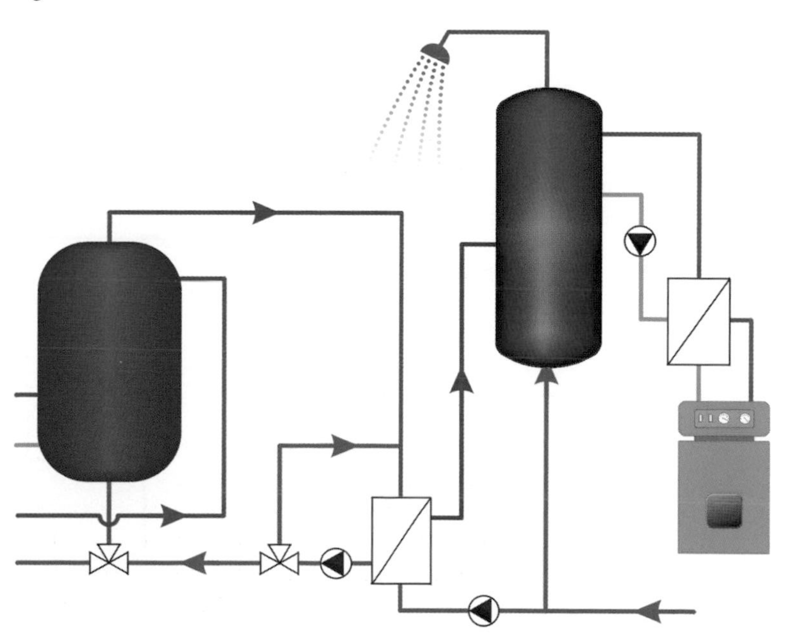

Figure 5.14.
Heat exchanger using the storage charging principle

5.4.3.1 STORAGE CHARGING PRINCIPLE WITH AUXILIARY HEATING IN THE STANDBY STORAGE

The following pairs of temperatures can be used for the basis of the design of the buffer discharge circuit heat exchanger (using the same example as before):

- Buffer tank circuit side:
  Inlet: 55°C
  Outlet: 17°C

- Domestic water side:
  Inlet: 12°C
  Outlet: 45°C

Owing to a large temperature spread in the discharge circuit of the buffer store, the highest possible cooling of the buffer is achieved – even with charge conditions at high temperature levels. This temperature spread is established on the basis of practical experience in which the domestic water entry temperature is preset by the domestic water mains. The volumetric flows for the calculation of the heat exchanger are obtained from the thermal energy to be transferred. The energy to be transferred by this store-charging system is designed according to the water consumption profile and the collector field efficiency.

In the same way as for solar stores in one- and two-family homes, the domestic water store (in the case where the store is charged through a buffer circuit) should be charged in the lower area if possible, which means it then has a dedicated solar storage area. This is then, apart from regular thermal disinfection by a boiler, only charged by solar heat from the buffer circuit. Auxiliary heating by the hot water boiler is then carried out only for the upper standby section of the domestic water tank.

In the following section, the heat exchangers for charging the solar area in the domestic water store are calculated.

5.4.3.2 ONCE-THROUGH-FLOW PRINCIPLE

With the once-through-flow principle, the heat exchanger (Figure 5.15) is matched to the maximum domestic water consumption peak so as to be able to discharge the buffer circuit at the moment of highest consumption. This is an essential precondition, as otherwise the heat in the buffer is not transferred sufficiently at the peaks and the tank can only be discharged partially. If only a small part of the overall consumption is withdrawn at peak times, the heat exchanger can be designed for example to 50% of the calculated or measured withdrawal peak.

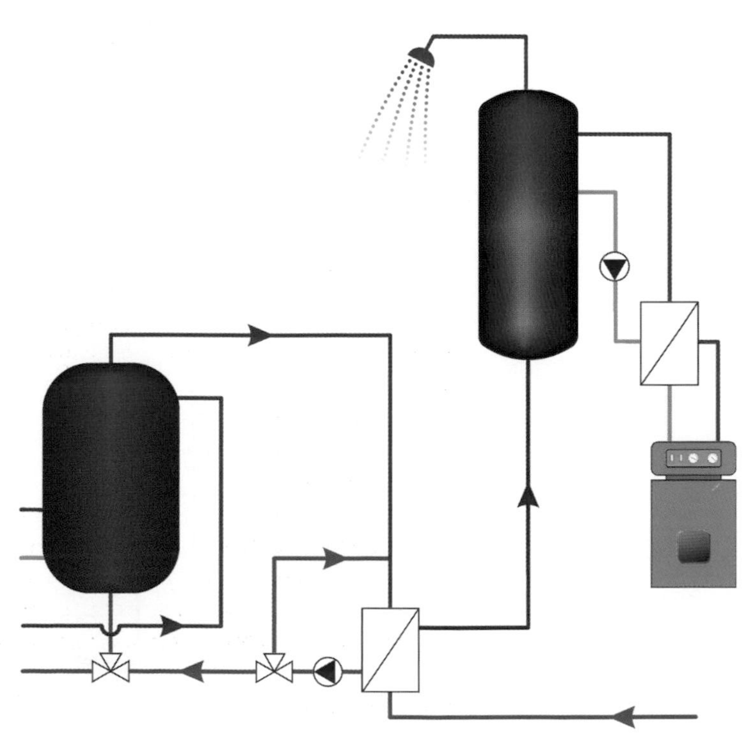

Figure 5.15.
Heat exchanger using once-through-flow principle

In practice, systems are designed to achieve an adjustable outlet temperature on the domestic water side (only for systems with high solar fraction or auxiliary heating in the buffer storage tank) or to the lowest possible return temperature in the buffer tank circuit.

The following temperature pairs are used for the design (again, using our example):

- Buffer tank circuit side
  Inlet: 55°C
  Outlet: 17°C

- Domestic water side
  Inlet: 12°C
  Outlet: 45°C

As the basis for calculating the required maximum performance, the highest withdrawal peaks must be established, for example with the help of the simultaneity model (estimation by multiplying the flow through a single consumption point by the square root of the number of consumption points). The performance is calculated from the mass flow of the withdrawal peak and the desired inlet/outlet temperatures. Calculation of the performance peak (in kW):

$$\dot{Q} = \dot{m} c_W \Delta\theta$$

with $\dot{m}$ from the simultaneity calculation for the withdrawal peak, $c_W = 1.16$ Wh/kgK and $\Delta\theta = 45°C - 12°C = 33$ K.

In this once-through-flow variant, it is essential to take account of the increased pressure loss (through the heat exchanger) in the domestic water system. At peak withdrawal times supply problems could otherwise occur (in the form of pressure and temperature variations).

However, not all of the heat exchangers designed for these peaks function satisfactorily in the partial load area. Very low or virtually no pressure losses in the heat exchanger calculation clearly show a heat transfer that is too low in the partial load area of the calculated operating case. The reason for this is the low k-value of the heat exchanger, caused by laminar flow in this performance range which will reduce the heat transfer rate per unit of transfer area. During design and installation the selected heat exchanger must therefore be recalculated in discussions with the manufacturer. This problem can be reduced or circumvented by designing the heat exchanger to, for example, a maximum of 50% of the peak withdrawal rate.

5.4.3.3 COMPARISON OF THE ONCE-THROUGH-FLOW PRINCIPLE AND STORAGE TANK LOADING PRINCIPLE

The greatest advantages of the once-through-flow principle are a (theoretical) very low return temperature to the buffer tank and simple integration into the existing system. The advantage of a lower return flow temperature can, however, be achieved in practice only with correct heat exchanger dimensioning and accurate installation of the control system. For solar systems with large buffer volumes and circuit dimensions, the system inertia should be considered in relation to the control of the once-through-flow principle. Considering the problems that currently still exist, the greatest care is needed when using the once-through flow principle.

The store-charging principle can also be used where very high performances have to be transferred with inexpensive heat exchangers and without any effects on the domestic water network (no pressure variations). With a suitable controller, its function is the same as the once-through-flow principle. If the connection of store charging is implemented as shown in Figure 5.15, the system behaves in its operation like a combined storage charging/once-through-flow system. If the amount of withdrawn hot water is low, the whole domestic volumetric flow is led through the heat exchanger. If no hot water is drawn off, then the system operates according to the storage charging principle by using the storage charging pump. If very high quantities of hot water are drawn off, cold water is led through the heat exchanger according to the volumetric flow of the charging pump. The rest flows into the store and can later be heated by solar energy.

5.5 Safety technology

5.5.1 Collector field

If the collector system is classified as steam or pressure equipment (such as the European Pressure Equipment Directive 97/23/EC, see section 12.2 and/or national regulations), strict additional requirements apply. For instance, in Germany, steam boiler regulations for a Group III steam boiler system (up to 50 l volume) (for limits see Appendix B 'Technical regulations'), stipulate that a larger collector field must be divided into partial fields with 50 l in each. The individual collector fields must then be provided with closing devices, which must be protected from unintentional closing, and safety valves (Figure 5.16). The respective safety valve must, however, not be closable with respect to the collector leg. The safety valves on the field section/collector leg must be authorized for temperatures up to 225°C and also have a common discharge to a collecting reservoir in the building. The blow-off lines must be made of copper (for corrosion resistance). Their exit cross-section and the cross-section of the safety valves must meet the steam regulation requirements; the discharge lines from the safety valves must thereby be steam-safe and discharge without danger to personnel. The dimensions for the safety valves can be taken from Table 5.3.

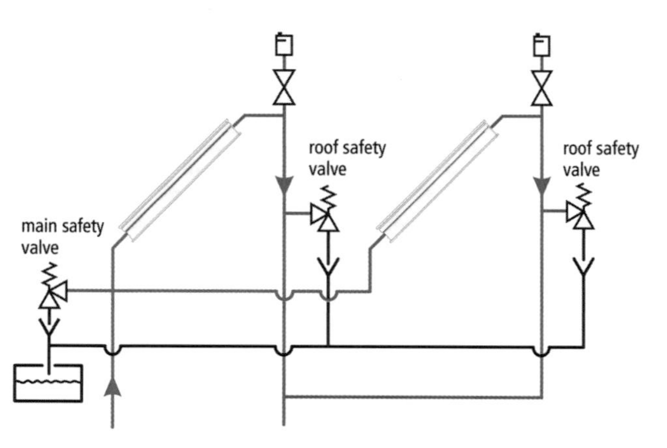

Figure 5.16. Arrangement of the safety valves when dividing the system into partial fields

Table 5.3. Safety valves (entry cross-section) for collector fields

Valve size (entry cross-section) DN	Collector surface area (m²)
15	50
20	100
25	200
32	350
40	600

If the solar energy system is not designed as a steam boiler system, according to the new formulation of the European Pressure Equipment Directive, it is still necessary to test the ability to shut off and correspondingly secure individual parts of the collector field, in order to be able to carry out any necessary maintenance to the collectors without too much expense.

If the path from the roof to the main safety valve is long, the discharge from the safety valves in the collector field should be into individual containers close to the field to save money. These containers must be temperature- and weather-resistant; galvanized steel should not be used, in order to avoid reactions with the heat transfer medium.

5.5.1.1 EVAPORATION IN THE COLLECTOR FIELD

With the high stagnation temperatures of modern collectors, boiling of the heat transfer medium simply through pressures significantly above 10 bar cannot be ruled out. As solar energy systems are normally fitted with standard components for up to

PN10, the manufacturers and designers limit themselves to a maximum pressure of 6 bar or 10 bar in the collector circuit. Vapour formation is thus entirely possible in the collector field at 6 bar. If the solar system is designed with a high degree of solar coverage, long and frequent stagnation times can be expected. In such a situation the following should be noted.

If such systems are operated at high pressure, then boiling in the collector cannot safely be avoided, but – because of the pressure – it takes place at a higher temperature. As a result the whole filled volume of the collectors is subject to a high thermal load. Conventional glycol mixtures can be subject to accelerated ageing because of this. Therefore systems in which long stagnation times are expected should be set at a maximum operating pressure of, for example, 4 bar; or a heat transfer medium should be selected with a high thermal resistance for flat and evacuated tube collectors.

Below the temperature and pressure conditions described in the vapour pressure chart for the heat transfer medium, no vapour phase will form. A glycol/water mixture of 40/60% by volume will vaporize above a temperature of about 170°C at a pressure of for example 8 bar in the collector circuit. Below this temperature, or with a higher pressure, the heat transfer medium will remain in the liquid phase.

The collectors of such systems are designed for the evaporation and resulting condensation of the heat transfer medium. This affects the whole content of the collector as well as a part of the connecting lines because of the temperature progression in the collector and the collector circuit. Evaporation may well affect the whole collector circuit (particularly with short lines, for example with roof central heating systems), depending on the collector type. This vapour volume must be taken into account when calculating the expansion vessels. It is essential to take into account the dangers from vapour escape during the installation of automatic bleeders or safety valves in the vapour formation zones. Automatic bleeders must therefore be closed or shut off. The exit openings on safety valves in the potential vapour formation zones must be fitted so that there can be no danger to personnel.

In drainback systems, the controller usually switches off the pump when the water in the store reaches the maximum allowed temperature. The collector field will then drain, and boiling of the collector fluid is thus avoided. The collectors will go into 'dry' stagnation. Although this is an elegant solution, refilling of the collector circuit while it is stagnating at very high temperatures may need special design attention.

5.5.2 Collector circuit

The (main) safety valves should be dimensioned as per Table 5.3, in which the nominal widths must correspond as a minimum with the values shown.

The safety valves must be matched to the heat performance of the collector leg and/or the collector field, and must also be able to discharge the maximum performance of the collector field in vapour form. In inherently safe systems (in which even at system stagnation the expansion vessel is capable of taking the increase in volume of the system content and the corresponding vapour volume in the collector field) discharge is also permitted in the liquid phase at the main safety valve. The main safety valve must be authorized for temperatures up to 120°C. The response pressure of the safety valve must correspond with the maximum operating excess pressure permitted for the collector. The closing pressure of the valve must be a minimum of 90% of the response pressure.

The response pressure of the main safety valve on the collector circuit should be designed so that it is significantly (1 bar) below the response pressure of the field safety valves. This ensures better monitoring of the system function (excess medium first escapes into the connection space). For systems with a static head of water of more than 10 m at the installation location of the main safety valve, it can have the same response pressure as the valves in the collector field. Owing to the higher system pressure at the installation location (field pressure plus static pressure) the main safety valve will be the first to open in the event of a pressure rise.

The blow-off line must be installed so that in the event of a vapour escape there can be no risk to personnel. The blow-off line is led into a collecting tank to catch the glycol/water mixture. The blow-off lines from the safety valves in the collector field should also end in this collecting tank. This must be temperature-resistant and must be able to take at least the liquid volume (= collector content + connecting lines) displaced by the vapour in the collector field. If possible, a capacity to take the whole system contents is sensible for filling and repair work. For refilling the plant, a manually operated pump should also be available.

5.5.2.1 REQUIREMENTS OF THE COLLECTOR CIRCUIT

A typical flat-plate collector may reach stagnation temperatures above 200°C; evacuated tubes reach temperatures up to 280°C. The high temperatures occur during system stagnation – that is, in the event of failure or the desirable switching-off point of the pump in the collector circuit. If the system starts up again after such a switching off at high temperatures, temperatures of up to 130°C can arise in the collector circuit.

In order to operate the system safely even at high temperatures, a maximum pressure of 6 bar is usually permitted in the collector circuit and for systems with high static pre-pressure 10 bar. The operating pressure is usually 0.5–0.8 bar above the static pressure of the system to safely prevent the penetration of air. In individual design concepts, an operating pressure of 4 bar has been selected to avoid evaporation in the collector circuit during complete system operation. Further descriptions about system pressure can be found in section 5.5.4.

5.5.3 Buffer store circuit

Steam laws and regulations vary widely between countries. Here, an example is given for Germany. The buffer store is classified as a pressure vessel according to Section 8 of the German Pressure Vessel Regulations:

- Containers in which the pressure is applied only by liquids, whose temperature does not exceed the boiling temperature at atmospheric pressure (that is, 100°C for water), belong to group V 'Pressure vessels with a permitted operating excess pressure p_{max} of no more than 500 bar (p_{max}< 500 bar)…'. Buffer storage tanks in Group V, which are made to correspond with the specifications in the AD Codes of Practice, can be started up without the acceptance of an expert or a competent person. The test confirmation and proper marking of the container are sufficient.
- If the maximum temperatures in the buffer storage tank are >100°C at atmospheric pressure it is classified as: Group II 'Pressure vessels with a permitted operating excess pressure p of not more than 1 bar and a permitted pressure contents product p_{max} × content in litres of more than 200 …'. These containers, according to Section 10, paragraph 2 of the German regulations, must be subject to repeated testing. This test must be carried out by a competent person (but not by an expert); the extent of the testing is established in Section 10.

In general it is recommended that the containers be limited to operating temperatures below 100°C and designed according to the requirements of Group V, in order to avoid the cost of testing.

As far as the buffer stores are concerned, temperatures should be restricted to the permitted maximum temperature of the selected pressure vessel group. This can be done by setting the maximum store temperature on the solar controller, or by means of an additional safety temperature limiter (STL). This equipment is also specified in the case of indirect heating for temperatures in the heating medium >110°C. At this point it must, however, be considered that temperatures in excess of 110°C in the heating medium occur only very briefly in the collector circuit, and the standard is clearly valid for heating media with constant temperatures in the heating medium >110°C. With some manufacturers the use of STLs in connection with solar storage is required only above a specified collector surface area per litre of tank content.

If a safety temperature limiter has been triggered by briefly exceeding the response temperature, further heating of the storage is stopped, as the pump and hence the solar system are permanently switched off. In such a case the system can only be started up again manually. To avoid unnoticed stoppages, it is essential for an STL to be connected to a fault alarm signal.

As this concerns use of the standards for conventional heating systems, which cannot be sensibly applied fully to solar systems, the supplier of the components should be asked for a written comment in the case of any uncertainties.

The storage should be safeguarded by a corresponding safety valve that is suitable for the size of the store and the thermal output of the collector field. The safety valves must thereby correspond to the requirements listed in Table 5.4.

Table 5.4.
Safety valves for buffer storage tanks in accordance with DIN 4753 P1

Nominal capacity of water container (l)	Maximum heating capacity (kW)	Maximum capacity corresponding to collector field size (m^2)[a]	Connection diameter of safety valve, min.
up to 200	75	125	DN 15
200–1000	150	250	DN 20
1000–5000	250	417	DN 25
over 5000			1.25 in

[a]Maximum capacity 600 W/m^2 of collector

5.5.4 Expansion vessels

If the buffer store is integrated into a closed system, installation of an expansion vessel is necessary. The standard calculation programs can be used for dimensioning. The following example is intended to provide an idea of the size of the expansion vessel.

EXAMPLE

The expansion vessel is to be calculated for a buffer store with a content of 8000 l, a filling temperature of 10°C and a maximum temperature of 90°C. The response pressure of the safety valve is 2.5 bar, and the filling pressure in the buffer tank 0.2 bar. The pressure information represents an excess pressure against atmospheric pressure. The calculation in this case results in a minimum volume of 543 l. According to the manufacturer's information two expansion vessels with a rated volume of 280 l each are selected to reduce costs.

At the maximum temperature of 90°C in the buffer tank, it must be ensured by means of a cooling length or an auxiliary vessel that a maximum temperature of 70°C is not permanently exceeded at the membrane of the expansion vessel.

5.5.5 Fittings on the domestic water line

The safety valves on the domestic hot water storage tank are dimensioned as described in Table 5.4 for the buffer. The safety valve should be installed into the cold water feed, and must not be capable of being shut off with respect to the domestic water. To avoid the safety valve responding to the expansion from the heating up of the water, installation of an expansion vessel is also recommended for the domestic water tank (Figure 5.17).

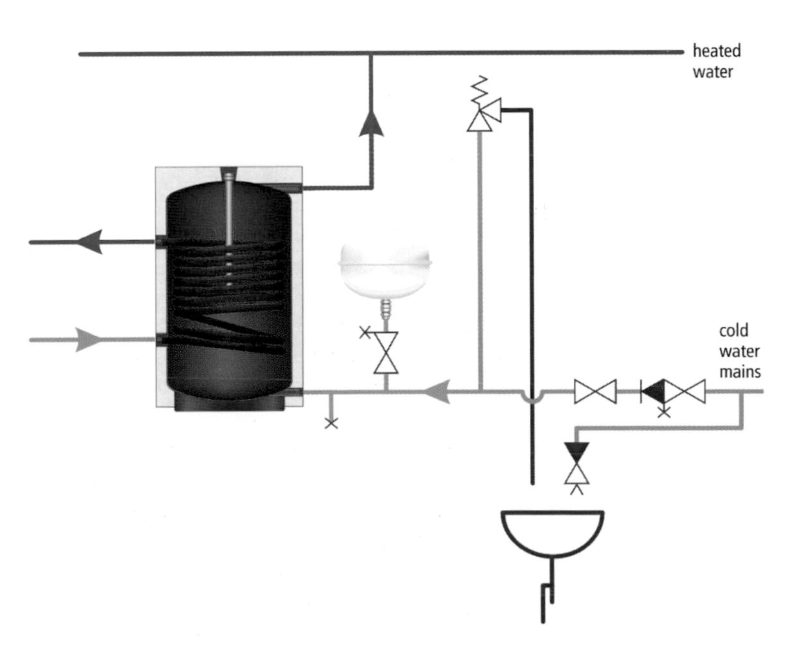

Figure 5.17.
Safety devices for domestic water heater

5.6 Economic considerations

The following considerations apply to the cost side. By determining annuities for the cost of the solar system and taking account of the energy yields of the solar system, the specific costs per kWh of solar-generated energy are calculated. These specific costs are called the *solar heat price*. If, for example, a system is planned for a multi-family home, it is possible to determine a mixed price from the proportionate costs for fossil and solar heat for calculation purposes. It is also possible in this way to make the costs for the individual dweller transparent in calculating the water heating costs, by converting these costs to the square metre of the respective dwelling unit.

5.6.1 Annuity method

In the annuity method, non-periodic and periodic payments with changing amounts are transformed into periodic constant payments over a considered period. The annuity is this periodic constant amount, and is made up of the interest and repayment portions for the repayment of the capital used. This method of calculation permits the types of payment arising in various periods to be summed directly. For example, interest and repayments for the investment are transformed with the help of the annuity factor a into average payments over a considered period T.

The payments are divided into two categories according to their temporal requirements: one-off or non-periodic payments, and regular payments. The individual costs are summarized under the general terms capital-related, consumption-related and operation-related. The considered period is taken as the service life of the short-lived and/or the more capital-intensive system components, so that the residual value has to be determined for the remaining system components. For solar energy systems consideration of the system as a whole has become the norm. After the considered period expires the whole system is regarded as being written off; no residual value is set.

The following types of costs are to be considered:

- capital-related costs, such as interest and repayments as well as servicing costs
- consumption-related costs: costs for the auxiliary energy (electricity for the pump)
- operation-related costs, such as insurance costs.

5.6.1.1 GENERAL PROCEDURE

For all types of costs the annuities are separately determined and added together:

$$AN_{tot} = AN_c + AN_s + AN_o$$

where AN_{tot} is the total annuity, AN_c is the annuity for capital-related costs, AN_s is the annuity for consumption-related costs, and AN_o is the annuity for operation-related costs.

The annuity for capital-related payments is given by

$$AN_c = A_0\,(1 - \mathrm{R})a + f_k A_0 ba_{SER} = AN_i + AN_{SER}$$

where AN_c is the annuity for capital-related payments (= €/a or \$/a or £/a); AN_i is the annuity for investment-related payments (€/a or \$/a or £/a) = $A_0(1 - R)a$; AN_{SER} is the annuity for servicing payments = $f_k A_o ba_{SER}$; A_0 is the procurement costs for the system;

$$a = \frac{q^T\,(q - 1)}{q^T - 1} = a(q, T)$$

T is the period considered in a (a = years, latin: annus); q is an interest factor which = $1 + (p/100)$, p is the interest (%); f_s is a factor for establishing the service costs in %/year of investment amount without grant; ba_{SER} is a price-dependent annuity factor for servicing payments = $b(T, q, r_s) \times a(q, T)$, and $b(T, q_s, r_k)$ is a cash value factor; r_s is the annual price change factor for servicing payments; and R is a component-specific residual value factor. The residual value is usually set to zero for solar systems.

The annuity for consumption-related payments is given by

$$AN_c = A_{c1} ba_c$$

The annuity for operation-related payments is given by

$$AN_o = A_{o1} b a_o$$

5.6.1.2 EXAMPLE CALCULATIONS ACCORDING TO THE ANNUITY METHOD

A solar energy system is considered with a price of €76,694/$76,694/£53,686 including VAT and planning. The system supplies a solar heat yield of 55,000 kWh/a and has a surface area of approximately 100 m². Table 5.5 shows the calculation sheet for € or $, and Table 5.6 the sheet for £.

Individual simulation programs such as T*SOL (see Chapter 11) offer the option of carrying out the economic calculations in parallel with the project-related simulation calculations. In this way the economics can be considered easily.

Table 5.5. Calculation sheet according to VDI 2067 for solar systems (values in € or US$)

	System				
	p = 6%, T = 20 yrs		p = 6%, T = 25 yrs		
	No grant	30% grant	No grant	30% grant	
Investment amount, A_0	76,964	53,686	76,964	53,686	
Annuity factor, a	0.087	0.087	0.078	0.078	
Investment-related payments	6672	4671	5982	4187	
Price change factor for servicing	1.05	1.05	1.05	1.05	Price-dynamic for calculation
Servicing factor, f_s	0.005	0.005	0.005	0.005	
Payments for servicing in first year	383	383	383	383	
Price-dynamic annuity factor	1.506	1.506	1.651	1.651	
Annuity for payments for servicing, AN_{SER}	578	578	633	633	
Price change factor for auxiliary energy, r_{aux}	1.05	1.05	1.05	1.05	Price-dynamic for calculation
Payments for auxiliary energy in year 1	141	141	141	141	
Price-dynamic annuity factor	1.506	1.506	1.651	1.651	
Annuity for payments for auxiliary energy, AN_{aux}	212	212	232	232	
Price change factor for insurance, r_{ins}	1.05	1.05	1.05	1.05	5000 m² building at €0.41/$0.41 total supplement €20.5/$20.5 per annum
Payments for insurance in year 1	20	20	20	20	
Price-dynamic annuity factor	1.506	1.506	1.651	1.651	
Annuity for insurance payments, AN_{ins}	31	31	34	34	
Total annuity (€/a or $/a)	7492.93	5490.71	6881.13	5086.50	
Solar yield (kWh/a)	55,000	55,000	55,000	55,000	
Heat price (€/kWh or $/kWh)	0.14	0.10	0.12	0.09	

5.6.2 Types of costs for solar energy systems

The following general conditions apply to the individual types of costs in solar systems.

The considered period for the annuity calculation is normally 20 years for solar systems. This figure is often seen as too conservative, as the main cost component (the collector field) can easily have a service life of 30 years. Individual considerations are thus often carried out with time spans of 25 and 30 years. Interest rates are always changing, but an interest rate of $p = 6\%$ has been used for the above calculation.

5.6.2.1 CAPITAL-RELATED COSTS

The capital-related costs are the largest proportion of solar system costs. They are made up of the investment amount, A_0, and – as a small proportion of this – the payments for servicing and maintenance. The investment amount, A_0, can be reduced by public subsidies.

Table 5.6. Calculation sheet according to VDI 2067 for solar systems (values in £)

	System				
	p = 6%, T = 20 yrs		p = 6%, T = 25 yrs		
	No grant	30% grant	No grant	30% grant	
Investment amount, A_0	53,876	37,580	53,876	37,580	
Annuity factor, a	0.087	0.087	0.078	0.078	
Investment-related payments	4670	3270	4187	2931	
Price change factor for servicing	1.05	1.05	1.05	1.05	Price-dynamic for calculation
Servicing factor, f_s	0.005	0.005	0.005	0.005	
Payments for servicing in first year	268	268	268	268	
Price-dynamic annuity factor	1.506	1.506	1.651	1.651	
Annuity for payments for servicing, AN_{SER}	405	405	443	443	
Price change factor for auxiliary energy, r_{aux}	1.05	1.05	1.05	1.05	Price-dynamic for calculation
Payments for auxiliary energy in year 1	99	99	99	99	
Price-dynamic annuity factor	1.506	1.506	1.651	1.651	
Annuity for payments for auxiliary, energy AN_{aux}	148	148	162	162	
Price change factor for insurance, r_{ins}	1.05	1.05	1.05	1.05	5000 m² building at £0.29; total supplement: £14.4 GBP per annum
Payments for insurance in year 1	14	14	14	14	
Price-dynamic annuity factor	1.506	1.506	1.651	1.651	
Annuity for insurance payments, AN_{ins}	22	22	24	24	
Total annuity (£/a)	5245	3843	4816	3561	
Solar yield (kWh/a)	55,000	55,000	55,000	55,000	
Heat price (£/a)	0,10	0.07	0.08	0.06	

The factor f_s for calculating the servicing and maintenance costs may be estimated as 0.3–0.5% of the investment cost (without subsidies). This factor is lower than for conventional heat generators, as the wear on a solar system is lower. The larger the system, the smaller the factor f_s.

This can be explained by the type and extent of maintenance work. This consists mostly of visual inspections in which the travelling time of the maintenance personnel constitutes a significant part. Moreover, the percentage of the investment costs for moving parts such as pumps, valves and expansion vessels becomes considerable as the system size increases.

5.6.2.2 CONSUMPTION-RELATED COSTS

In solar systems these cost components are restricted to the payments for auxiliary energy – in other words mainly the electricity costs for controller and pumps. Standard figures for the level of this consumption portion are in the range 2–5% of the solar yield.

5.6.2.3 OPERATION-RELATED COSTS

In conventional systems for energy generation the costs of operating personnel, boiler operatives etc. are included in this cost component. These personnel are not required for the solar system considered. Instead, the costs for insuring the solar energy system are inserted. According to the type of insurance (building insurance/solar system insurance) different costs arise, which relate to the size of the building for an extension of the building insurance, or to the investment in the solar system for the solar system insurance. In practice solar systems have not generally been insured.

5.7 Solar contracting

An economic alternative to the self-operation of a heating system is contracting. In contracting, an energy supply company, also called a *contractor*, supplies the building completely with room heating and hot water. Solar thermal systems are ideal for incorporating into contracting arrangements, where the contractor sets up, finances and operates the conventional hot water heating system and the solar thermal system (see Figure 5.18).

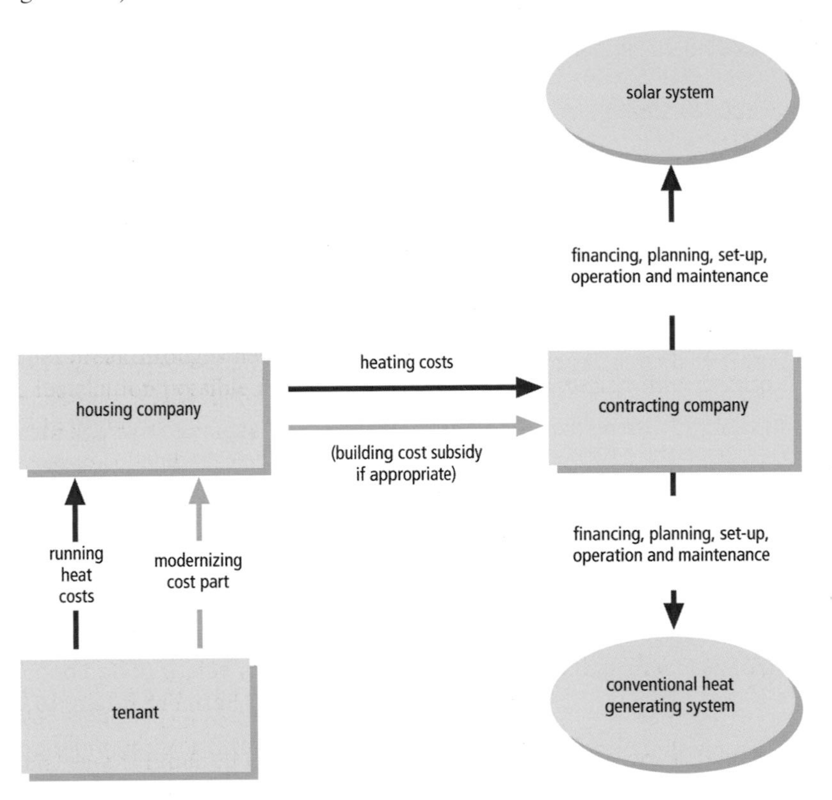

Figure 5.18. Heat supply model, solar contracting

The twin-sourced generated heat is sold to the heat consumer. The heat consumer can be either the housing company or the direct tenant. The contractor is therefore interested in the best possible operation of the overall system including the solar thermal system. The optimum operation of the solar thermal system is reduced to its running operating costs. The investment and operating risk is transferred to the contractor. The housing society improves the renting potential on the current market by practising environmental protection and the use of advanced technology.

The supply of solar heat by means of solar contracting contributes to an improved market penetration for large solar thermal systems in the building of multi-storey flats for rent. The contractor (heat supplier), through his specialist expertise helps to overcome the existing market reluctance, such as lack of know-how or lack of confidence in the functionality of solar thermal systems.

The remaining additional costs for solar thermal systems under the present financial boundary conditions can be further reduced by the use of standardized large systems and larger quantities in contracting.

Solar contracting may also be eligible for grants within the scope of solar subsidy programmes. The contractor knows about the various solar grants that are available, undertakes management of the grant, and thus relieves the building owner of the responsibility.

Of course, building owners can also independently install conventional heat generation systems and supplementary solar thermal systems at their own expense and thereby at their own risk. As a rule, however, contracting solutions are economically more advantageous from the customer's point of view. The sample calculations in Tables 5.7 (for € or $) and 5.8 (for £) should make this clearer.

Tables 5.7 and 5.8 compare the apportionment costs for self-operated building heating by the building owners with the equivalent costs for district heating, conventional contracting and solar contracting. The calculation is based on a new building project in Hamburg of two multi-storey blocks of flats with 70 dwelling units with 5000 m^2 of total living area using a low-energy building method. The thermal connection rating is 250 kW, and the total heat requirement for room heating and hot water is 400 MWh.

Table 5.7. Cost comparison: self-operated natural gas heating boiler, district heating, conventional contracting. Solar contracting, actual market situation in Hamburg (gross), prices in € or US$

Position	Unit	Self-operated gas heating boiler	District heating (coal/gas fired)	Conventional contracting	Solar contracting
Central heating systems	€ or $	57,520		53,686	53,686
Solar thermal systems	€ or $				51,129
Planning	€ or $	6902		6442	10,481
Hamburg solar programme	€ or $				15,339
BAW grant	€ or $				12,782
Capital costs (8%)	€/a or $/a	7527		7025	10,184
Fuel/district heating costs	€/a or $/a	18,410	28,379	16,748	14,878
Operating, maintenance	€/a or $/a	1519	623	1626	2124
Mixed heat price	€/MWh or $/MWh	68.5	72.6	63.4	68
Apportionment costs	€ or $ per m² per month	0.45	0.48	0.42	0.45

Table 5.8. Cost comparison: self-operated natural gas heating boiler, district heating, conventional contracting. Solar contracting, actual market situation in Hamburg (gross), prices in £

Position	Unit	Self-operated gas heating boiler	District heating (coal/gas fired)	Conventional contracting	Solar contracting
Central heating systems	£	40,264		37,580	37,580
Solar thermal systems	£				35,790
Planning	£	4831		4509	7589
Hamburg solar programme	£				10,737
BAW grant	£				8947
Capital costs (8%)	£/a	5269		4918	7129
Fuel/district heating costs	£/a	12,887	19,865	11,724	10,415
Operating, maintenance	£/a	1063	436	1138	1487
Mixed heat price	£/MWh	48.0	50.8	44.4	47.6
Apportionment costs	£/m² per month	0.32	0.34	0.29	0.32

The conventional heat generator consists of a natural gas boiler. The gas price corresponds to a special customer tariff, and the district heating price corresponds to the standard tariff. The existing purchasing advantages on the part of the contractor for system technology and primary energy have been taken into account. The solar thermal system has an absorber surface area of 100 m² with a total yield of 45,000 kWh/a.

The solar contracting solution achieves approximate price parity with heating apportionment costs of €0.45/$0.45/£0.32 per m² per month compared with self-operated heat generation. In comparison with the conventional contracting solution, the solar contracting solution must compete with the additional monthly apportionment costs of about €0.03/$0.03/£0.02 per/m². In comparison with the district heating solution all other solutions investigated provide price advantages. This shows that a solar contracting solution lies within the range of the normal apportionment heating costs for heat supply. The heating costs for buildings, for example in Hamburg, currently lie within a range of €0.51–1.31/$0.51–1.31/£0.36–0.92 per m² per month. On top of this the erection of a solar thermal system increases the attractiveness and hence the rentability of the flat.

The cost relation between a conventional self-operated heat generator and solar contracting essentially depends upon the fuel prices and the investment costs for the solar thermal system. For an energy price increase of 5% per annum the graph shown in Figure 5.19 arises for three of the investigated cases (self-operated heat generation, conventional and solar contracting).

The solar contracting solution shows a lower dependence on the natural gas price increase than the conventional contracting solution by the solar-generated part. Both contracting solutions have cost advantages over self-operated heat generation.

The investment costs of solar thermal systems depend heavily on the size of the system. Large surface area collectors can be made and installed more economically. Systems with collector surfaces of 30 m² or more are suitable for solar contracting.

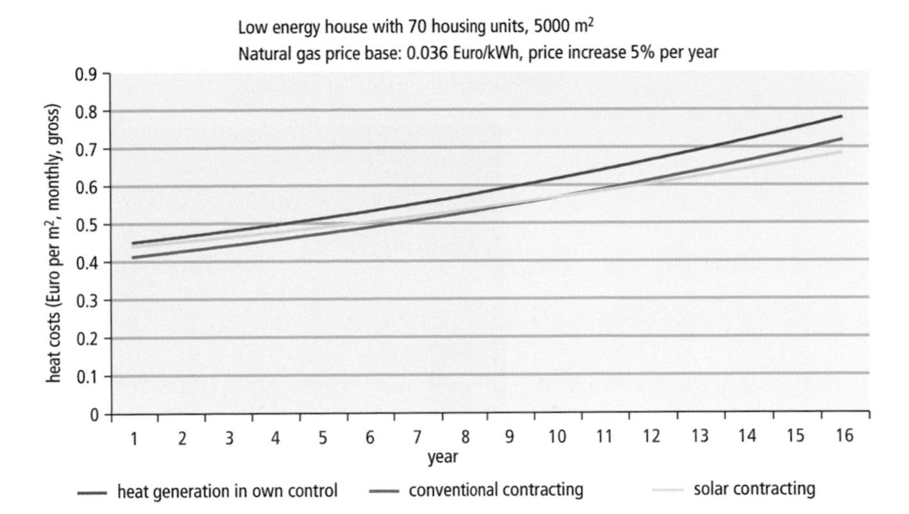

Figure 5.19. Development of apportionment heating costs based on a natural gas price increase of 5% p.a.

The precondition for successful implementation of solar contracting projects is the suitability of the building and of course the existing interest of the building owner in contracting the heat generation out to a third party. The solar thermal system must be incorporated at an early stage in the planning. In the case of the building, the roof slope (structural strength, restoration requirements, alignment, pitch and shading), and the work of incorporating the hydraulics (solar circuit piping, space available for buffer tank in the heating room) must be taken into account.

As a rule the contracting company offers a standard heat supply contract in which, during the contract period, the heat price, price adaptation, maintenance and repair service, calculation methods, liability and insurance of the heat generating plant are accurately controlled.

If all the above preconditions are met, solar contracting offers the private building owners or the housing co-operative, without any additional costs (or with only a low level of such costs, depending upon the local subsidy situation), a calculable, solar-supported heat supply without having to carry great risks themselves. This takes place to the benefit not only of both contract partners, but also to that of the environment.[38,39,40,41]

5.8 Solar district heating

Here we are talking about local supplies of heat for heating and hot water to small housing estates, or to building complexes such as hospitals (Figure 5.20).

In solar-supported district heating systems the heat is fed into a local heating network from one or many collector fields, and from there it is distributed to the connected households. A short- or long-term storage tank is coupled into the heat distribution network.

The concept of solar-supported district heating systems has come into consideration mainly in connection with the building of new housing estates or large building complexes that have been designed as low-energy buildings, as an environmentally friendly supply variant. It is important in such projects that, from the outset, the best possible technical preconditions are created for the use of solar energy by means of integrated energy planning. These include:

- a plan of the development with the orientation of the buildings that are favourable for active and passive use of solar energy (south alignment)
- planning of the estate or buildings according to solar architectonic criteria
- increased thermal protection in the buildings (low-energy building method)
- low-temperature heating systems, which permit a low network return temperature and hence a higher solar yield
- central heating plants and storage tanks arranged centrally to minimize distribution losses
- sufficient space for the heat storage tank.

Solar-supported district heating systems represent an inexpensive and ecologically sensible concept for the utilization of solar thermal energy if the technical conditions are suitable.

Figure 5.20.
This 3000 m² collector surface area system in Ry, Denmark, supplies the estate in the background by a district heating network

5.8.1 Solar energy systems with short-term heat storage

In solar district heating systems with short-term storage, buffer storage of several cubic metres of capacity is integrated into the network. The design of such systems is done on the basis of a solar fraction of 80–100% of the domestic water heating requirement in the summer months. This gives a solar fraction of 40–50% of the annual domestic water requirement – that is, around 20% of the total requirement for space heating and hot water generation.

In comparison with small local systems in individual buildings, large solar energy systems, which feed heat into a central heat supply, can be installed for less than half the investment costs:

- Small systems for one-family homes with about 5 m² of collector surface cost around €960/$960/£670 per m².
- Large systems with over 100 m² of collector surface cost around €500–600/$500–600/£350–420 per m².
- Local area heating systems with several hundred to several thousand m² of collector surface cost about €360/$360/£252 per m².

5.8.2 Solar systems with long-term heat storage

A solar-supported local heating system with long-term storage supplies larger housing estates of more than 100 dwelling units with both hot water and space heating. The time offset between the solar radiation available in the summer and the maximum heating requirement in the winter is compensated for by the seasonal thermal storage system. The pilot systems in Germany are designed for a solar fraction of about 50% of the total heat requirements. Solar district heating systems with long-term thermal storage are currently still in the development phase.

5.8.3 Guide values for the design of solar district heating systems

The design of solar-supported district heating systems must be carried out by means of detailed calculations taking account of the size of the housing estate, the thermal insulation standards, and the solar fraction to be achieved.

Table 5.9 contains guide values for the rough dimensioning of a solar district heating system.

Table 5.9. Guide values for rough dimensioning of a solar district heating system

System type	Solar district heating with short-term thermal storage	Solar district heating with long-term thermal storage
Minimum system size	From 30–40 housing units or 60 people	From 100–150 housing units (each with 70 m² living area)
Collector surface area	1–1.3 m²/person	1.4–2.4 m²/MWh per annum 0.14–0.2 m² per m² living area
Storage tank volume	70–100 litres/m² collector surface	1.5–4 m³/MWha 1.4–2.1 m³ per m² collector area
Useful solar energy	350–500 kWh/m²a	230–350 kWh/m²a
Solar fraction	Hot water 50% Total 10–20%	Total 40–70%

5.8.4 Components of solar district heating systems

5.8.4.1 STORAGE TANKS

As short-term storage tanks, either steel buffer tanks in the standard sizes (up to 6 m³), in which the desired total volume can be achieved by connecting several individual tanks in succession, are used, or special custom-made designs are arranged. Different principles are used in the selection of long-term heat storage systems.

EARTH RESERVOIRS

This storage system is designed as a concrete container that is either partially or completely submerged in the earth. It is lined to seal it against vapour diffusion, and is thermally insulated. The storage medium is water (Figure 5.21).

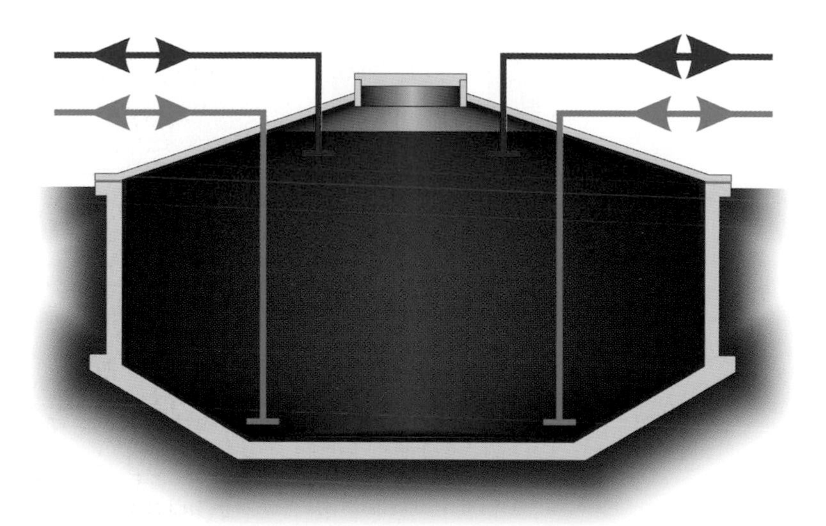

Figure 5.21. Earth reservoir

Practical German examples include:

- Hamburg-Bramfeld: Tank structure with nozzles; thermal insulation in wall area 20 cm, in the cover area 30 cm; volume: 4500 m³.

- Friedrichshafen-Wiggenhausen: Self-supporting concrete structure, thermal insulation in wall area 30 cm, in the cover area 40 cm; volume: 12,000 m^3; approximately €120/$120/£84 per m^3 installed volume; internally lined with a 1.2 mm thick stainless steel foil vapour diffusion seal.

Another variant, the gravel/water storage tank, is constructed more simply. A pit lined with a water-tight plastic film is filled with a gravel/water mixture as the storage medium. The storage tank is thermally insulated at least at the side and the top.
Practical example:

- Chemnitz: Storage tank with a gravel/water filling that takes on the structural function of the walls; storage volume 8000 m^3 with 5300 m^3 water equivalent; €140/$140/£98 per m^3 water equivalent; good seal is achieved by a 2.5 mm thick HDPE (high density polyethylene) film.

EARTH PROBE STORAGE SYSTEM

For this type of storage system, heat exchanger pipes are laid horizontally in the earth or vertically into drilled holes (U-tube probes), and are thermally insulated up to the surface. The surrounding soil is used directly as the storage medium, and heats up or cools down. Practically any size of storage volume is possible. The soil characteristics however, play an important role (Figure 5.22).
Practical examples:

- Neckarsulm: Storage volume around 175,000 m^3.
- Arnstein: Approximately 3000 m^3 storage volume; utilization about 36%.

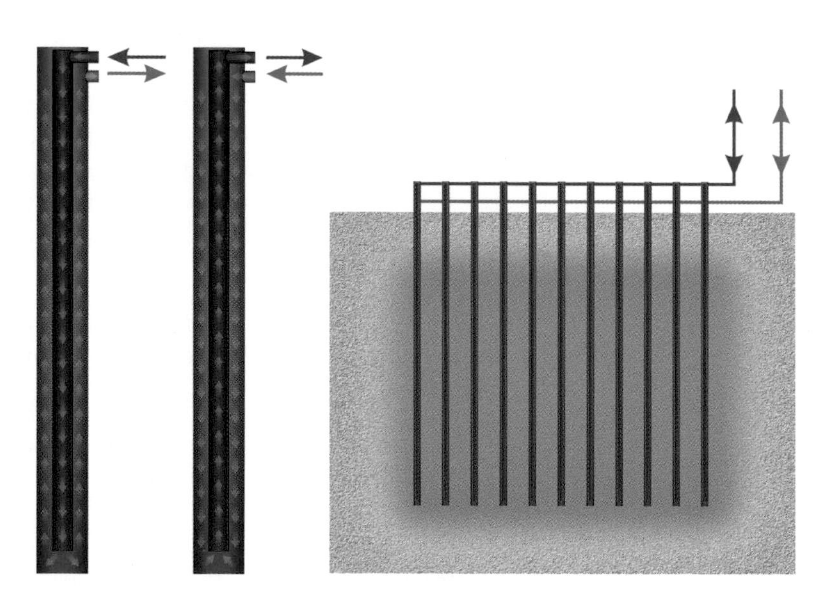

Figure 5.22. Earth probe storage system

AQUIFER STORAGE SYSTEM

Holes are drilled in pairs into water-bearing earth layers to depths of 50–300 m. Warm water is pumped into the soil, which serves as the storage medium, by means of a borehole (well); the water is subsequently discharged through another borehole. There is no need for thermal insulation. Soil formations with low groundwater flow speeds are necessary.
Practical example:

- Berlin: Within the scope of the reconstruction of the Reichstag (Government building) and the neighbouring buildings, two aquifer storage systems were implemented for cold storage at 60 m deep and hot storage at 300 m deep (in combination with a heat pump and a vegetable oil: CHP).

5.8.4.2 DISTRICT HEATING NETWORKS

Various different network structures can be used for distributing the heat:

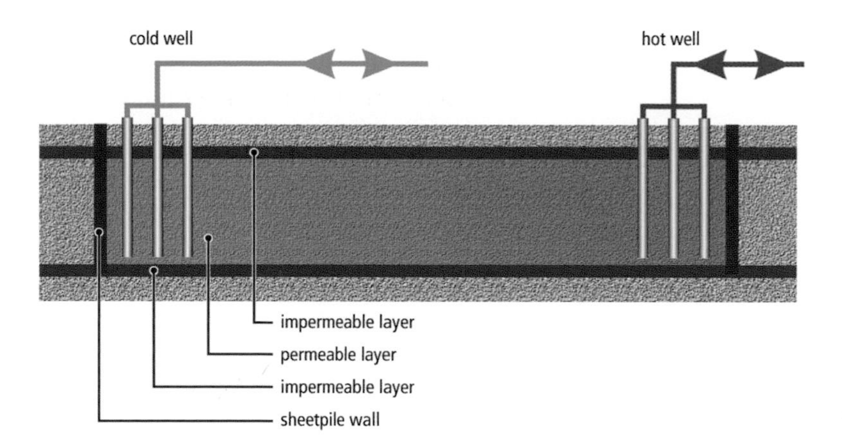

Figure 5.23. Aquifer storage system

- *2+2 conductor network*. This comprises two conductor lines for the heat supply to the buildings (domestic water heating takes place locally in the individual buildings), and an additional two conductor lines for the solar circuit.
- *4+2 conductor network*. This comprises four conductor lines for the heat supply: two lines each for domestic hot water (DHW) and space heating, plus an additional 2 lines for the solar. It permits central domestic water heating, but results in higher circulation losses.
- *3 conductor line network (two variants)*. An (older) variant uses a separate supply for the space heating and water heating together with a common return. In connection with solar district heating a system supply, a common return and a solar supply are now frequently used (newer variant).

5.8.4.3 HOUSE CONNECTING STATIONS

We can differentiate here between two main connecting options to the building: direct and indirect integration of heating.

Direct connection, without an intermediate heat exchanger, is cheaper and involves lower heat losses. *Indirect heating* decouples the district heating network through a heat exchanger from the lines in the building. This variant is usually preferred by house-owners.

5.8.5 Practical experience

The pilot systems with short- and long-term heat storage systems implemented so far have shown that this new – and still unfamiliar – technology brings with it a high requirement for information as well as minor problems. However, all systems have been put into operation without great difficulties. The solar systems operate well and reliably, and the long-term heat storage systems fulfil expectations. The main problems involve their combination with conventional system technology, together with their dimensioning and implementation.

In almost all systems, return temperatures that were too high occurred in the heat distribution network. Often it was simply that heaters that were too small were installed and operated, which made low return temperatures impossible. In one example this was due to faulty regulation of the internal house heating system by an average of 10–15 K higher than the value established by load simulations. The high return temperatures of the heat distribution network (monthly average up to 55°C), on the one hand, led to high storage temperatures and hence to low collector efficiencies; on the other hand the heat storage could not be discharged to the pre-calculated low temperatures. In this way, part of the heat content was not used in the storage. Since then, return temperatures have been reduced to a value of maximum 45°C.

More information on operating experience can be found in *Professional Experience with Solar District Heating in the German Project Friedrichshafen / Wiggenhausen-Süd*.[42]

In general, the initial operating results have shown that through the solar energy system and the linked detailed instrumental monitoring of the whole plant, previously unrecognised operating problems involved in conventional system technology have been made evident. These difficulties cannot therefore be set against solar systems, as they occur in many buildings without solar energy systems and have often not been detected. On the contrary, the use of a solar energy system with the corresponding instrumentation leads to the finding and rectification of these problems. Optimization of the solar energy system yield hence always requires optimization of the conventional system technology.

6 Solar concentrating systems

The temperature level that can be achieved with non-concentrating solar thermal collectors is limited. At the most a level of approximately 200°C can be reached with high-end vacuum tube collectors. A further increase of temperatures beyond this level is technically hardly achievable. In addition, above 100°C the collector efficiency decreases significantly. However, looking at the required temperature levels in the market segments of process heat or electricity generation by thermal processes, far higher temperatures are needed. Such temperatures can only be generated by the concentration of sunlight.

6.1 Concentration of solar radiation

Concentration of sunlight for large-scale applications is commonly done with reflecting concentrators; lens systems cannot be used owing to their high price and limitations in size. Instead, a parabolic-shaped reflector concentrates the solar radiation either on a focal line or on a focal point. The concentrator needs to track the sun, so that its incident rays are always perpendicular to the aperture area.

In principle, the main choice is between one-axis and two-axis tracking systems (see Figures 6.1a–6.1c). Systems with one-axis tracking concentrate the sunlight onto an absorber tube in the focal line of the concentrator, whereas two-axis tracked systems focus the rays of the sun onto a round-shaped absorber at the focal point.

The theoretical upper value of concentration is 46,211; it is limited by the fact that the sun is not a point source of radiation. By concentrating solar radiation a maximum temperature of 5500°C – the temperature of the sun's surface – can be achieved. However, in practice these maximum values have never been reached, and in general they are not needed in any case. With a rising concentration ratio, the theoretical temperature limit also increases. In practical applications, operating temperatures commonly do not even come close to the theoretically possible temperature (see Table 6.1). There are two main reasons for this: first, it is not possible to manufacture or install to the ideal and, second, heat is carried away, which continuously reduces the temperature. However, if the heat removal is interrupted, the temperature in the absorber can rise dramatically.

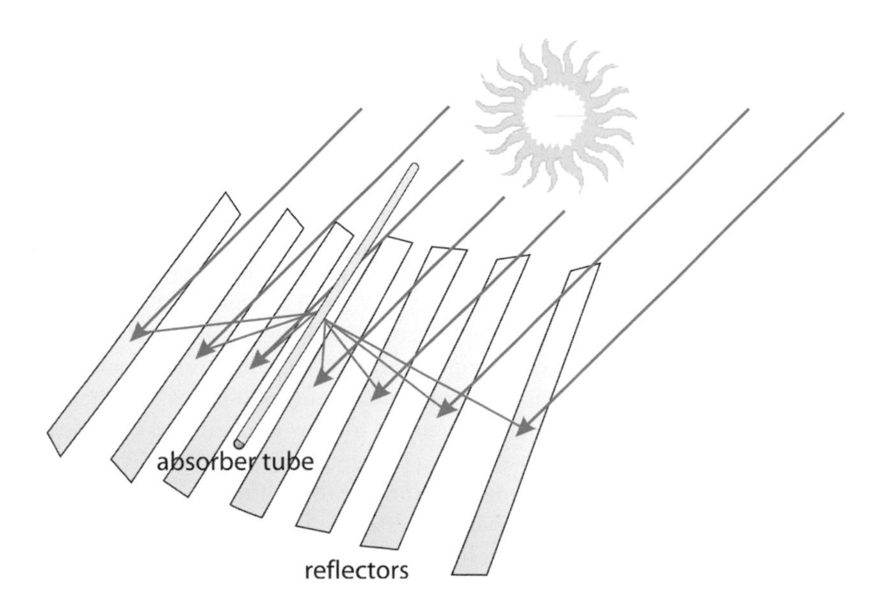

Figure 6.1a.
Concentration of solar radiation: reflectors with one-axis tracking

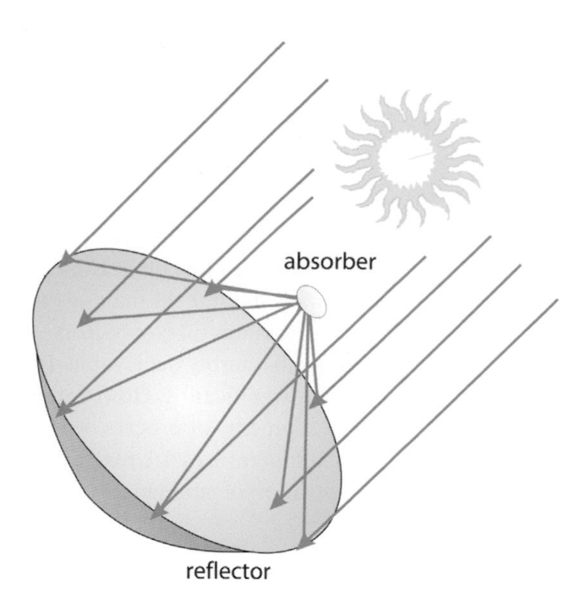

Figure 6.1b.
Concentration of solar radiation: single reflector with two-axis tracking

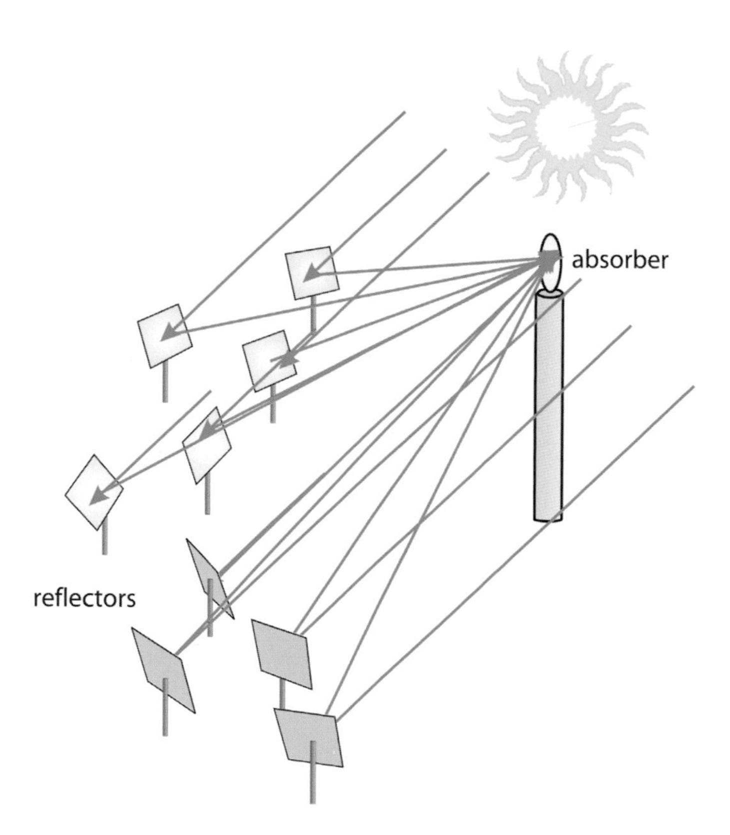

Figure 6.1c.
Concentration of solar radiation: multiple reflectors with two-axis tracking

6.2 Concentrating systems providing process heat

The generation of energy- and cost-efficient heat at temperature levels above 150°C is possible only with solar systems that make use of concentrating collectors, in contrast to non-concentrating systems or collectors with low concentration ratios (such as compound parabolic concentrators). Commonly the demand for process heat is strong in the temperature ranges 80–250°C and from 900–1500°C. Of particular interest is the market segment for low-temperature process heat from 80–250°C. Applications operating in this temperature range have used approximately 300 million MWh annually in the EU (8% of the end energy demand). The chemical industry, the pulp and paper and textile industry and the food processing industry consume large parts of this heat. The areas of application are numerous and include not only such widely varying processes as the heating of baths (for example for electroplating or cleaning),

Table 6.1. Concentration ratios of various systems

Collector type/system	Concentration ratio	Operating temperature (°C)	Theoretical temperature limit (°C)
Parabolic trough collector LS-3 and EuroTrough	82	~400	910
Solar tower plant with REFOS-pressurised receiver	~500	~1100	1590
EuroDish (dish/Stirling) system	2500	650	2510

drying, chemical processes (thermal separation, for example), melting or boiling, but also the generation of low-temperature process steam and the supply of heat to drive absorption refrigeration systems (solar cooling).

A selection of concentrating collectors was presented in section 6.1. With regard to the economic aspects of the provision of process heat, *parabolic trough collectors* are of particular interest. These concentrate the incident direct solar radiation linearly by means of parabolic curved reflectors onto a black-coated absorber tube. The aperture span of these collectors usually ranges between 2 and 4 m. The absorber tube takes up the solar radiation and converts it to heat, which it transfers to the heat transfer fluid that flows in the tube. Commonly used heat transfer fluids are water (hot water or steam), air and also thermal oils. Generally the latter re-transfer the heat to water or air in a heat exchanger at a further stage of the line.

Compared with glazed flat-plate collectors, these systems exhibit fewer heat losses. Among other aspects, this results from the small absorber surface area compared with the aperture (solar radiation collection area) and the selective coating of the absorber tube, which reduces the emittance of infrared radiation and thus radiation losses. In addition, a glass envelope that sits around the absorber tube reduces convective losses. In order to achieve even higher collector efficiencies, and thus make it possible to reach higher operating temperatures, the space between the glass envelope and the absorber tube is evacuated.

A tracking device – commonly a motor and a transmission device – enables the parabolic trough collector to follow the sun on one axis. Parabolic trough collectors are usually installed with a north–south orientation. This orientation results in a higher annual energy yield than an east–west orientation. However, the latter orientation shows a more even distribution of the annual energy yield. In the case of two-axis tracking the requirements for construction, control and maintenance are higher and so therefore are the costs, so that the one-axis-tracked parabolic trough has proven itself as the more reliable and more efficient system. Further developments in parabolic trough technology aim at improving optical efficiencies while using less material.

A further significant difference between tracked concentrating collector systems and the conventional solar thermal systems such as glazed flat-plate collectors or evacuated tube collectors is shown in the operation and safety concepts.

In conventional collector systems, if a standstill occurs (for example if no heat is removed, or if a component fails) then in order to prevent stagnation temperatures being exceeded, various substantive measures (such as expansion vessels) have to be incorporated. In contrast, tracked collectors make use of safety control routines that defocus the collector to avoid exceeding stagnation temperatures. However, suitable safety devices, such as a flow indicator, are needed to ensure the flow of heat transfer fluid through the absorber, and thus the removal of heat, before the collector is focused, in order to avoid overheating of the system.

Some evacuated tube collectors integrate compound parabolic concentrators (CPCs) into the glass envelope (see Figure 6.2) in order to enhance the aperture. Also simple flat absorber stripes are used for that purpose.

Figure 6.3 shows the efficiency of various collector types by plotting the annual energy yield per square metre of collector surface area against the mean flow temperature. Whereas the energy yield for non-concentrating collectors decreases significantly with increasing flow temperature up to a maximum temperature of 100°C, the decrease is considerably less for evacuated tube collectors with CPCs and parabolic trough collectors. The common non-tracking glazed flat-plate collectors and evacuated tube collectors use the entire solar radiation (that is, the direct irradiation and the diffuse sky irradiation), in contrast to concentrating solar systems, which use

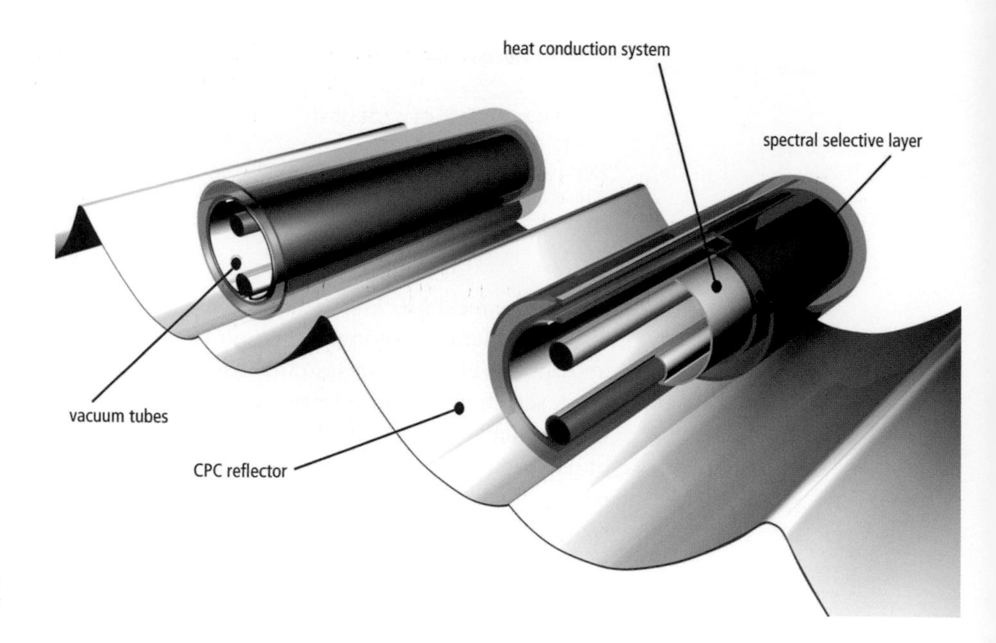

Figure 6.2. Schematic structure of a CPC tube collector. Source: Consolar

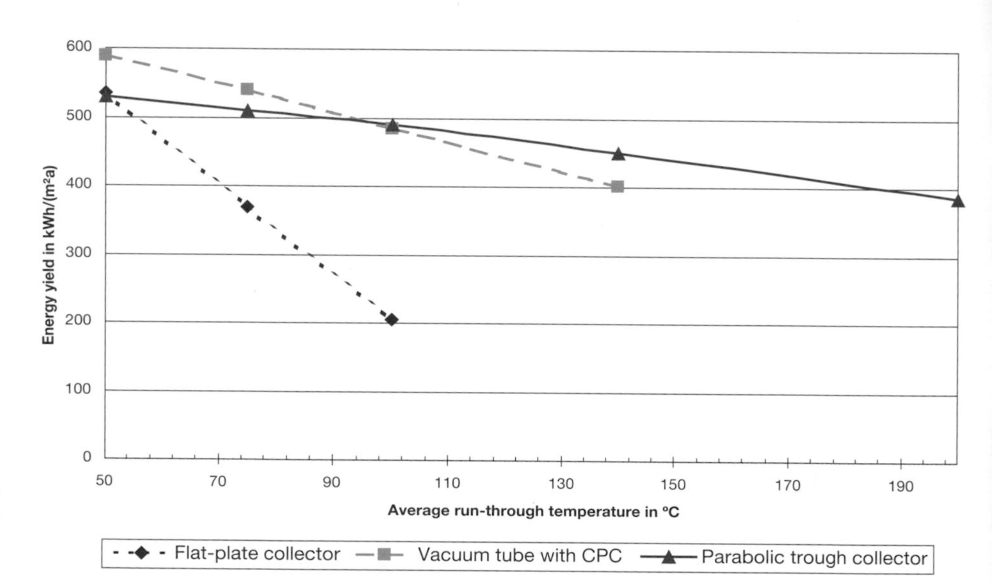

Figure 6.3. Energy yield against mean flow temperature for various collector types at Würzburg, Germany. Source: Klaus Hennecke

only the direct irradiation. The energy yield in Figure 6.3 was calculated for southerly oriented collectors with an inclination angle of 40°, located in the city of Würzburg, Germany.

The energy yield of a collector system depends not only on the construction of the collector but also on the annual amount of direct irradiation. Würzburg receives a mean annual direct irradiation of 1066 kWh/m^{2}a. Comparison of parabolic trough collectors of simple construction with conventional solar thermal collectors shows that, even in a mid-European climate, the parabolic trough collector has a number of advantages, in particular when higher operating temperatures are needed. At southern European locations with a higher share of direct irradiation the advantages become even larger.

The system integration of concentrating collectors to provide process heat does not differ much from conventional systems that supply process heat. The centrepiece of the system is the collector field: that is, arrangement of the collectors on the ground or on the roof of a building. The heat transfer fluid (HTF) circulates through the field. By measuring the temperature of the HTF at the outlet of the collector, a controller regulates the flow rate in harmony with the irradiance. The gained heat is then transferred to a heat exchanger, where it is either used directly in a process (for example to heat a bath or to preheat feed water or combustion air; see Figure 6.4) or stored in a short- or long-term heat store.

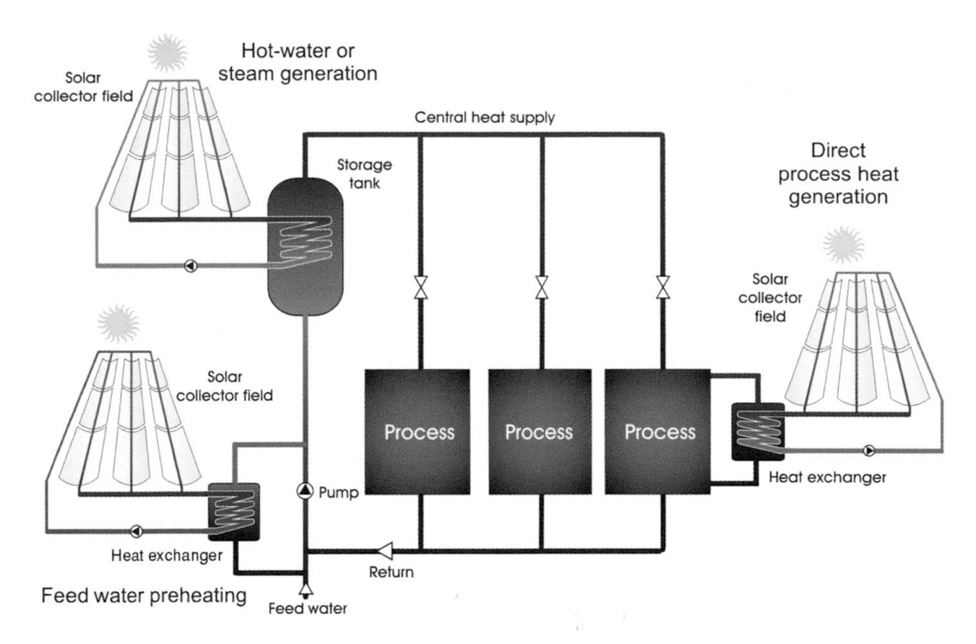

Figure 6.4. Schematic of various options for integrating solar systems into a conventional heat supply. Source: DLR

The simplest and most cost-effective integration is the direct input of solar heat into the process (see upper picture in Figure 6.5). This variant only makes sense if the respective process runs continuously and the heat demand is larger than that being provided by solar energy. The schematic shows an indirect system in which the collector circuit is separated for freezing and corrosion reasons from the application process by a heat exchanger. For economic motives the dimensioning of the collector field should ensure that the maximum solar energy yield does not exceed the heat demand at any time.

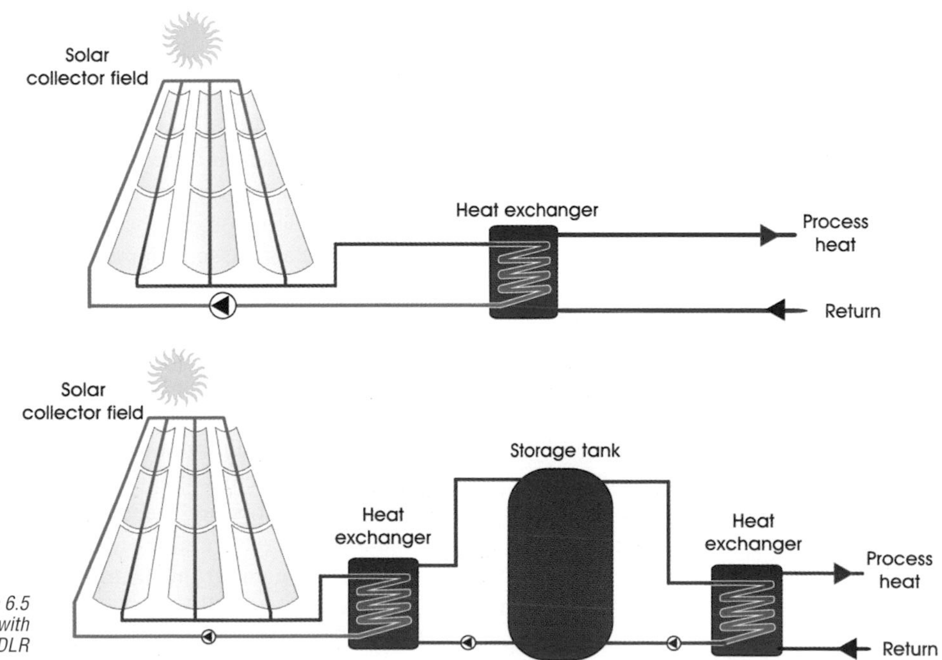

Figure 6.5 Solar system without (above) and with storage (below). Source: DLR

More common are processes that are operated for only 5–6 days a week, or processes with frequent interruptions. Such situations require the use of a store for any excess heat. The required heat can then be supplied to the user whenever it is needed. In this case the dimensioning of the store and the collector field depends not only on the required heat demand, temperature level and system design, but also on the desired heat capacity of the storage, and thus the amount of time for which heat can be supplied. In general a distinction is made between: short-term storage with a capacity of a number of hours, which covers daily fluctuations; storage systems with a capacity for several days; and seasonal storages. The store has to be dimensioned according to the specific demand.

In contrast to solar thermal power plants for electricity production, solar thermal process heat systems should be installed directly at the location where the heat is required in order to avoid losses caused by the need to transport the heat. Hence sufficient space for the collector field and a relatively high annual irradiation need to be available at the specific location.

Typical investment costs for a solar thermal process heat system are given in Table 6.2.

Table 6.2. Investment costs for various collector types (indicative; market prices fluctuate and differ per country)

Collector type	Costs (€/m²)
Flat-plate collector	250-300
Compound parabolic concentrator (CPC)	300-350
Parabolic trough collector	300-400
Evacuated flat-plate collector	400-600
Evacuated tube collector	400-600
Evacuated tube collector with CPC	400-600

The maintenance costs (cleaning etc.) for conventional collectors are of the order of €2.5/m²a; for parabolic trough collectors they amount to approximately €5/m²a.

Considering the costs for the total system and the single components, about 80% of the investment cost relates to the collector field (for field sizes of >1000 m²), including the erection, support structure and piping. The remaining 20% relates to the heat exchangers, pumps, control system and planning. Whereas the integration of short-term storage does not influence the investment costs significantly, long-term storage tanks can be far more costly, reaching shares of 10–20% of the total investment cost.

The total investment costs for solar thermal systems for process heat are about €250–1000/$250–1000/£165–600 per kW_{th} (kilowatt–thermal) of installed capacity. Accordingly the resulting energy costs are about €0.02–0.05/$0.02–0.05/£0.014–0.035 per kWh_{th} for low-temperature heat supply, and €0.05–0.15/$0.05–0.15/£0.035–0.10 per kWh_{th} for mid-temperature heat.

For the near future it is important to demonstrate the integration and reliability of solar thermal process heat systems in suitable applications. One obstacle for such systems is the availability of space to erect the collector field. Higher specific costs compared with those of conventional process heat systems are another barrier. However, cost reduction potentials are seen in modifications of the collector designs, and in smaller, modular collector units that are suitable for erecting on a rooftop or for roof integration. Further potential is expected from increasing automated operation, thus reducing the operation and maintenance (O&M) costs. Mass production, the reduction of O&M costs and the improvement of system design and collector efficiencies are expected to cut costs by half by the year 2010.

Until now the European market did not offer concentrating systems for process heat applications. However, several current European projects are aimed at the development of commercial collectors, and some companies in Israel and the USA already offer commercial systems. Also, a number of systems are operating in the USA, with collector field sizes ranging from 200 to 3000 m².

In addition, with the 'Campaign for Take-Off' the European Commission is pursuing the aim of having 2,000,000 m² of solar thermal collectors installed for industrial process heating and solar cooling. By reaching this goal, it is expected to realize a saving of prime energy of about 2 million MWh/a. In 2001 solar thermal collectors with a total area of 10,000 m² were operating in the field of industrial process heat in Europe. In the near and mid-term future, concentrating collectors could play an important and growing role in the spreading of supplying solar process heat.

6.3 Concentrating solar thermal systems for electricity generation

The use of approximately 1% of the surface area of the Sahara for solar power plants would be sufficient to meet the entire global electricity demand. In particular solar thermal power plants, apart from photovoltaics, offer the opportunity to produce solar electricity in the tropics at low cost. These power plants do not use the photo effect like photovoltaic systems, but apply thermal processes to generate electricity. There are three different types of solar thermal power plant:

- parabolic trough plants
- solar tower plants
- dish/Stirling systems.

6.3.1 Parabolic trough plants

The first solar thermal power plants were developed in the USA in 1906. The first demonstration plants were erected and successfully tested in the USA and near Cairo, Egypt, still a British colony at that time. Amazingly, these systems looked almost like the systems of today. However, problems with materials and other technical difficulties put an end to the first attempts at large-scale solar electricity generation in 1914, shortly before the outbreak of the First World War.

In 1968 the USA laid the basis for the renaissance of solar thermal electric technology. The US public electric utilities were obliged by the power of the Public Utilities Regulatory Policy Act to buy electricity from independent power producers at a clearly defined tariff. After a doubling of the electricity costs in only a few years owing to the oil crisis, the Californian electric utility Southern California Edison (SCE) offered long-term conditions for the feed-in of electricity from renewable energy systems. In combination with tax incentives such as an exemption from paying property tax for solar power plants, the development of solar thermal power plant projects started to become financially interesting. In 1969 the company LUZ was founded, which concluded a feed-in contract for solar thermal electricity over a period of 30 years with SCE in 1983. The first commercially operated solar thermal power plant with parabolic trough technology was erected in 1984. From then on new solar thermal power plants with increased size and improved technology followed each year (see Table 6.3). In the mid 1980s electricity prices went down again. After the abolition of tax exemptions at the end of 1990, LUZ went bankrupt just before starting the erection of its tenth solar thermal power plant.

Table 6.3. Characteristics of different parabolic trough collectors

Collector type	LS-1	LS-2	LS-3	EuroTrough
Year of first installation	1984	1986	1988	2001
Concentration ratio	61	61	82	82
Aperture width (m)	2.5	5.0	5.66	5.66
Collector length (m)	50	48	99	150
Aperture (m^2)	128	235	545	825
Absorber tube diameter (mm)	42.4	70	70	70

Although electricity generation from solar thermal power plants is significantly cheaper than from photovoltaic systems, no further commercial solar thermal power plants have been built since 1991. However, a number of new solar thermal power plant projects are currently being developed. The World Bank has allocated US$200 million for financial support of the erection of combined solar thermal and natural-gas-fired power plants in developing countries such as Egypt, Mexico, India and Morocco. A higher feed-in tariff for electricity from solar thermal power plants is being paid in Spain, and other southern European countries are preparing similar support measures.

Up to now solar thermal power plants with parabolic trough technology have been the only commercially operating plants. After the oil crisis nine parabolic trough plants were erected from 1984 to 1991 in the Mojave Desert in California on a surface area of more than 6 km^2. They are called *SEGS plants* (solar electric generation systems) (see Figure 6.6 and Table 6.4). More than a million mirror elements with a total aperture of

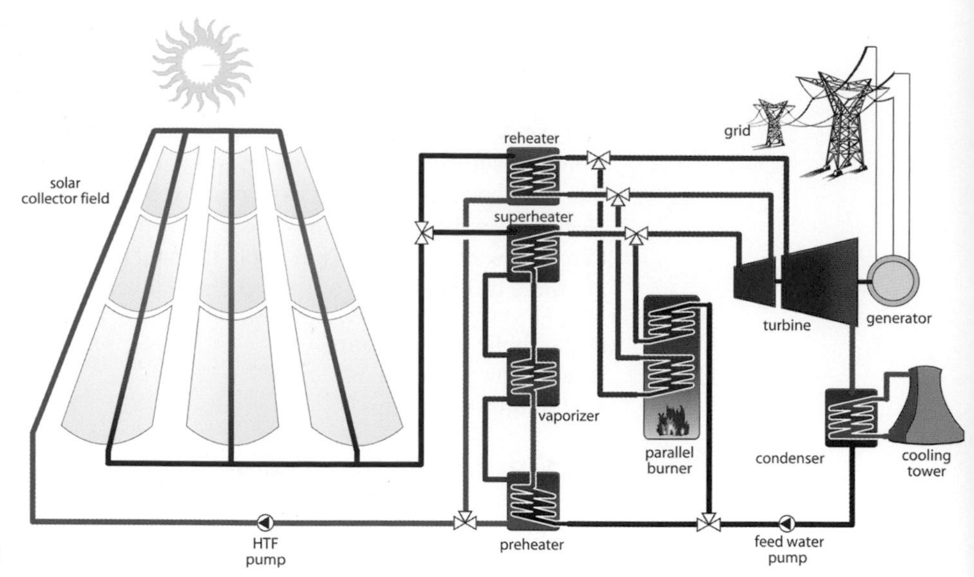

Figure 6.6. Schematic of an SEGS plant

Table 6.4. Technical data for parabolic trough-based SEGS plants in California

Plant	I	II	III	IV	V	VI	VII	VIII	IX
Year of commissioning	1984	1985	1986	1986	1986	1988	1988	1989	1990
Net capacity (MW)	13.8	30	30	30	30	30	30	80	80
Land use (1000 m²)	290	660	800	800	860	660	680	1620	1690
Aperture (1000 m²)	83	165	233	233	251	188	194	464	484
HTF outlet temp. (°C)	306	321	349	349	349	391	391	391	391
Efficiency (%)									
– Steam turbine (solar)	31.5	29.4	30.6	30.6	30.6	36.6	36.6	36.6	36.6
– Steam turbine (gas)	–	36.3	36.3	36.3	36.3	39.5	39.5	36.6	36.6
– Solar field (thermal)[a]	35	43	43	43	43	43	43	53	50
– Solar-to-electric (net)[a]	9.3	10.6	10.2	10.2	10.2	12.4	12.3	14.0	13.6
Specific invest. costs (US$/kW)	4490	3200	3600	3630	4130	3860	3860	2890	3440

[a] Design

2,300,000 m² focus the sunlight in these plants, which have an electrical capacity of 354 MW. Each year the SEGS plants generate about 800 million kWh of electricity, enough to cover the demand of 60,000 Americans. Eight of the SEGS plants can also be operated with fossil fuels, so that electricity can be supplied during night-time or periods of bad weather. However, the annual share of fossil fuel, in this case natural gas, is limited by law to 25% of the entire annual thermal input. The total investment for the SEGS plants amount to more than US$1.2 billion. Up to now these plants have supplied more than 10 billion kWh of electricity to the grid. The levelized electricity costs have decreased from system to system, with US$0.26/kWh for the first SEGS plant down to US$0.12–0.14/kWh for the plants that were erected last.

The principle of operation of the parabolic trough plants is easy to understand. Large reflectors, arranged in the form of a trough-shaped collector, focus the sunlight onto a focal line. Several collectors are joined together in rows of 300–600 m (Figure 6.7). Again, a number of parallel rows then form the entire solar field.

Each single collector can be turned around its longitudinal axis to track the sun. The sunlight is concentrated up to 80 times onto an absorber tube that is located in the centre of the focal line. A glass envelope is placed around the absorber tube in order to reduce heat losses. Heat radiation losses are minimised by applying a special high-temperature-resistant selective coating to the absorber tube. In the Californian plants a specific thermal oil flows through the absorber tubes and is heated to temperatures of up to 400°C.

A parabolic trough collector can also be designed according to the Fresnel principle, as shown in Figure 6.8. A prototype that applies this principle has been erected in Belgium. Heat exchangers transfer the solar heat from the heat transfer fluid (HTF) to a water/steam cycle. The feed water is first brought up to higher pressure

Figure 6.7.
Parabolic trough power plants in California. Source: KJC

Figure 6.8.
Parabolic trough collectors. Source: Volker Quaschning

before it is preheated, evaporated and superheated by the HTF. The superheated steam drives a conventional steam turbine generator set to generate electricity. In two-stage turbines with high- and low-pressure stages the steam is reheated between the two stages. Once leaving the turbine the steam is expanded and is then condensed to water before reaching the feed water pump again. In cases of bad weather or at night-time the steam cycle can also be operated with a parallel fossil-fuel-driven boiler.

In contrast to photovoltaic systems a solar thermal power plant can guarantee a daily security of supply. This aspect increases the attractiveness of solar thermal power technology in a power plant portfolio. A useful supply of electricity can either be realized by hybrid operation with fossil fuels or, if solar only, CO_2-emission-free operation is desired, by using thermal storage (Figure 6.9). A well-proven principle in storing high temperature heat makes use of two-tank storage with, for example,

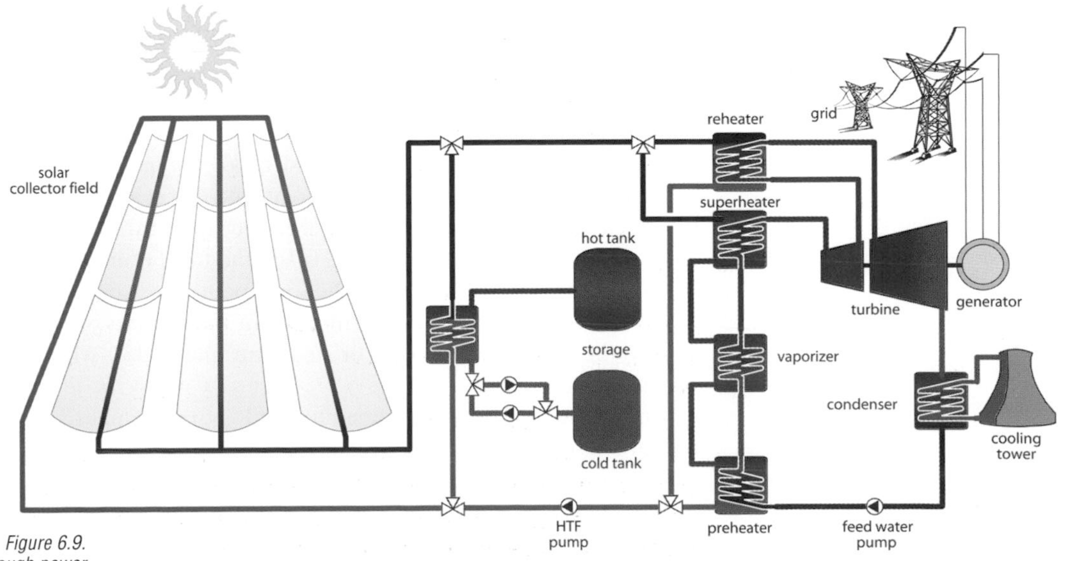

Figure 6.9.
Schematic of a parabolic trough power plant with thermal storage

molten salt as heat carrier. In the case of excess solar heat this is transferred via a heat exchanger to the molten salt, which is pumped from a warm to a hot store. In periods with less solar radiation the hot molten salt can be pumped back to the warm tank, heating up the HTF, which then drives the steam cycle.

The parallel steam generator (boiler) can also be fuelled by biomass or hydrogen (produced from renewable electricity). This is another option for generating electricity without emitting additional CO_2.

Current technological developments aim at further improvement of the efficiency and thus reduction of costs. For example, in Southern Spain close to the city of Almería the direct generation of solar steam is being demonstrated (Figure 6.10). Here the parabolic trough collectors directly heat and evaporate water under high-pressure conditions to a temperature of 400°C. Steam at such conditions can directly drive a steam turbine, so that such plants could dispense with the HTF and the heat exchangers.

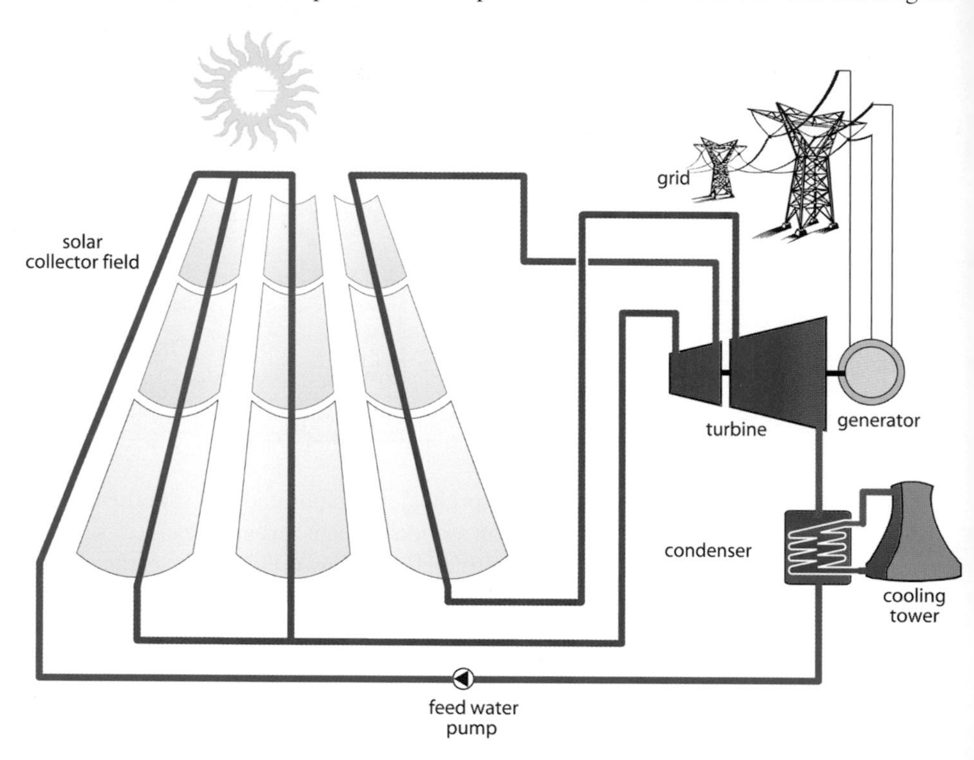

Figure 6.10. Schematic of a parabolic trough plant with solar thermal direct steam generation

The largest potential cost reduction is expected from large-scale production of solar thermal power plants, thus making use of the economies of scale. On a long-term basis cost reductions from €0.15 down to €0.05 per kWh are seen as possible. This would drive costs down to a similar order of magnitude as those of conventional fossil-fuelled power plants, but without the emission of harmful greenhouse gases.

6.3.2 Solar tower plants

Solar tower plants offer another option for producing solar thermal electricity. In this system several hundred, or even several thousand, reflectors are positioned around a central tower. Each of the reflectors, also called a *heliostat* (Figure 6.11), tracks the sun under computer control in order to focus the direct sunlight to the central receiver located at the top of the tower. The accuracy of the tracking is very important to ensure that the sun's reflected rays reach the focal point.

An absorber is positioned at the focal point. The concentrated sunlight heats up the absorber to temperatures of more than 1000°C. Air or molten salt transfers the heat to the power cycle – a gas or steam turbine cycle – where the heat is then converted into electricity.

In contrast to parabolic trough technology, no commercial tower plants are in operation, so far. In Almería (Spain), Barstow (USA) and Rehovot (Israel) pilot plants are operated in which system configurations are being optimized and new components are being tested (Figure 6.12). Also in Spain, the first commercial solar tower plants are in an advanced planning stage.

Figure 6.11.
Heliostats. Source: Volker Quaschning

Figure 6.12.
Solar tower test installations in Almería, southern Spain. Source: Stefan Franzen, CIEMAT

The tower concept with open volumetric receivers (Figure 6.13) works as follows. A fan sucks the ambient air into the receiver upon which the heliostats focus the sunlight. Commonly wire mesh, ceramic foam or a metallic or ceramic honeycomb structure is used as the receiver material. This structure is heated by the solar radiation and transfers the heat to the airflow. The ambient air cools the front part of the absorber, whereas very high temperatures arise in the rear part of the absorber material. Thus radiation losses are minimised. The heated air with temperatures between 650°C and 850°C is impelled to a waste heat boiler where water is

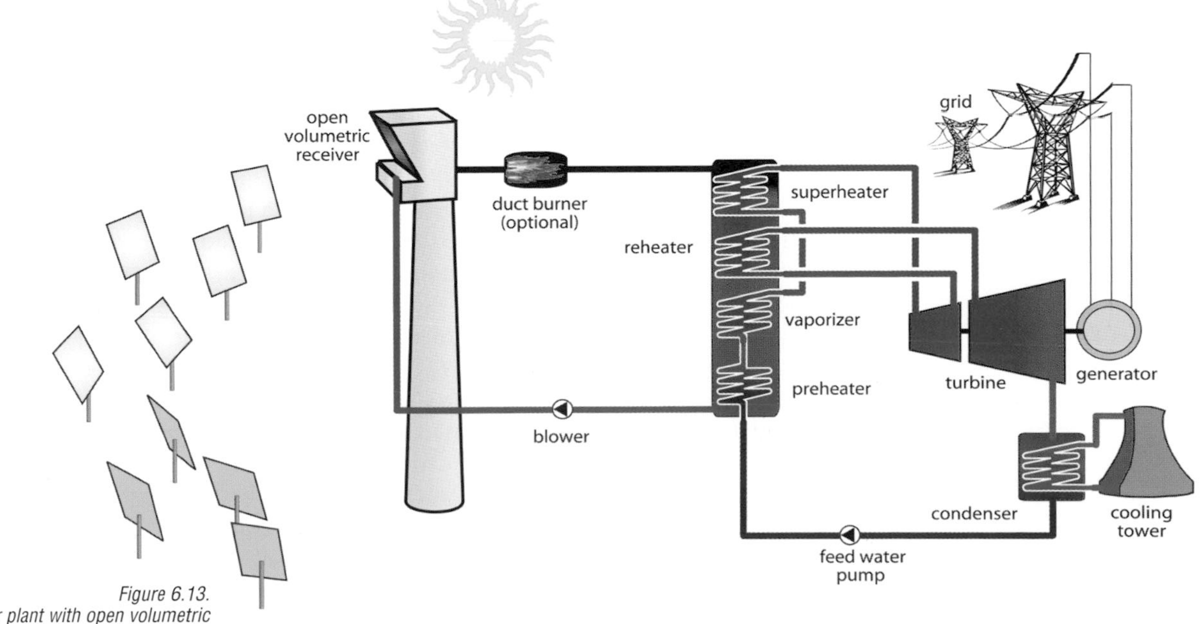

Figure 6.13.
Solar tower plant with open volumetric receiver

evaporated and superheated. Again, the superheated steam drives a steam turbine generator set to produce electricity. This power plant variant can make use of other fuels, for example by means of a duct burner.

Another tower concept that was developed from that just described offers promising options for the mid-term future: the closed volumetric receiver (REFOS concept; Figure 6.14). Here a transparent cupola-shaped silica glass separates the ambient air from the absorber. The air is heated in a pressurised air receiver at a pressure of approximately 15 bar to temperatures of up to 1100°C. With air of this temperature level a gas turbine is driven. Downstream a steam cycle makes use of the waste heat of the gas turbine.

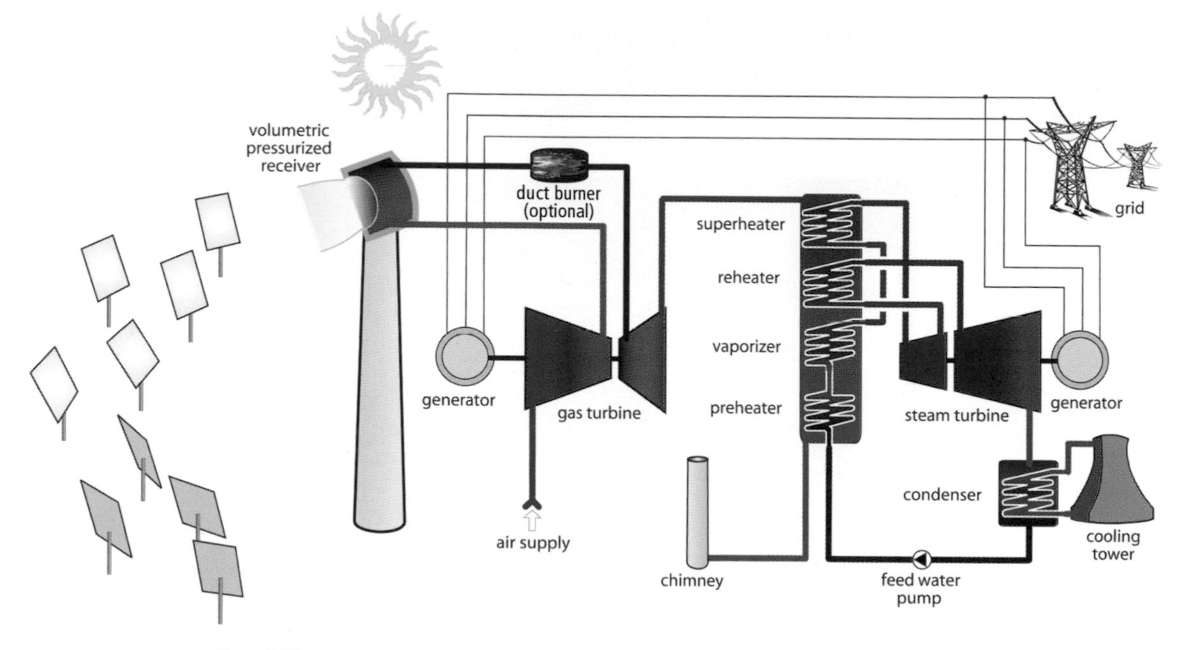

Figure 6.14. Solar tower plant with a pressurised volumetric receiver for solar operation of gas and steam turbines (REFOS)

In general, electrical efficiencies of combined cycles are higher (in the range of 50%) than steam cycle efficiencies (about 35%). The integration of solar thermal energy into combined cycle processes allows solar-to-electric efficiencies of more than 20%. This improvement and its technological prospects justify the increased technological effort and the higher costs of this receiver technology.

The levelized electricity costs are expected to be slightly higher for the first commercial solar tower plants as, in contrast to parabolic trough power plants, there are no series-produced solar tower plants in operation. However, owing to the higher efficiencies that are possible, lower costs are expected in the medium term.

6.3.3 Dish/Stirling systems

Whereas the erection and commercial operation of solar parabolic trough and tower plants is economically viable at electrical capacities of several megawatts, dish/Stirling systems can be also used in smaller units, for example as stand-alone systems to supply remote villages with power.

In these systems a concave mirror in the shape of a large bowl concentrates the solar radiation to a focal point, where the receiver is located. In the receiver the solar radiation is converted into heat and transferred to the heart of the system, the Stirling engine. This engine converts the heat directly into kinetic energy, which drives a generator to produce electricity. In order to direct the solar radiation to the receiver it is necessary to move the concave mirror in two axes, hence tracking the sun.

Solar heat is not the only form of heat that can drive the Stirling engine; heat from any combustion process can also be used. Combinations with a biogas burner can enable the dish/Stirling system to produce electricity also during night-time or at periods of bad weather. The use of biogas in such a system is also entirely CO_2-neutral.

Some prototype dish/Stirling systems have been erected and operated in Saudi Arabia, Spain and the USA (Figure 6.15). The levelized electricity costs are still relatively high in comparison with those of solar tower or parabolic trough plants.

Figure 6.15. Dish/Stirling demonstration systems in Almería, Southern Spain. Source: Volker Quaschning

Table 6.5. Technical characteristics of the EuroDish dish/Stirling system shown in Figure 6.15 Source: Schlaich Bergermann und Partner, Stuttgart

Concentrator diameter: 8.5 m	Reflectivity: 94%
Aperture: 56.6 m^2	Working medium: Helium
Focal distance: 4.5 m	Gas pressure: 20–150 bar
Average concentration ratio: 2500	Receiver-gas temperature: 650°C
Electrical gross capacity: 9 kW	Max. operating wind speed: 65 km/h
Electrical net capacity: 8.4 kW	Survival wind speed: 160 km/h

However, a dramatic cost reduction is thought to be possible when such systems are produced in large numbers and in volume production.

6.3.4 Economics and outlook

Concentrating solar systems use only the direct part of the solar radiation, whereas non-concentrating solar systems such as photovoltaic systems also make use of the diffuse part. Tables 6.6 and 6.7 show that the direct normal irradiation increases with decreasing latitude towards the equator more rapidly than the global horizontal radiation.

Table 6.6. Direct normal irradiation and global horizontal irradiation values for various European capitals

	London	Berlin	Paris	Rome	Madrid	Lisbon
Latitude (°N)	51.5	52.5	48.9	41.9	40.5	38.6
Direct normal irradiation (kWh/m^2a)	690	686	842	1565	1593	1269
Global horizontal irradiation (kWh/m^2a)	956	993	1088	1561	1582	1686

Table 6.7. Values for direct normal irradiation and global irradiation of various regions that are interesting for the installation of solar thermal power plants

	Bari (Italy)	Tabernas (Spain)	Oujda (Morocco)	Cairo (Egypt)	Luxor (Egypt)
Latitude (°N)	41.1	36.1	34.2	30.1	25.4
Direct normal irradiation (kWh/m^2a)	1884	2180	2290	2350	2975
Global horizontal irradiation (kWh/m^2a)	1659	1832	1995	2093	2438

It is true that solar thermal power plants can also be operated in regions with rather low direct irradiation, but the economic viability decreases significantly there. Suitable regions show an annual total direct normal irradiation in the order of 2000 kWh/m^2a or above. In Europe the most suitable regions can be found in southern Spain, southern Italy, Greece and northern Africa.

Today, at good locations, levelized electricity costs of the order of €0.15/kWh can be realized (see Figure 6.16 overleaf). Volume production and technical improvements could lower these costs to below €0.10/kWh. Technically, solar thermal electricity can also be transported from northern Africa to central Europe. Given future transport costs of €0.01–0.02/kWh, a significant part of the electricity supply in central Europe could be covered by environmentally friendly solar thermal electricity in the long-term future.

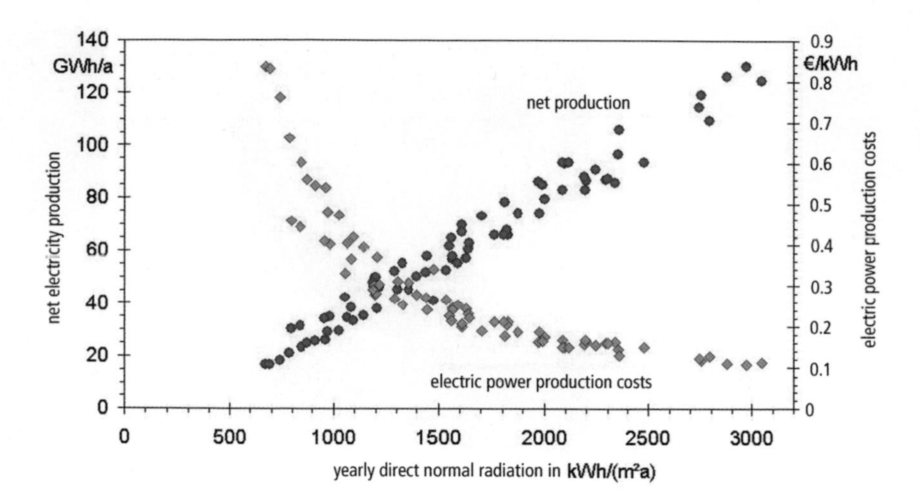

Figure 6.16.
Net electricity production and levelized electricity costs of a 50 MW_{el} parabolic trough power plant for 65 different worldwide locations plotted against the annual direct normal irradiation

7 Solar heating of open-air swimming pools

7.1 Introduction

Solar heating of open-air swimming pool water has some decisive advantages over other methods of using solar energy thermally:

- *Temperature level.* The required temperature level is comparatively low, at 18–25°C. This permits the use of less expensive polypropylene absorbers.
- *Solar radiation and time of use.* The swimming season coincides with the time of the highest solar radiation. Commonly at latitudes in central Europe open-air pools are operated from the beginning or middle of May until the middle of September. During this period approximately 65–75% of the annual solar radiation occurs. At lower latitudes, the swimming season can be longer. Because of higher air temperatures the need for swimming pool heating may decrease, but with a smaller collector high efficiencies can be reached.
- *Simple system design.* The pool water flows directly through the absorber, powered by the filter pump. The storage tanks normally required for solar energy systems are not required, as the pool itself takes over this function.

Solar heating for open-air swimming pools has been used for several decades now, and is a well-established technology. However, this does not mean that this application of solar thermal energy has reached its limits yet.

According to statistics in *Sun in Action II*[43], the updated overview of European solar heating markets by the European Solar Trade Industry Federation (ESTIF), about 3000–4000 m^2 of unglazed collectors were erected yearly in the 1990s. The estimated production and sales for 2000 and 2001 were 10,000 m^2 yearly.

If we look at the developments over recent years, heating of the pool is too costly for most swimming pool owners. Existing older conventional heating systems are, however, often replaced either by absorber systems, or the owners do without heating altogether.

If conventional fuels are used for heating pools and spas they are most likely to use natural gas; however, some heat pump pool heaters have emerged in the US markets.

This chapter describes in detail the components, planning, installation and economics of swimming pool absorber systems. The emphasis is on the solar heating of swimming pools (open-air pools in the private area with pool sizes of up to 250 m^2) and public open-air pools (local authority or privately operated). We shall also discuss combination options with systems for domestic water heating and for room heating support. The various options for making use of solar thermal energy in the indoor pool area were described in more detail in Chapter 5.

7.2 Components

7.2.1 Absorbers

Solar open-air pool heating uses absorbers to collect the energy (Figures 7.1 and 7.2). The collector design is characterized by the lack of either a transparent cover and housing or thermal insulation. This simple construction is possible as the systems operate with low-temperature differences between the absorber and the surroundings and with relatively uniform return temperatures (10–18°C).

The swimming pool absorber is generally made from plastic, utilizing a specially stabilized polypropylene polymer as the plastic extrusion. As an alternative, some

Figure 7.1.
Solar absorber in an open-air pool.
Source: Lange GmbH, Telgte

Figure 7.2.
Swimming pool absorber system for a private pool

manufacturers use EPDM extrusion, as it is more flexible and can be fixed directly to roof shingles. Pool absorbers are generally drained in the winter to prevent frost damage to the absorbers and piping.

Because of the risk of corrosion of thermal collectors with copper absorbers, these can be operated in solar systems for swimming pool heating only if a separate solar loop is installed (that is, an indirect system).

7.2.1.1 EFFICIENCY AND YIELD

The use of unglazed and uninsulated absorbers for solar open-air pool water heating has some advantages, owing to the special operating conditions.

In the typical operating range, with a temperature difference, $\Delta\theta$, between the outside temperature and the mean absorber temperature of 0–20 K, absorbers often operate with a higher efficiency than glazed collectors. This is because the optical losses (normally about 10–15% with respect to the amount of solar radiation) through a transparent cover do not arise, and the thermal losses are not so significant because of the low temperature difference. These thermal losses increase with operating temperatures, but this rarely occurs because of the moderate absorber temperatures found under normal operating conditions. The wind speed is the decisive factor that causes losses and hence has a negative effect on the efficiency of the absorber (Figure 7.3).

7.2.1.2 DESIGNS

Apart from a few special designs, plastic absorbers can be subdivided into two groups:

- tube absorbers (small tube absorbers)
- flat absorbers.

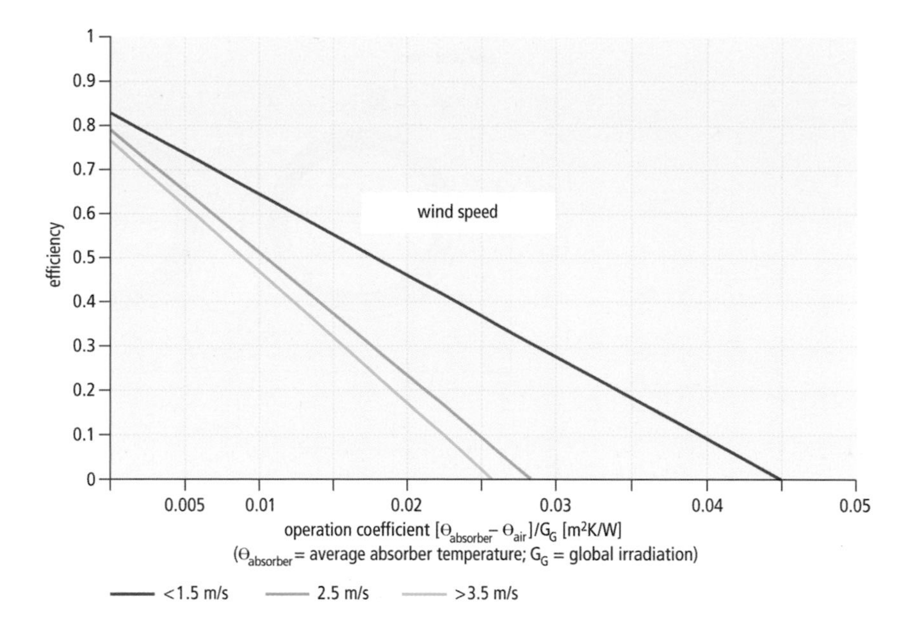

Figure 7.3. Characteristic curve as a function of wind speed

The *tube absorber* (Figure 7.4) is the simplest design. A number of smooth or ribbed tubes (small tubes) are arranged in parallel and, according to the design, are connected together either with intermediate webs or by retainers at a given spacing. Absorber lengths of up to 100 m can be achieved, and obstructions such as chimneys or rooflights can easily be circumvented.

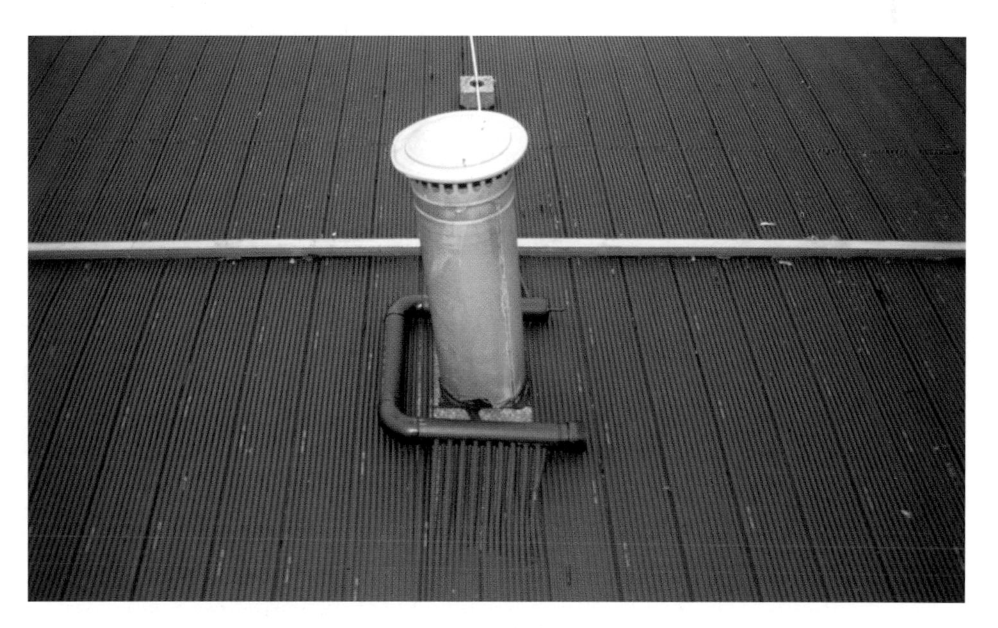

Figure 7.4. Chimney bypass with an EPDM tube absorber. Source: DGS

In the case of *flat absorbers*, sometimes also called *plate* or *cushion absorbers*, the channels are linked together structurally. This produces plates of different dimensions with a smooth surface. This has the advantage that there are no grooves in which dirt or leaves can accumulate and solidify. The self-cleaning effect during rain is also better.

The influence of the design form on the conversion factor with different inclination angles can be measured, but it is minimal. Variations of the angle of incidence lead to small differences in the conversion factor only for flat collectors. They lead to larger variations with ribbed tube absorbers than with normal tube absorbers.

All absorbers are very easy to handle (see also Chapter 5 on installation): thus for example all common types can be walked on.

Figures 7.5 and 7.6 show respectively a summary of the absorbers available on the market, and the different methods of connecting the absorber to the collection and distribution pipes.

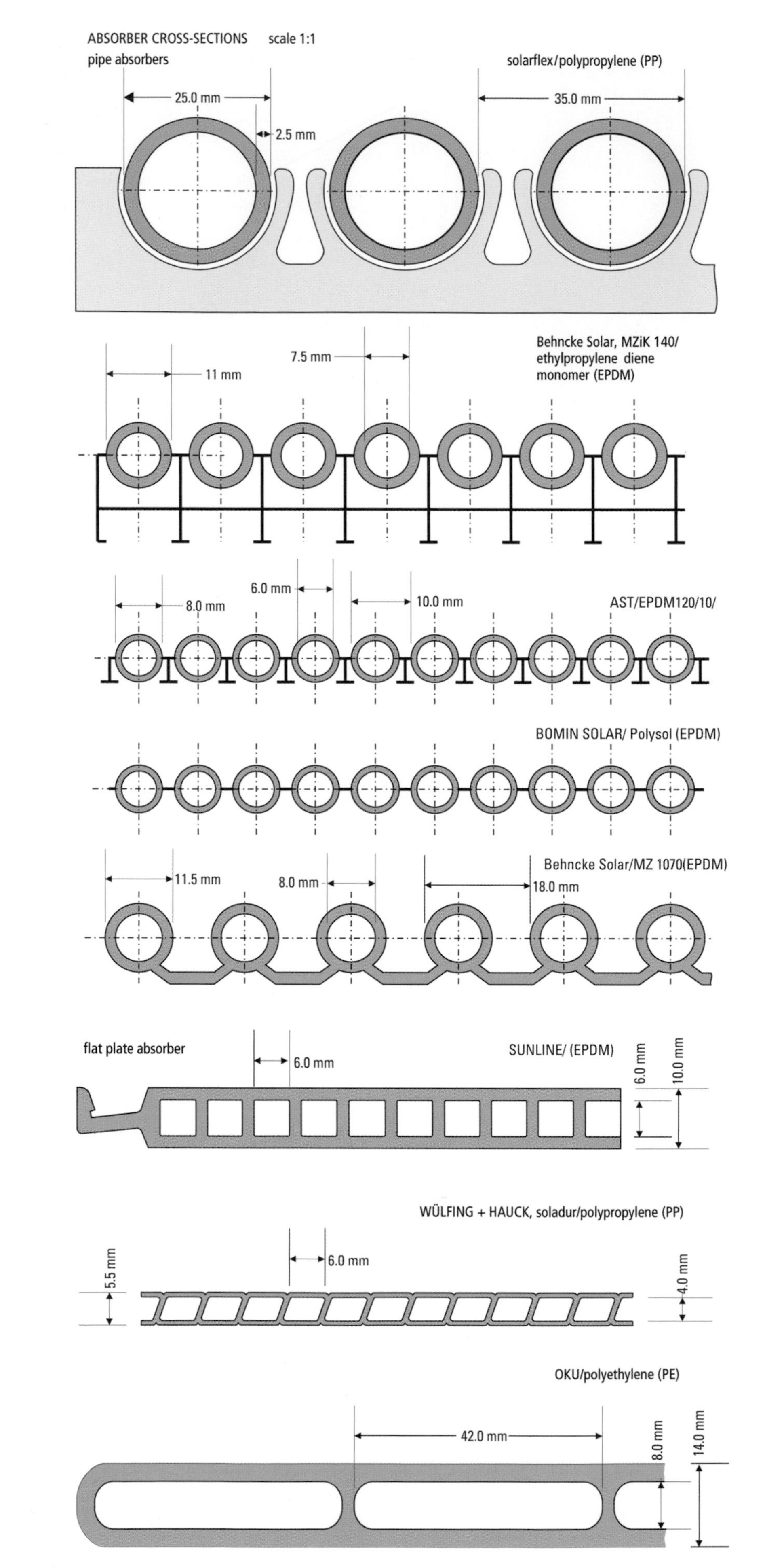

Figure 7.5. Different designs of absorber in cross-section

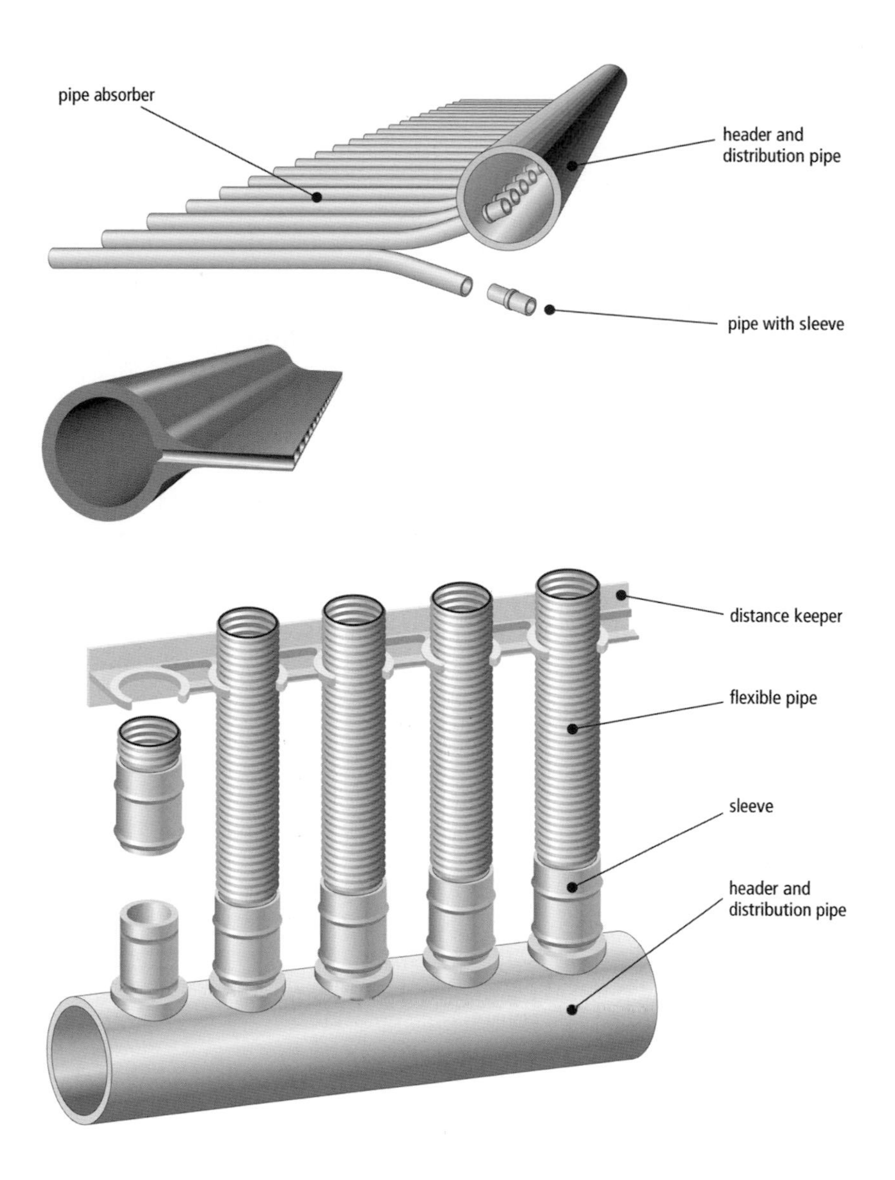

Figure 7.6.
Different methods of connecting the absorber to the collecting/distributing pipes

7.2.1.3 PRESSURE LOSS

It is important to consider the pressure loss of the absorber during planning. As the pressure loss for the whole system defines the pumping power, the objective of the designer here should be to achieve the lowest possible system power requirement with the highest possible thermal yield.

In the following we shall therefore discuss briefly the very different pressure losses of the individual absorbers. In general it is true that the maximum permissible excess operating pressure is quite low (0.5–1.5 bar). Only the finned tube absorber has a higher value (3.0 bar). The tube cross-section or diameter has the greatest influence on the pressure loss. When using absorbers with small tube diameters, lower maximum lengths are possible than for absorbers with large diameters. This must be taken into account during system planning.

7.2.2 Piping and header pipes

In principle the same components as used in the solar circuit for solar heating can be used for open-air swimming pool construction. Copper or steel pipes cannot be used here because of the risk of corrosion; plastic pipes are the only ones that can be used.

The header pipes are generally made especially for the particular absorber type and are offered by the manufacturer together with it. Sometimes they are actually integrated directly in the absorber.

According to the system, end-stoppers, sleeves, socket ends and pipe connections are usually offered together with the collector and distributor pipes, which can be fixed by gluing, by welding or – as shown in Figure 7.7 – with clamps. In the case of long straight pipelines very high temperature-related length changes must be taken

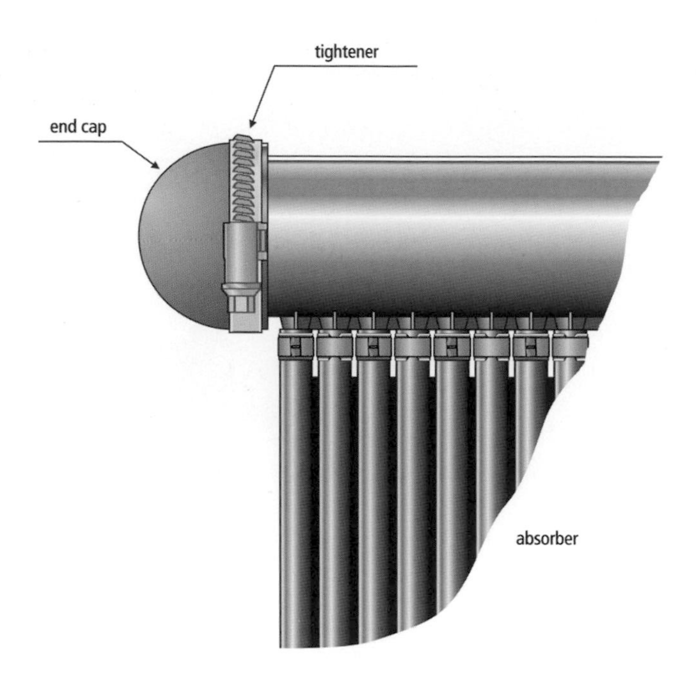

Figure 7.7. End cap of a collector pipe fixed with clamps

into account for plastics (coefficient of thermal expansion up to 0.2 mm/mK). Appropriate expansion bends must be installed with anchor clamps, which are fixed so that they can slide in the direction of the pipeline (see also section 7.5.2.2).

7.2.3 Materials

Solar absorbers are exclusively made from plastic. They can be hard and rigid or soft and flexible according to the plastic mixture. The use of plastic permits operation of the solar system with chlorinated swimming pool water. It is, however, necessary to consider the chlorine content. A high dose (from about 5 mg/l) can damage the absorber. The exact limits from which damage can occur depend on the composition of the plastic.

Plastics are also used for pipelines. These are made from rigid materials, however. The following plastics are basically the ones that can be used:

- EPDM (ethyl propylene diene monomer)
- PP (polypropylene)
- PE (polyethylene)
- ABS (acrylnitrile butadiene styrene copolymer)
- PVC (polyvinyl chloride – hard or soft).

Because of their good properties, two absorber materials are market leaders in spite of their higher costs: EPDM and PP. Table 7.1 gives a summary of the typical properties of the designated materials for absorbers and pipelines. Both require specific formulations to avoid degradation in sunlight.

7.2.4 Pumps, heat exchangers and other components

7.2.4.1 PUMPS

The materials used for pumps also have to fulfil the requirements for corrosion protection. Because it is normally not possible to use pumps without metal, corrosion-resistant materials should be used. The impeller, for example, is usually made from cast bronze, and the shaft from chrome nickel steel. The housing usually consists of grey cast iron, but plastics may also be used. Some manufacturers also offer swimming pool pumps made completely from plastic, such as glass-fibre-reinforced PP or POM (polyoxymethaline). Pumps can also be obtained on the market in which the pump shaft does not come into contact with the swimming pool water owing to the design. If the capacity of the existing filter pump is not sufficient to pump the swimming pool water through the additional absorber system, one or more supplementary pumps

Table 7.1. Properties of plastics used as absorber or pipe materials

Material	Properties	Deployment temperatures	Lifetime	Deployment
EPDM	– artificial rubber – Flexible, also frost resistant when filled with water	– 50 °C to 150 °C	> 30 years, manufacturer's guarantee often 10 years	Absorber
PP	– Polyolefine – Polymer from Propylene – Pipes weldable – Frost resistant (when not filled with water)	– 30 °C to 120 °C	> 20 to 30 years, manufacturer's guarantee often 10 years	Absorber, piping
HDPE	– High density polyethylene – Weldable – Mostly pipes, rigid – UV resistant through soot – Frost resistant (when not filled with water)	– 30 °C to 110 °C	> 30 years	Piping, absorbers
Soft PVC	– Can be glued – Deteriorates when softeners escape	– 20 °C to 65 °C	Strongly dependent on application	Piping
Hard PVC	– Can be glued, UV resistant	– 5 °C to 100 °C	> 20 years	Piping
ABS	– Polymer	– 10 °C to 80 °C	> 20 years	Piping, distribution piping

must be used. Owing to the large volumetric flow in comparison with domestic water solar systems, and the resulting pipe diameters, the pumps are correspondingly dimensioned and have power settings of several kW, or even more in the case of very large systems (see Figure 7.8).

Figure 7.8. Circulating pump for solar loop, cast iron housing, chrome nickel steel shaft, cast iron or gunmetal impeller. Manufacturer: Herborner, Herborn

7.2.4.2 HEAT EXCHANGERS

Standard solar systems for open-air pool heating have a simple system construction, in which no heat exchanger is necessary. If, however, another type of heating is required, heat exchangers are necessary. The heat exchanger must naturally meet the same material requirements as on the swimming pool water side. Stainless steel (V4A or St.1.4571) is generally used here. All sorts of heat source, such as heat pumps or gas heating boilers, can be connected and a temperature sensor positioned for control purposes. Certain system configurations (see section 7.3), however, require the use of heat exchangers, which are described in more detail in sections 2.4.4 and 5.4.

7.2.4.3 OTHER COMPONENTS

According to the system connection, pumps and/or partly motor-controlled valves (Figure 7.9) are necessary for the operation of the system. Plastics are also used here for the valves. Such commercially available fittings for swimming pool systems can be obtained in both PVC and PE or similar materials. As the flow is regulated by motor-

Figure 7.9. Motor-controlled three-way valve made of PVC. Manufacturer: Resol, Hattingen

controlled valves, some manufacturers of control equipment also offer such fittings for swimming pool system operation. In addition, non-return valves, shut-off valves or slide valves and ventilators are required. These fittings are also standard accessories for swimming pool technology, and are available in correspondingly appropriate materials.

7.2.5 Controllers

A swimming pool absorber system uses the well-known principle of *temperature difference control*. However, the temperature differences that lead to a switching procedure are significantly smaller than in the case of domestic water heating systems, for example. Thus the starting of the solar pump or the positioning of the three-way valve takes place at 2–4 K, whereas at 0.5–1 K the pump is switched off again, or the three-way valve is switched over.

When the pool temperature exceeds a given value the solar system is switched off again. This value can, for example, be approximately 28°C. The maximum temperature should be carefully selected. On the one hand a reduced refreshing effect for the swimmers at higher temperatures plays a role; on the other hand the selected maximum temperature should not be too low, as the pool can act as 'buffer storage' for less sunny days.

The important thing for any form of control is the correct positioning of the temperature sensors. For switching on the absorber circuit pump, the absorber temperature is compared with the pool temperature. However, the pool temperature is not acquired within the pool itself but in the filter circuit. For the most accurate control of the system the on and off signals for the solar systems are separated from one another. This means that the absorber temperature is not used for switching off; instead the feed temperature is compared with the pool temperature. Accurate sensor elements increase the thermal yield of the solar system. As a rule Pt1000 sensors are used. It is, however, also important that the sensors are compared in pairs. It is of secondary importance whether the absolute temperature is accurately displayed.

It has also been found to be useful, especially for large solar systems, to acquire the absorber temperature not by means of clip-on or immersed sensors, but rather by means of a separate non-filled reference absorber section. In this way the effect of the relevant variables of radiation, air temperature and wind is transferred to the output with a very low control delay time.

Controllers for large swimming pool absorber systems can usually also record other variables such as irradiated power and volumetric flow, so that balancing and determination of the efficiency of the solar system can take place. There are also controllers that can operate different pools with different temperature levels. A paddling pool for small children has a much higher set temperature than a swimming or diving pool. If there is low irradiated power and hence available heat from the absorber systems, this can be supplied for example directly to the children's pool, as the thermal output from the absorber is sufficient for this but not for the larger pools.

Figure 7.10 shows a standard control scheme for solar open-air pool heating. Correct positioning and suitable sensors play a decisive role here. Some controllers are also able to control the auxiliary heating. In order to use energy rationally it is particularly important to consider the desired maximum temperature in the case of auxiliary heating. If one chooses a high target temperature the losses increase and hence the energy consumption also increases. If the temperature is raised, for example, from 25°C to 25.5°C the energy consumption increases by up to 10%.

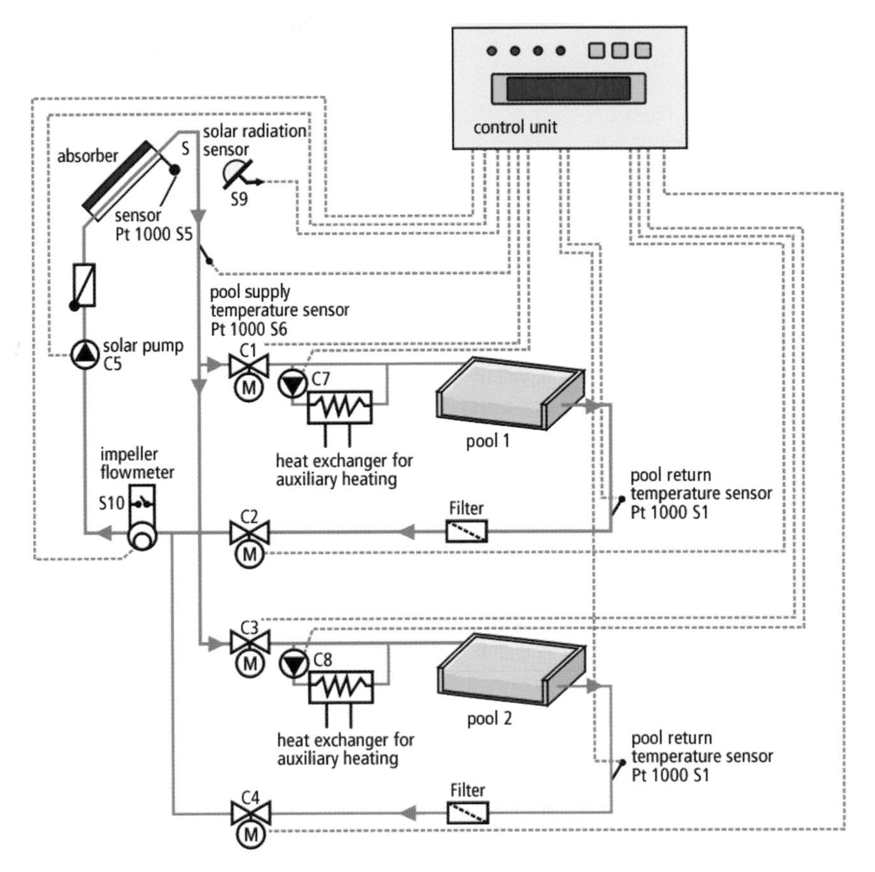

Figure 7.10. Example of a control scheme for a solar open-air pool heating system

NOCTURNAL COOLING

In hot and sunny weather, normal solar radiation striking the surface area of the pool may result in pool temperatures exceeding the point at which the pool is refreshing. A simple switch can be employed to simply reverse the sensors and trick the controller into engaging the pump and circulating pool water through the collectors. If a clear sky is present, thermal radiation from the pool water can effectively reduce water temperatures and ensure a refreshing pool. In large pools where swimming competition requires temperatures within a several degree threshold, this feature can be very useful for outdoor competition.

7.2.6 Covering of the swimming pool

The heat losses from a swimming pool occur mainly from the surface of the water through evaporation, but also through convection and radiation (see also Figure 7.16). To prevent overnight reductions in the pool temperature a cover over the water surface with a suitable thermal insulating effect is recommended. This prevents up to 100% of the overnight evaporation of the pool water from taking place, and the losses through radiation and convection are significantly reduced too.

A cover is particularly efficient in temperate climates, if the pool water is to be held at a high temperature (> 25°C) or if the pool is in an unprotected position. Here it is possible to save between 30% and 50% of the energy according to the location and position of the pool, or the temperature can be held at a higher level. Because of the considerable extra costs for covering the large pool surfaces in municipal open-air pools these covers are usually not installed.

In the case of private swimming pools, simple and cheap covers can always be installed. The absorber surface can be kept smaller because of the savings due to the covering, and the total investment costs can thus be reduced in spite of the extra costs for the cover. Just like the absorber, the covers should be UV- and temperature-resistant. They also are made of plastics such as, for example, closed cell PE foam, PE bubble-foils or PVC cellular profiles (Figure 7.11). The heat can be significantly better maintained in the pool with PE foam covers than with PVC profiles.

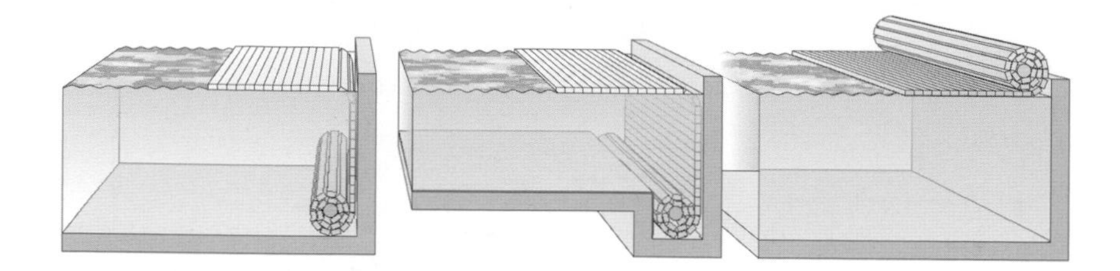

Figure 7.11. Different covering variants for swimming pools to reduce heat losses

7.3 Systems

7.3.1 Solar private open-air pool heating

In the case of private swimming pools the pool surface is seldom larger than 100 m^2. Commonly only simple filter circuits are installed here. Conventional auxiliary heating systems are being installed less and less. As for domestic water heating systems, solar systems are now being offered as complete packages with all necessary components. The absorber surface area is dimensioned according to the size of the pool (see planning section) and is thereby offered in different sizes. There are several methods of implementing the hydraulic circuit and the operation of the absorber circuit. The two most sensible and most frequently used systems are described in detail in the following sections.

7.3.1.1 SYSTEMS WITH THREE-WAY VALVES

The absorber circuit is integrated into the existing filter circuit with a three-way valve (Figure 7.12). This means that the filter pump must be suitably dimensioned in order to overcome the additional pressure loss in the absorber circuit. As a rule this is the case in private swimming pools. It depends strongly on the height difference between the pool surface and the absorber. If this is greater than about 5 m, an additional absorber circuit pump is normally required (see section 7.3.1.2).

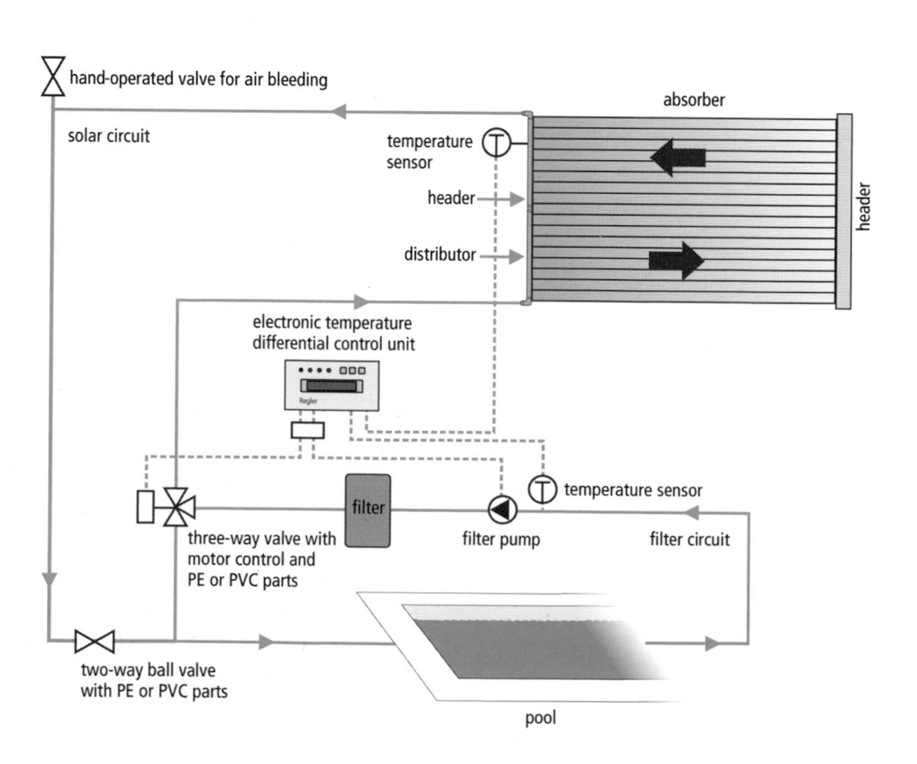

Figure 7.12. Circuit diagram of absorber system with three-way valve

If a suitable temperature difference (2–4 K) exists between the absorber and the pool water, the solar controller starts the pump operation. The motor-controlled three-way valve is set so that the water flows through the absorber, heats up, and is then led back to the pool.

There are even simpler solutions in which the three-way valve is operated manually instead of by the controller. However, these are rarely used.

7.3.1.2 SYSTEMS WITH ADDITIONAL ABSORBER CIRCUIT PUMP

In these systems an additional pump operates the absorber circuit. In the conditions described above, this is switched on in addition to the filter pump. A non-return valve should be installed in the feed to the absorber circuit to prevent the absorber field from running empty after switching off the pump. This can be operated electrically or pneumatically. In such a case a non-return valve should also be installed in the filter circuit to prevent incorrect flows (Figure 7.13).

In a variant without a non-return valve, a set of fans or a ventilation fitting should be installed. If the absorber runs empty when the pump is stationary, a vacuum can occur, which would damage the absorber. A simple ventilating device allows air to flow into the absorber. When the pump is switched on, the air in the absorber is forced out through the absorber and filter circuit into the pool. A ventilating device ensures that the air can immediately escape again from the absorber when the pump is switched on.

In a system with an additional absorber circuit pump there is also an alternative operating mode. If the pool filter circuit is not permanently operated, the absorber circuit can be connected independently of the filter circuit. The pool water is removed in front of the filter system and pumped through the absorber by the absorber circuit pump. For this purpose the absorber circuit pump should have a fine filter connected upstream.

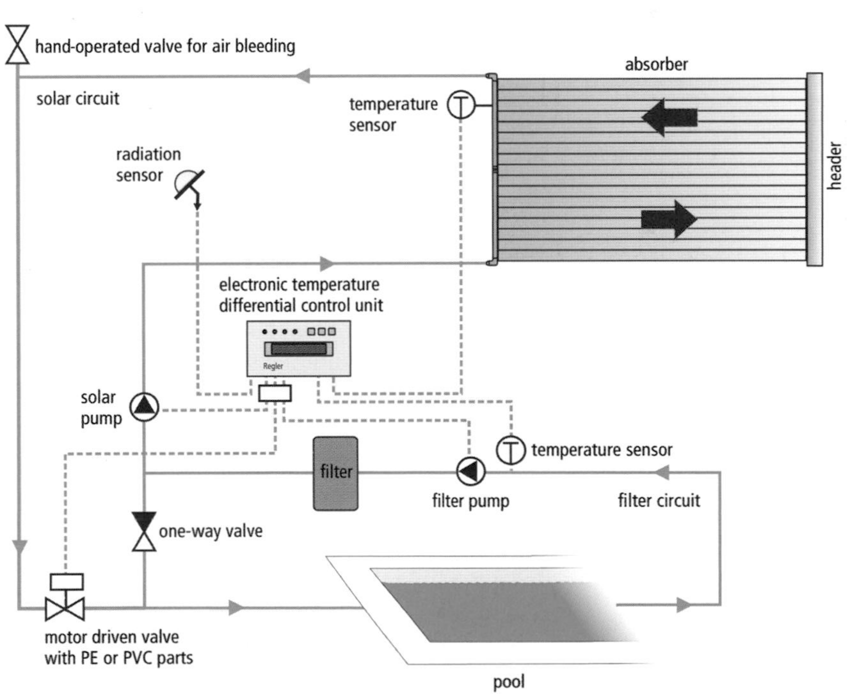

Figure 7.13. Circuit diagram of absorber system with additional solar pump

7.3.1.3 INTEGRATION OF AUXILIARY HEATING

Information on the integration of an auxiliary heating system can be found in the corresponding section for public open-air pools. This also applies to private swimming pools.

7.3.2 Solar heating of public open-air swimming pools

A further application for solar open-air pool heating is that of public pools, which are mostly operated by local authorities, although some are privately operated. Here, according to the type of pool complex, one or more pools are heated by the absorber system. How many, and which pools are supplied with solar heat, depends on the system configuration and the surface available for installation of the absorbers. In large open-air pools an absorber surface area of several hundred square metres can easily be required.

7.3.2.1 HYDRAULIC CIRCUIT

Solar circuits in public open-air pools are normally operated with a separate solar circuit or absorber circuit pump. The hydraulic construction is much more complex than for private swimming pools because of the hygiene requirements.

A system in a large open-air pool functions according to the following principle. The waste water is led from the pool into a central water storage tank. This tank acts as a 'water-level display' for the whole swimming pool water circuit. Evaporated water is replaced here by fresh water. The water is pumped through the filter from the water tank. One or (according to the design of the filter system) several parallel-connected filter pumps are used for this. After this the water is returned to the pool via the water treatment system.

In front of the water treatment system, the absorber field is connected to the circuit in a bypass system. The solar loop pump diverts part of the volumetric flow and pumps it through the absorber field. The size of the partial volumetric flow depends on the size of the absorber field. The solar-heated water is led to the main flow again after the diversion and finally arrives back in the pool.

A motorized valve should be installed in the absorber circuit feed line and a non-return valve after the solar pump. These two fittings prevent the absorber field from running empty when the system is not in operation.

Before the water reaches the pool the hygiene parameters are set. Chlorine and chemicals are introduced to regulate the pH value as necessary. The chlorine injection point should always be integrated behind the absorber field diverter, as the chlorine concentration in the absorber circuit must not exceed 0.6 mg/l. If there is a surge of chlorine (under certain circumstances up to 10 mg/l) the absorber may be damaged.

OTHER CIRCUIT VARIANTS

The hydraulic circuit described above is the most simple and economic. It does, however, mean that all pools must be operated via the same filter system.

If each pool of an open-air complex has its own filter circuit, the absorber system must be integrated in other ways. One option is to connect the absorber field hydraulically to several filter circuits. However, a partial flow must always be diverted from one filter circuit and heated by the absorber system. In this way each pool can be solar-heated successively. For example, at first the swimming pool can be heated, and if additional solar heat is still available this can then be provided to the diving pool.

Other arrangements of the hydraulic circuit allow the supply of the solar-heated water to only one, several or all pools according to the current thermal output of the absorber unit.

When realizing such a circuit, the position of the pools and their filter circuits as well as their distance from the absorber field must be considered for both hydraulic and financial reasons.

7.3.2.2 CONNECTING THE ABSORBER FIELD

With solar-heated open-air pools there is often the difficulty that the absorber field has to be installed on several roof areas. In addition, it is rarely possible for each absorber field to be the same size, and also the pipeline lengths are all different. A circuit conforming to the Tichelmann connection (also called Z-connection, see Chapter 2) is worth striving for, but is usually not feasible. It is therefore even more important for all the absorber fields to have a uniform flow through them. This entails careful arrangement of the pipelines and absorber field pumps. Ball valves should be installed at suitable points to guarantee simple emptying and filling of the absorber fields.

7.3.2.3 HEAT RECOVERY FROM THE REVERSE FLOW FILTER WATER

The filter system must be cleaned at regular intervals, or the filter itself may have to be cleaned after contamination. This is done by a reverse flow of fresh water. The water flushed through the filter must be drained to the sewer, which means that heat is inevitably lost. The fresh water that replaces the water drained into the sewer can be preheated via a heat exchanger. Coaxial heat exchangers with automatically circulating cleaning pellets have proved themselves to be useful for this.

As for other applications, the economics of heat recovery from the reverse flow filter water must be carefully considered. As a rule this is worthwhile mainly for pool temperatures over 23°C. Installations carried out within the scope of a complete restoration of the pipeline system and a long swimming season can also improve the economic viability.

7.3.2.4 INTEGRATION OF AUXILIARY HEATING

Conventionally operated auxiliary heating is necessary if the pool water has to be maintained at a constant temperature. Some open-air pools like to offer their visitors warm swimming pool water independently of the sunshine, which requires auxiliary heating when the solar radiation is insufficient.

Auxiliary heating is operated by means of a conventional system (preferably gas heating) and an additional heat exchanger. In a dual-heated system the auxiliary heating should always follow the solar heating. If the water is not of the required temperature after recirculation to the filter circuit, the auxiliary heating covers the residual heat requirement (Figure 7.14).

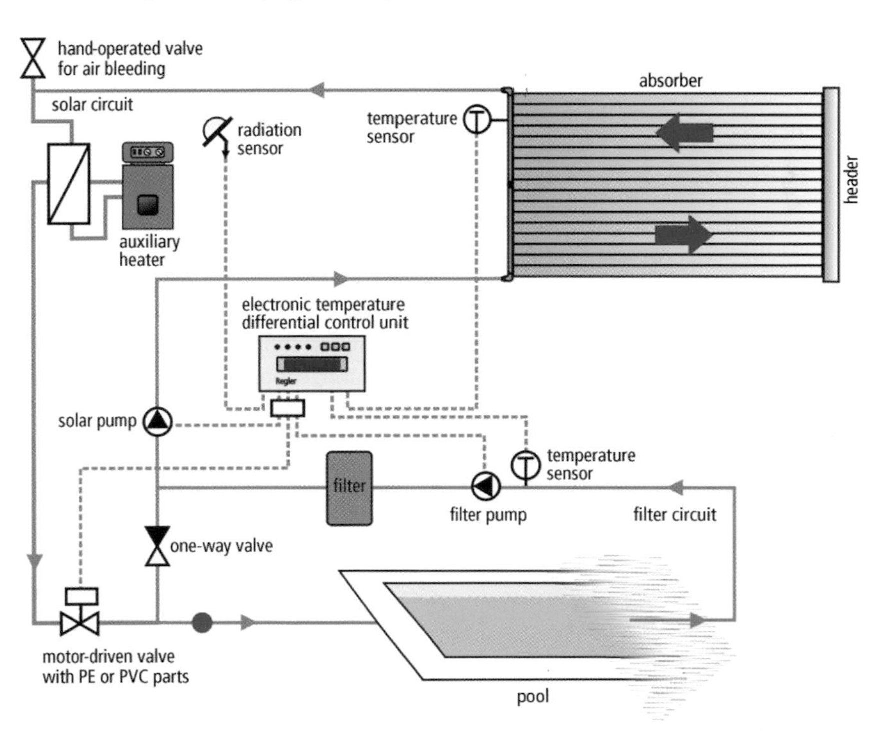

Figure 7.14.
Circuit diagram of large absorber system with additional heating

7.3.2.5 RATIONAL USE OF ENERGY IN PUBLIC OPEN-AIR BATHS

Much of the total heat requirement is basically determined by the level of fresh water required. The amount of fresh water is very high in most open-air pools: 10–20 m^3 per square metre of pool surface per season are typical figures.[44] Of course, these figures depend on the pool usage. This amount of water has to be heated from the cold water temperature to the pool temperature. In central Europe, this means that about 15–20 kWh of heat is required to heat 1 m^3 of fresh water to pool temperature. Whether the amount of fresh water added goes beyond the required level is easily established on the basis of the annual quantity of water consumed if the water is taken from the public drinking water network.

With old filter technology the required quality can be achieved only by the addition of significantly more fresh water, which in turn must be brought up to the required pool temperature. Together with the increased energy requirement of old pumps and the greater flow resistance of older technology, this can represent half of the energy consumption of an open-air pool complex. So by replacing old equipment a significant part of the energy can also be saved.

7.3.2.6 INTEGRATION INTO EXISTING SWIMMING POOL TECHNOLOGY

The inclusion of a solar system into an existing open-air pool does not normally pose special problems. At a suitable point part of the volumetric flow of the pool water is led through the absorber. According to the design this might even be the whole volumetric flow. As the same technology is used for the pipelines in solar open-air pool heating, the connections do not represent any problems. The most favourable position for the connection of the absorber bypass is after the filter system and before the water treatment station.

The auxiliary heating should always be installed after the return flow of the absorber bypass into the filter circuit so that it can be simply controlled and operated effectively.

If such integration is not possible for structural reasons an alternative must be found. However, it is always necessary to consider the disadvantages, for example connection of the absorber bypass after the chlorine injection point (see section 7.3.2.1).

7.3.3 Combined solar domestic water heating, open-air pool heating and room heating support

A solar system for domestic water heating and for solar room heating support can be sensibly combined with solar open-air pool heating. These systems are designed for heating support in the transition months. In the summer, however, no room heating is required, and hence excess heat is available during this time that can be transferred to the swimming pool water. Additional heat exchangers are required for this, as such solar systems are not operated with plastic absorbers, but rather with glazed flat-plate collectors or evacuated tube collectors.

The system is similar to a solar system with two storage tanks and priority switching, whereby the first tank is a combined tank and the role of the second tank is taken on by the swimming pool (Figure 7.15). Charging of the storage tank has the priority. When this has reached its maximum temperature, or if there is insufficient solar radiation to heat the tank, the system is switched over to the swimming pool in the summer. In the winter the collector surface area, which has been increased for heating support, is completely available to the storage tank. This can both heat the domestic water and be used for room heating support (combined storage tank). With this system it is necessary to take account of the special material requirements for the solar loop/swimming pool water heat exchanger. This must be made from suitable materials (for example stainless steel or possibly special copper alloys) in order to prevent corrosion.

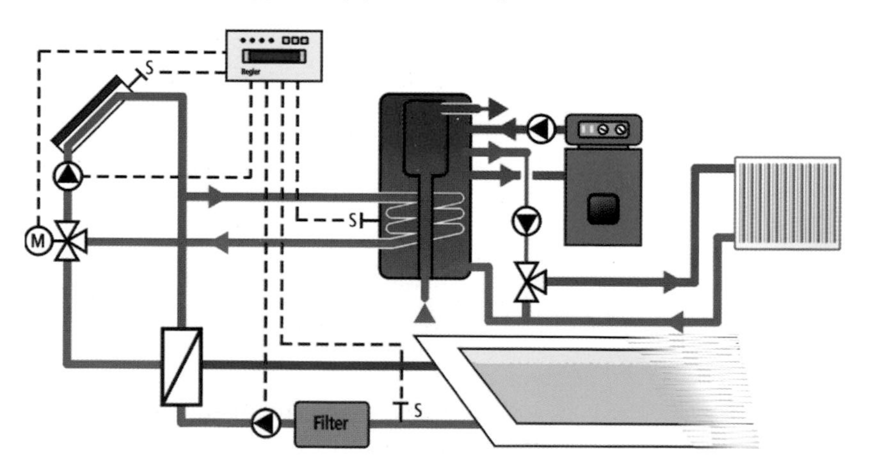

Figure 7.15. Control diagram for a solar system for domestic water heating, room heating support and open-air pool heating

7.4 Planning and dimensioning

7.4.1 Fundamental considerations

As for solar thermal systems for domestic water heating, the prevailing conditions of solar radiation and heat consumption are of great significance in the planning of solar swimming pool heating systems. The heat consumption of a swimming pool is in turn determined by the size of the pool surface area, the depth of the water and the colour of the pool, the desired water temperature, and the ambient meteorological conditions (air temperature and wind speed).

In open-air pools the pool water can be heated to a solar fraction of up to 100% with corresponding system dimensioning. The resulting slight temperature variations in the pool water do not normally affect the visitors' enjoyment, as more visitors come in sunny weather than in periods of poor weather. In such a single-source solar pool water heating system the water volume acts as the storage medium, which cushions the temperature variations. In lengthy periods of poor weather the few regular swimmers must, however, expect and accept a lower pool water temperature compared with a conventionally heated pool.

Table 7.2 gives a summary of the heat gains and losses arising in a swimming pool, together with the variables upon which these depend.

When planning a solar system for pool water heating, the aim must be to balance the thermal losses from the swimming pool against the solar gain as far as possible, so that there is little need for auxiliary heating. However, in the following section we shall also discuss the design of large open-air pools with conventional auxiliary heating.

Table 7.2. Heat gains and losses in a swimming pool, and the influencing variables

Heat gain	Heat loss	Influencing variable
Direct solar radiation into the pool	Evaporation	Pool surface area
Conversion of pump energy	Convection	Pool water temperature
Addition of heat (solar or conventional)	Radiation	Climatic conditions at site
	Surrounding earth	Wind conditions on the pool surface
	Exchange of filter flushing water for fresh water	Number of swimmers in pool
		Groundwater conditions

7.4.2 Approximation formulae for establishing absorber surface area, volumetric flow and pressure loss

7.4.2.1 PRIVATE OPEN-AIR POOLS

The size of the absorber field of the solar system is geared to the size of the pool surface. For private pools the range of possible designs – according to which the average temperature level is to be achieved – is very wide, as here, unlike public open-air pools (see section 7.4.2.2), economic aspects tend to be more in the background.

In principle, the design ratio between absorber surface and swimming pool surface area can be defined as follows:

- Temperate climates: absorber surface area = 0.5–1.0 × the pool surface area.
- Subtropical climates: absorber surface area = 0.3–0.7 × the pool surface area.

Because of the significantly smaller pool surface, private swimming pools should be equipped with relatively cheap covers. If a cover is used, a smaller design ratio can be selected.

For private swimming pools, unlike public open-air pools, the absorber field is often installed on the pitched roof of the house. With a roof pitch >15° the absorber should have an orientation between –45° and +45° to the equator. The less favourable the alignment of the absorber surface, the greater the design ratio that should be selected.

7.4.2.2 MUNICIPAL OPEN-AIR POOLS

Figure 7.16 shows the heat losses from an open-air pool. The evaporation losses on the pool surface represent more than 60% of the total losses. However, these figures depend strongly on the circumstances, such as solar irradiation, pool temperature, air temperature and wind speed.

A pool cover can significantly reduce the evaporation losses. However, for public open-air pools with large pool areas this requires very high investment costs, and so such covers are seldom found. In certain circumstances part of the evaporation losses can be compensated for at less expense by increasing the absorber surface area rather than by covering the pool.

Apart from the heat losses over the pool surface, the climatic conditions (radiation, temperature and wind speed) must be taken into account during

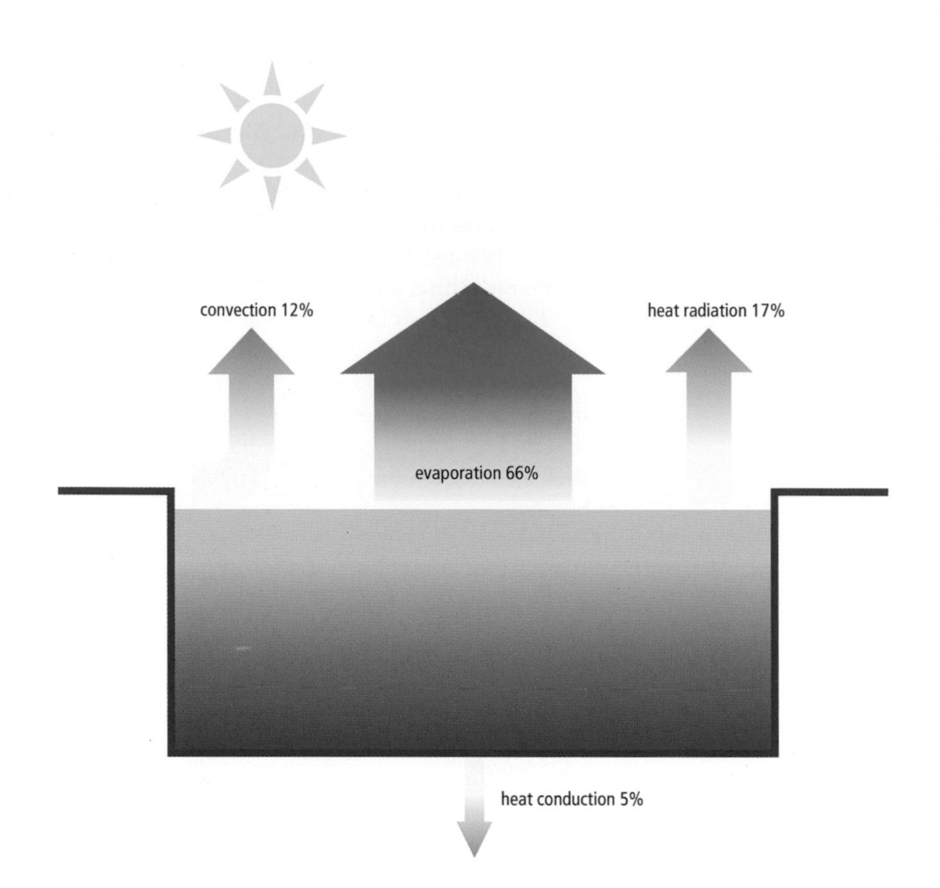

Figure 7.16. Heat losses from an open-air pool[45]

dimensioning. Given the direct relationship between pool surface area and heat loss (and the heat required to compensate for this), the ratio between absorber surface and pool surface is the decisive criterion for the design of the absorber field.

SINGLE-SOURCE HEATING

Many years of experience of operating solar open-air pool heating have shown that this ratio should be in the following range for temperate climates:

Absorber surface area = 0.5–0.7 × Pool surface area

In subtropical climates, depending on the comfort requirements, 0.3–0.5 × Pool surface area may be used.

With this design, under standard conditions an average pool temperature of 22–23°C is established.

The exact design depends on other general conditions. If a high pool temperature is desired, the absorber surface area and hence the design ratio must be increased. In practice it is often the available space for the absorber field that defines the ratio. The alignment of the absorber is not so important for the design. This is largely because the absorber system is normally installed on flat roofs for municipal open-air pools.

DUAL-SOURCE HEATING

If the pool water is maintained at a temp of 21–22°C by conventional auxiliary heating, the design ratio of the absorber to the pool surface is selected in a similar way as for single-source heating. The average pool water temperature is significantly higher than for single-source heating. In spite of this, the energy used for auxiliary heating is not reduced to the same extent, as the solar yield increases with larger absorber surface areas. This is due to the increasing losses because of the higher average pool water temperature

Dual-source heating must always be examined critically from the ecological point of view. For a typical design ratio of 0.6, dual-source operation increases the average pool water temperature by 1 K compared with single-source operation. For this comparatively low increase in comfort, a relatively large amount of conventional energy has to be used, which leads to increased operating costs.

CONSIDERATION OF VISITOR BEHAVIOUR

Consideration of the behaviour of the swimmers can help in selecting the system (single-source, dual-source) for heating a pool. Investigations have shown that during periods of poor weather – for example when the air temperature falls below 17°C – the number of swimmers falls for both types of heating. In a heated open-air pool the number of swimmers may well be several percentage points higher, but in considering the economics of the operation, the pool operator must understand that this high level of comfort involving constant pool water temperatures of, for example, 21°C is used by only a few swimmers. High visitor numbers are generally achieved only on warm sunny days, on which single-source heating functions well through the solar system.

7.4.2.3 ABSORBER FLOW AND PRESSURE LOSS

Uniform flow through the absorber surface is the most important criterion for high yields, particularly for large absorber fields in public open-air pools.

Furthermore, surface-related volumetric flow is important for efficient heat transfer. The volumetric flow should lie in the range 80–110 l/h per m^2 of absorber surface. This is a generally recognized value, but the manufacturer's information should always be referred to. With this volumetric flow, and global solar irradiance of 800 W/m^2, a temperature difference of 6–8 K is established between the feed and return lines.

In principle, the thermal performance of an absorber field increases with increasing volumetric flow. However, from a specific volumetric flow of approximately 80 l/hm^2 the thermal performance approaches a limiting value. At the same time the required electrical auxiliary energy increases. In order to achieve uniform flow through the absorber surface, the following points should be taken into account:

- $\Delta p_{absorberfield} > 0.5\ \Delta p_{system}$. The pressure loss along the absorber should be at least double the level of the sum of the pressure losses of the field pipework.
- In order to maintain this, long narrow absorber fields and large distributor pipes with a slow flow speed should be used.
- Pipework conforming to the Tichelmann principle should be striven for, but on its own is not sufficient.
- In the case of several individual fields the installation of throttle valves has proved worthwhile. With variable-area flowmeters the flow through the individual circuits can be monitored and matched.

As a rule, 0.2 bar is sufficient to cover pressure loss in the absorber lengths to ensure uniform flow, if the distributor is dimensioned correctly. If the collector mats are relatively short, the pressure loss can be correspondingly increased by series connection or by partial series connections in order to achieve uniform flow.

For good heat transfer between the absorber and pool water the speed should be designed sufficiently high so that the flow in the absorber channel is turbulent. The difference between the flow speeds in the distributor pipe and the absorber channel should, however, not be too great, or flow disturbances such as fringe eddies and suction effects can arise.

If the conditions described above are fulfilled, the optimum flow is achieved with minimum pump energy consumption. The consumption of electrical energy for the pumps should not be more than 2% of the available solar energy.

7.4.2.4 PUMP DESIGN

The design of the absorber circuit pump uses the same calculations as described in Chapter 3 for domestic water heating systems. The total volumetric flow of the system is obtained after determining the necessary absorber surface and taking account of the specific volumetric flow described in section 7.4.2.3. In addition, the pressure loss through the absorber, pipeline and all the fittings in the absorber circuit, and the height difference between the pool surface and the absorber field must be determined. Once the total volumetric flow and pressure loss are established, suitable pumps can be found in the manufacturer's literature for absorber pumps.

7.4.3 Computer-aided system dimensioning

Simulation with a suitable program is also possible for solar open-air pool heating, and in particular for estimating the yield. However, separate calculations have to be performed for the correct design of the absorber field with respect to uniform flow, as the common simulation programs cannot show the flow relationships in the different absorber fields.

Pure swimming pool heating systems (indoor and outdoor) can be simulated with several programs, including T*SOL and the Canadian EnerPool. See Chapter 10.

7.5 Installation

7.5.1 Absorbers

Absorbers can be installed on inclined or flat roofs with different designs. It is also possible to install them at ground level. As, even in temperate climates, the main times of use are in the warmest months where the elevation angles of the sun are greater than 50°, installation on rigging to optimize the yields is not necessary. If an inclined surface is to be used, then an alignment facing the equator in the range –45° to +45° is advantageous.

The installation of the absorber is strongly dependent on the type (tube or flat absorber) and the properties of the background. The structural strength of the roof plays only a minor role here. The absorber itself, when filled, has only a low surface weight of between 8 and 12 kg/m^2 according to the design. Concrete slabs are used to secure the absorber field against wind forces. For this the surface load can be considerably higher than the absorber field, so that in this case the structural strength of the roof should be checked.

Whenever roof penetrations are created, it is essential that they are closed again afterwards and made weatherproof. Fixings that do not require the roof skin to be penetrated are preferable, such as flat roof installation on concrete slabs as already mentioned.

7.5.1.1 MECHANICAL FIXING OF THE ABSORBER

In many applications the absorber is installed on a flat roof or a slightly inclined roof. Mechanical fixings for the absorber in the form of webbing or steel rails are suitable here (Figure 7.17). The key aspect here is to secure the absorber field against wind forces. In the case of public open-air pools, installation on flat roofs with wind security provided by concrete slabs has been found to be the simplest and cheapest solution. The concrete slabs must have component protection mats underneath them to protect the roof skin. As shown in Figure 7.18 the installation system can be mounted on the concrete slabs. The absorbers are then fixed between two concrete slabs using aluminium rails or webbing. Webbing has been found to be the cheapest and most versatile system for this. The absorbers are clamped with one upper and one lower webbing belt perpendicular to the direction of flow. The distance from one belt to the next should be no greater than 1.5 m. At regular intervals (also about 1.5 m) the upper and lower belts should be fixed with UV-resistant cable binders.

Figure 7.17. Mounting rails: Behncke Solar System

The mounting of absorbers with tension belts is also suitable for pitched roofs (Figure 7.19). Here, however, another type of point fixing must be chosen, which will vary according to the type of roof. This is also true for steel rails, which are more frequently used for pitched roof installation. With tiled roofs this point fixing can be made, for example, with roof hooks, as used for on-roof collector installation.

7.5.1.2 FIXING THE ABSORBER BY BONDING

The surface area of the most common types of roof covering (tiles, sheet metal, bitumen etc.) is smooth enough for it to be possible to bond the absorber to the

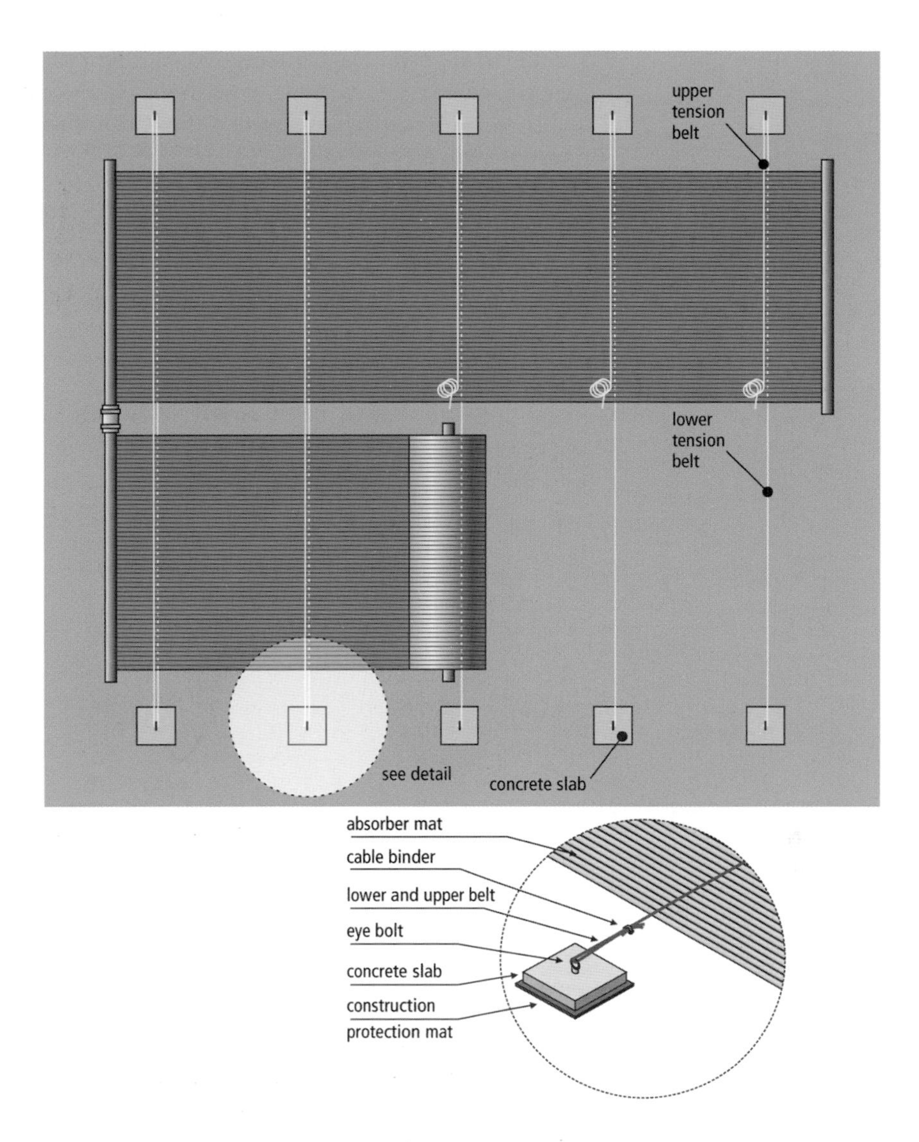

Figure 7.18.
Fixing the absorber with tension belts and concrete slabs on a flat surface

background with a special adhesive, which is usually available from the manufacturers. According to the manufacturer's system and the roof pitch, the adhesive layers should be applied at spacing between 30 and 100 cm. Some types of absorber have crosspieces on the back (see Figure 7.5), which not only create an insulating air cushion but also provide an additional seating surface for bonding purposes. With inclinations greater than 30° it is necessary to ensure that the absorbers do not slip

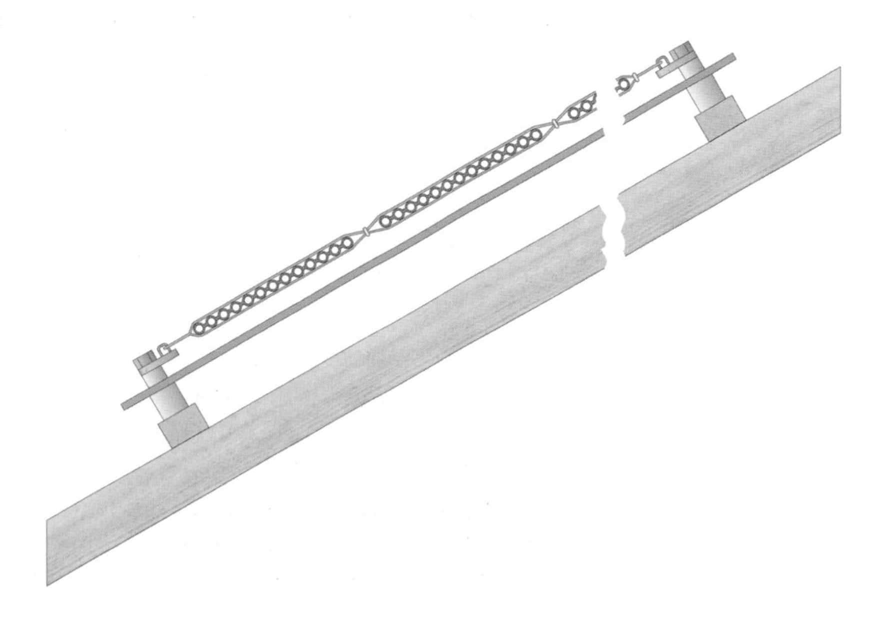

Figure 7.19.
Clamping absorbers with belts on an inclined roof

until the adhesive has hardened. For this purpose a double-sided adhesive absorber-mounting tape can be fitted close to the adhesive mass, for example. When bonding absorbers it must be ensured that the background is always dry, grease-free, and not contaminated (see Figure 7.20).

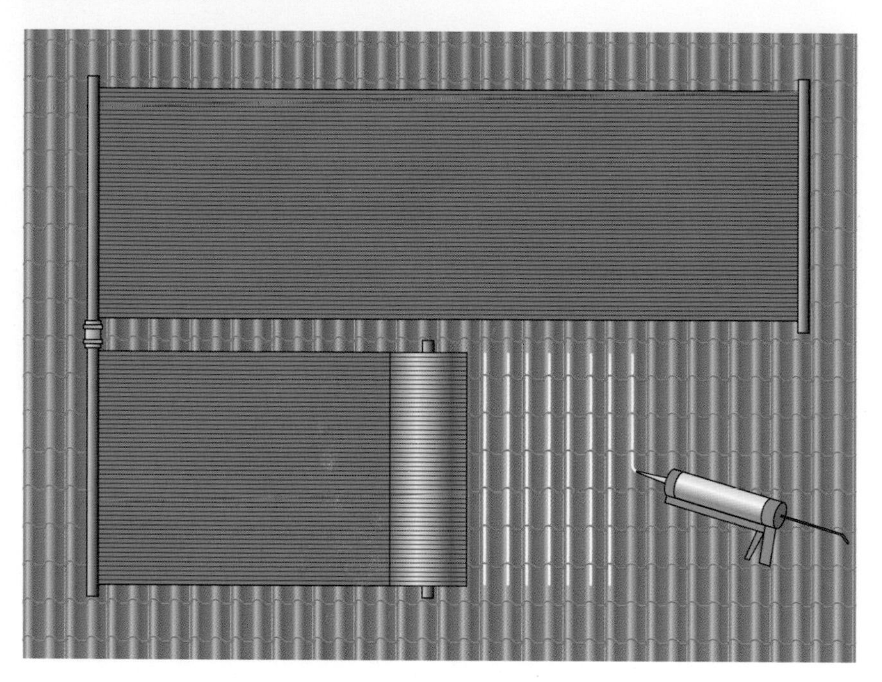

Figure 7.20. Bonding of absorber lengths on a pitched tiled roof

7.5.1.3 FURTHER ASPECTS OF ABSORBER INSTALLATION

For pitched roofs, the tube absorber should preferably be installed along the ridge/eave axis. In this way dirt that has gathered between the tubes can be flushed away by the rain. The absorber lengths can be very simply installed here by 'suspending' them from the collector and distributor pipes. They can again be prevented from slipping by means of belts. However, if transverse installation is selected for hydraulic reasons or because of the often-larger width of the roof surface, the absorber surface should be cleaned at least once per season (at the start of season) by spraying. This is also necessary for absorber fields mounted on flat roofs, as the dirt cannot be flushed away naturally by the rain.

If the roof has a pronounced profile (for example box section sheeting with large gaps at the top), a plastic pipe can be laid in the gap to prevent sagging of the absorber mats (see Figure 7.21).

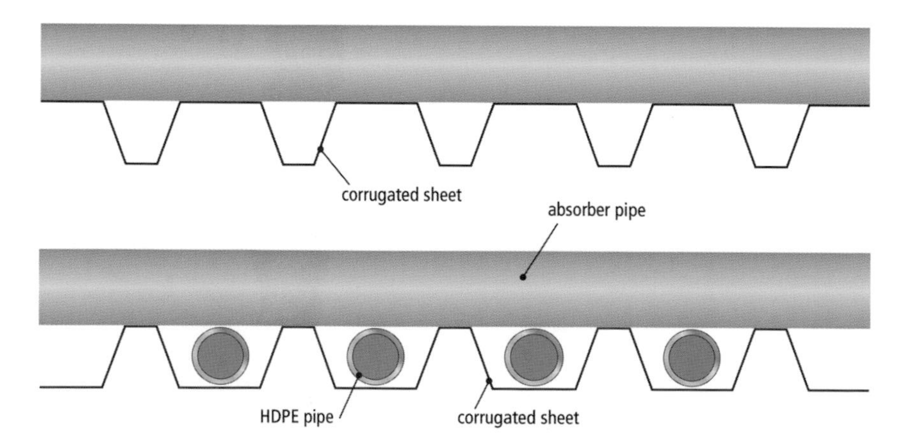

Figure 7.21. Trapezoidal sheeting with plastic pipe inserted to prevent sagging of the absorber

7.5.1.4 GROUND LEVEL INSTALLATION

As has already been indicated in the design section of the chapter, it may be that insufficient roof surface is available – particularly for public swimming pools, where several 100 m^2 of absorber field are required. In this case, it may be possible to install the system at ground level. Security from wind forces can be provided in the same way as for flat roof installation. It is also essential that the absorber field is protected from plant or weed growth. To do this, the flattest possible surface should be prepared while still allowing rainwater to seep through easily and preventing plant growth over the long term.

For additional protection the absorber area can be fenced off. Low hedging at a suitable distance protects the absorber field from wind and hence reduces heat losses by convection.

7.5.2 Pipeline installation

The relevant technical regulations (such as EN 805 in Europe) should be used for installation work. In addition the pipe lengths between absorber field and swimming pool should always be chosen to be as short as possible. Because of the low operating temperatures, thermal insulation is not normally used for the pipes. Special attention should be paid to length expansion of the plastic pipes due to the varying temperatures. If pipes are not to be insulated they should be covered with a black protective coating, which will protect them from UV radiation and help them to absorb additional heat from the sun.

7.5.2.1 COLLECTION AND DISTRIBUTION PIPES

In general the type of fixing that is used depends very much on the absorber system and the type of roof skin. The collection and distribution pipes are fixed to the roof with either screw clamps or anchors similar to those used for the on-roof assembly of flat collectors.

Since the absorber collection and distribution pipes are usually made of HDPE or PVC, the temperature-related expansion must be taken into account here. Individual pipes can, for example, be linked by rubber sleeves (see Figure 7.22). If the pipes to be connected do not directly butt against one another the temperature-related expansion can be compensated for by the rubber sleeve.

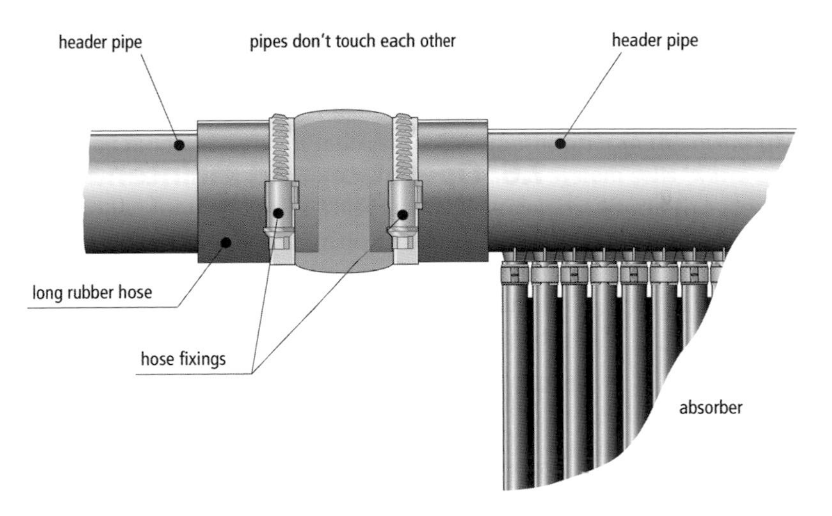

Figure 7.22. Rubber sleeve connectors for collector pipes to compensate for temperature-related expansion

7.5.2.2 SOLAR CIRCUIT

The piping for the solar circuit consists of plastic pipes. PVC pipes are often used but, if possible, PE or PP pipes should be used for environmental reasons. The pipes are fitted to buildings with screw clamps, and between the building and the swimming pool the pipes should be laid underground if possible. PVC pipes are connected together with a suitable adhesive. The common grey PVC pipes must not be subject to UV radiation and should therefore be laid under the roof or underground, or be given a protective coating. Black PVC pipes and their corresponding fittings are, however, permanently UV-resistant.

PE and PP pipes can be connected together by heating element stub welding (previously called butt welding). As this method is significantly more expensive than gluing PVC pipes and requires the relevant specialist knowledge, it is used mainly in municipal swimming pools. The ends of the pipes must be smooth and perpendicular to the pipe axis, so that a permanent leak-proof connection can be achieved. The heating element, in the form of a thin disc, is guided between the pipes to be joined,

Figure 7.24.
Private solar system for pool heating

7.8.2 Solar heating of a municipal open-air pool complex

The summer pool at Mariendorf belongs to the Berlin swimming pool authority. It is situated in the Tempelhof area in the south of Berlin, in the suburb of Mariendorf.

The open-air complex has one swimming pool of about 1000 m^2 and a learners' pool with the same pool surface area. Previously the pool water was heated to the required set temperature of 24°C during the season with a gas boiler. The boiler had to be shut down in 1999, and the technical department of the Berlin swimming pool authority considered replacing the conventional heating with a solar system for heating the pool water. Because of the total pool surface area of about 2000 m^2, an absorber surface of about 1400–1600 m^2 was required, according to design guidelines. The flat roof surfaces of the shower and changing room buildings were available for the installation of the absorber. These have a gross roof area of approximately 1700 m^2. With the corresponding circuit arrangements and pipeline installation it was possible to install an absorber surface area here of 1589 m^2 (Figure 7.25).

Figure 7.25.
Solar system for heating open-air pool water in the Mariendorf summer pool, Berlin. Source: DGS

Because of the position and shape of the roof surfaces the absorber field was divided into two areas, which were each operated with their own pump:

Field 1: 1041 m^2	Pump 1	Field 1(a) 722 m^2
	Field 1(b) 319 m^2	
Field 2: 548 m^2	Pump 2	Field 2(a) 422 m^2
	Field 2(b) 126 m^2	

Fields 1(a) and 1(b), and 2(a) and 2(b) have parallel flow.

All the swimming pool water is conditioned by means of a central filter system and a pure water pool. From the pure water pool, both swimming pools are fed in parallel with filtered and solar-heated water. This arrangement therefore does not allow one of the pools to be supplied as a priority with solar-heated water.

The absorber fields are hydraulically separated from one another. A solar pump is installed for each field, which is started up in the event of sufficient solar radiation. Control of the solar system is carried out by the controller for the complete swimming pool water conditioning system. A radiation sensor determines whether sufficient solar radiation is available to start the solar pump. Then the absorber fields are fed with filtered water from the pure water pool. The fresh water feed to replace the evaporated surface water can take place directly via the solar circuit and therefore permits lower feed temperatures and high efficiencies.

The considerable experience of the absorber manufacturer (AST) was available to be used during the planning of the system. The system was installed by a Berlin heating installation company. Because of the tight financial situation of the Berlin Senate an attempt was made to find a contracting partner for this system. The installation company took on this part, with the following conditions:

- running time: 10 years
- heat price: with a minimum consumption of 500,000 kWh, €0.023/kWh; with a consumption of more than 500,000 kWh, €0.013–0.023/kWh.

In the season of 1999 the system had a solar yield of about 520,000 kWh. The system data can be called up by means of long-distance data transmission from the controller. The investment cost of the system was about €128,000 (€82.00/m^2 of absorber surface); the subsidy from the Investment Bank Berlin and the financial authorities was 43%.

Visitors to the pool are informed about the solar yield and the pool temperature by means of a large display in the entrance area.

Some system data:

- collector surface area: 1589 m^2
- pool surface area: 2000 m^2
- started up: 1999
- operator: Berlin swimming pool authority, contact partner, Mr Thoma
- installation: GSM Heizung Sanitär, Berlin
- planning: GSM and AST, Austria
- yield: Over 500 MWh/a
- degree of system utilization: approximately 50%.

losses from the house, in connection with heat recovery, can lead to even lower energy consumption. By means of a sensible combination with solar air systems, it is also possible to input solar gain in addition to the energy savings. As both ventilation systems and solar air systems operate with the same medium, it is possible to integrate such a system without great expense. But also when no mechanical ventilation system is installed, it is possible to make use of a solar air heating system.

In passive houses, however, the use of solar air collectors is critical for economic reasons, as the heat energy requirements in spring and autumn are mostly low, owing to the high passive gain. This leads to a requirement for thermal storage, for which the simple and inexpensive solar air systems are less suitable. See Figure 8.1 for an example of a solar air system.

Figure 8.1. Example of a solar air system. Source: Grammer, Amberg

8.2 Components

8.2.1 Collector types

Solar air collectors can be differentiated according to the type of absorber flow pattern, or the type of collector cover.

8.2.1.1 DIFFERENTIATION ACCORDING TO TYPE OF ABSORBER FLOW PATTERN

Solar air collectors can be classified into three types of construction, depending on the way in which the heat transfer medium (air) is brought into contact with the absorber (see Figure 8.2).

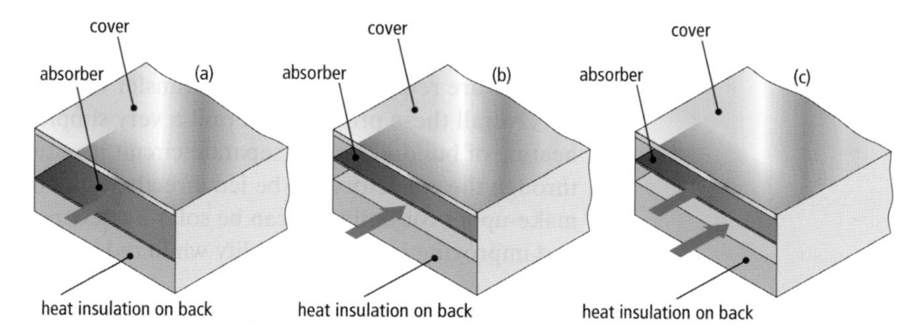

Figure 8.2. Designs of solar air collectors (a) flow over absorber (b) flow under absorber (c) flow on both sides of absorber

The fundamental design is the same for all three basic types. On the back of the collector there is thermal insulation in order to minimize the heat losses to the surroundings. The housing is closed on the front by means of a transparent cover.

In type (a) the absorber lies directly on the thermal insulation. and the air to be heated flows over the top. Type (b) has an air channel between the absorber and the thermal insulation through which the airflows. This prevents the flowing warm air from coming into contact with the cover, which significantly reduces the convective heat losses from the front. Type (c), with all-round flow, is in principle the same design as type (b), but the absorber has the airflow around both the upper and lower sides. The heat transfer from the absorber to the air is improved in this way, but heat losses can again occur on the front side. These three basic variants are covered collectors.

As described in the introduction, the heat transfer from the absorber to the air is poorer than for collectors using liquid media. Starting with these three basic types, many different modifications and further developments of solar air collectors are known. Optimization of the heat transfer is usually the main focus of the development. In assessing these developments, however, it must be borne in mind that the pressure losses should be minimized in the solar air collector. This development leads to optimization problems with the two contrasting objectives of improvement of heat transfer, and minimization of pressure losses.

8.2.1.2 CLASSIFICATION ACCORDING TO COLLECTOR COVER

As for liquid-based collectors, solar air systems also have collector variations in which the most simple and hence inexpensive design is aimed for. A non-covered air collector has a greater heat loss owing to the lack of a transparent cover, particularly with higher absorber temperatures. Its efficiency is thus lower. This design can, however, be of interest if working with lower absorber temperatures. Because the material costs are reduced and the production is simplified, with this type of collector and with suitable boundary conditions very low heat-production costs can be obtained. This type of solar air collector is used mainly for fresh air preheating, for the reasons described (see Figure 8.5). For applications in which higher operating temperatures are needed in the solar air collector, covered solar air collectors are selected.

8.2.1.3 STANDARD AIR COLLECTORS

A standard collector corresponding to type (b) above is shown in Figure 8.3. It has a frame, thermal insulation on the back and sides, a transparent glass cover, and an absorber. The absorber consists of a coated aluminium sheet that is designed as a U-profile. Placed side by side, these profiles produce a ribbed profile for the transfer of the heat to the air that is flowing through.

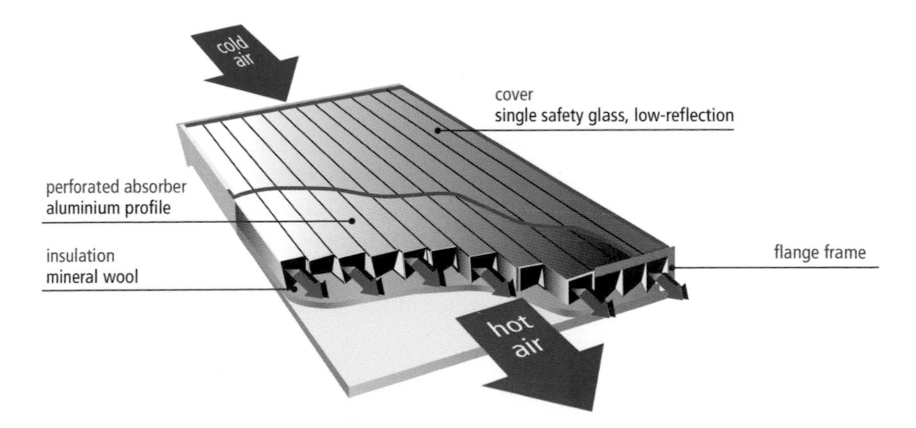

Figure 8.3. Schematic construction of an air collector with under-flow. Source: Grammer, Amberg

The collectors are available in different versions. For example, different depths are offered, depending on the system size. Increased airflow is permitted by the increased depth in larger systems.

In addition, because of the flow-specific series connection of the collectors, there are different modules available. The middle collector is connected on its narrow side to the end collectors by means of flanged connections. The end collectors have an integrated air connection. This connection is either designed as a pipe connection, for example to lead the waste air from a building through the collectors, or it is fitted with an air filter integrated into the rear opening in order to lead fresh air directly through the collectors.

8.2.1.4 ROOF INTEGRATION

For new buildings, or during refurbishment, if it is planned to integrate the collector into roof, the collector area is arranged and installed in modular form. Absorber troughs, whose height is derived from the required airflow (110–170 mm), without a transparent cover are used. They are offered in flexible dimensions of 400–1200 mm by 1000–2500 mm. The collector surface is bordered by the roof-covering frame. For each row of collectors two air-connecting parts are required on the rear wall of the trough. A diverting module is used to provide a side airflow through two collectors. Finally a single pane of safety glass is placed on the absorber troughs. Collectors can also be mounted on flat roofs (see Figure 8.4).

Figure 8.4. Air collectors placed on stands on the roof of a factory building. Source: Grammer, Amberg

8.2.1.5 OTHER VARIATIONS OF AIR COLLECTOR

To provide independent operation of the air circulation in buildings without a separate power supply (allotments, mountain huts etc.), collectors with an integrated photovoltaic (PV) module can be used. This supplies the electrical energy required to drive a d.c. fan when the sun shines. This variant can, however, also be used on top of dwellings that possess a power supply. The system is then operated by the PV module according to the weather, and no additional costs arise for fan operation.

Regulated back-ventilation of PV modules can also be achieved. The air collector is provided partially or completely with the modules instead of solar safety glass on the front side. The PV modules generate power, which is fed into the network, and the air heated by the waste heat from the modules is used for ventilation or heating purposes. This synergy effect leads to a higher efficiency in the PV module owing to the lower module temperature. Such systems are usually designed as large systems and, for example, are placed on stands on flat roofs. The PV system takes up only part of the surface of the solar air system, and it is not only used for generating the fan power.

8.2.1.6 FAÇADE COLLECTORS

Another interesting application for air-operated solar systems is their integration into facades. Here the collectors can, for example, completely replace the façade in a stud-bolt design, or they can be placed on an existing façade.

As the heating of a building normally takes place in winter and in the transition periods – that is, with low solar elevation angles – vertical arrangement of the collectors has some advantages. For economic reasons, too, façade integration of solar air collectors can be of interest, as during the new building or restoration of a façade costs can be saved in the façade construction. These can then be set against the costs of the solar air collector façade. For the inexpensive integration of solar air collectors into a façade, as well as from the energy point of view, integral planning is of great significance.

8.2.1.7 THE SOLARWALL™ – SYSTEM, TRANSPIRED AIR COLLECTOR

Apart from the façade collector system with transparent glazing, the uncovered collector, which has already been mentioned, is the one that is mainly used (Solarwall system, see Figure 8.5). Here a perforated, dark-coated metal absorber sheet is used as the outside jacket for a façade. By means of regulated suction of the solar-heated air boundary layer on the outer side of the metal sheet, the heat is collected and led to the ventilation system for heating purposes. This simple, direct method can be configured to preheat fresh ventilation air for buildings, thus improving indoor air quality while reducing energy cost. In lower latitudes this same vertical wall can be applied above a roof to capture heat incident on the inclined surface equally well. Applications for drying agricultural crops such as coffee, cocoa and tea have yielded excellent results in many countries.

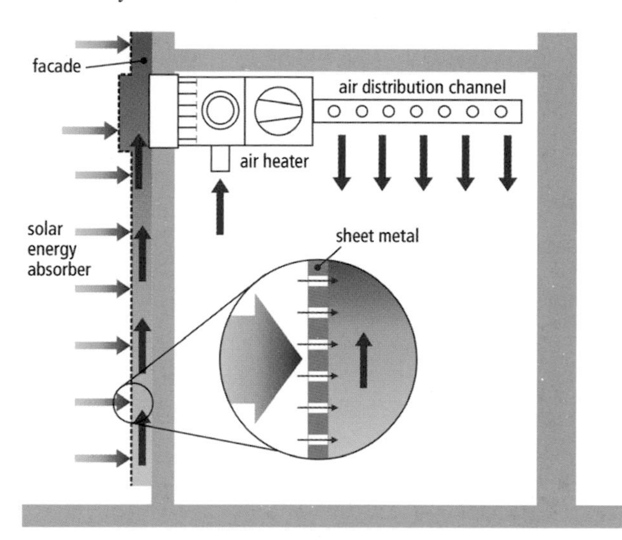

Figure 8.5. Construction of a solar air system with an unglazed aluminium absorber. Source: Solarwall International Ltd, Göttingen

8.2.2 Fans, blowers

The fans that are normally used in solar air systems are standard trade components as used in normal ventilation systems. In addition to the radial fans that are most frequently used, there are also axial and cross-flow fans. Axial fans can best be used in round pipelines, but they are mainly used in exhaust air systems. Cross-flow fans operate particularly quietly, but have a low air delivery rate and are mostly used only in special cases. Because of its flexible connecting options and its higher delivery rate capacity, the radial fan is preferred (see Figure 8.6). By means of different designs of blade (curved, forward or backward), optimum adaptation to the existing volumetric flows and pressure differences can be achieved.

For solar air systems in factory buildings with very high flow rates, the fans must have power settings of several kilowatts. With an annual average of 2000–2500 operating hours the power consumption required here is significant. In evaluating the additional energy requirement, the pressure loss in the solar air collectors and in the

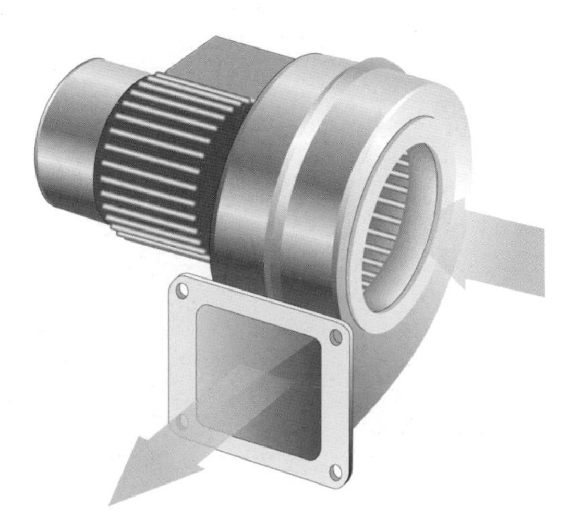

Figure 8.6. Radial fan

necessary pipes is crucial because of the need to integrate the solar air collectors. With careful planning and design of the system it is possible to reduce the additional electrical energy required to approximately 2–5% of the thermal gain.

8.2.3 Piping

The piping of solar air systems corresponds with the requirements for a normal ventilation system. They must be made from non-combustible materials, and must also be corrosion resistant. The materials mostly used are steel and stainless steel plate. Under certain conditions aluminium is also used. For hypocaust systems (see section 8.3.1.3), apart from the normal wrapped spiral-seam tubes, plastic tubes made of PE or PP are also used for the air passages through the building parts.

Circular cross-sections are mostly used for smaller channels; the pipes are made in a wrapped spiral system. For large air channels square or rectangular sections are normal. The pipelines are mostly available in lengths up to 10 m. They are connected together at the butt joints with gate valves or rebates and clamps; in individual cases they are also welded. Flexible tubes are used to cope with angles that circumvent inaccessible passages and deviate from standard values (connection of collector at inclinations ≠ 45°). Collector connections are additionally provided with thermal insulation. For pipes led outside the building, UV- and weather-resistant heat insulation is recommended. Inside the building this is not necessary.

In order to reduce noise, the use of sound-reducing components is recommended for solar air collector systems. This is particularly true for systems with high user requirements for comfort. High levels of noise occur particularly with high airflow speeds, and also at curvatures because of air friction and because of the use of low-friction wrapped spiral-seam tubes. Therefore, a silencer (see Figure 8.7) is usually installed in front of the room inlet vents in order to reduce the noise load.

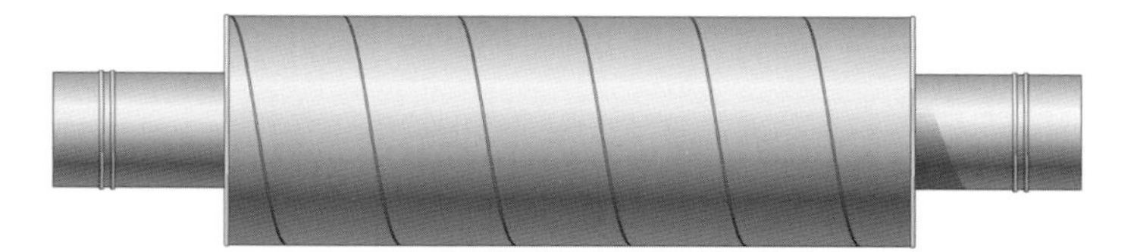

Figure 8.7. Silencer

8.2.4 Heat exchangers, heat recovery units

In (solar) air systems mainly *recuperators* or *plate heat exchangers* (recovery heat coefficient normally about 65% for large flow rates) or *rotary heat exchangers* are in use (see Figure 8.8). These can achieve recovery heat coefficients of up to 90% and simultaneously permit moisture exchange, which can be used for air-conditioning.

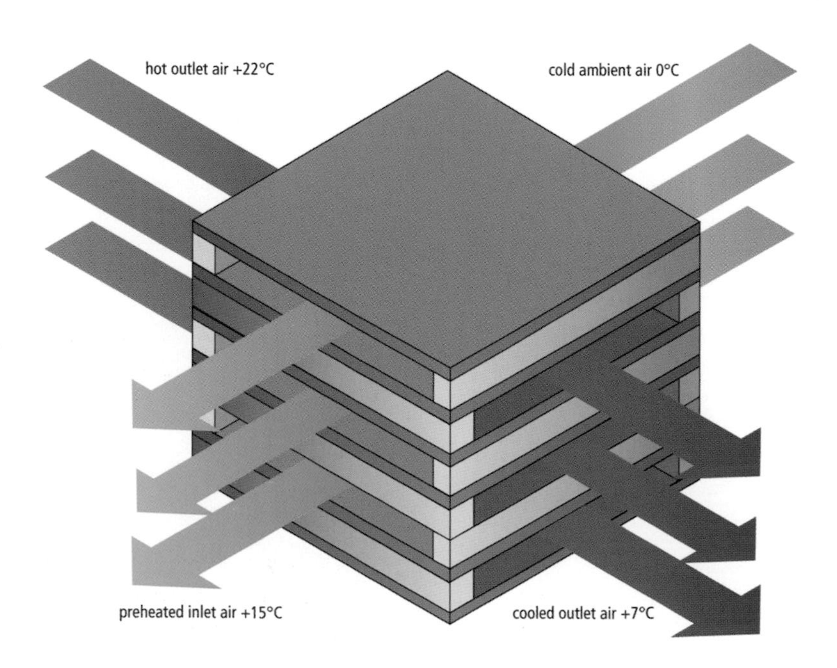

Figure 8.8. Schematic outline of a recuperator with example temperatures

The recovery heat coefficient gives the temperature changes that can be achieved by heat recovery with respect to the outer air or exhaust air side. In the case of devices with simultaneous moisture transfer, there is similarly also a moisture recovery coefficient.

In order to use the excess heat of a solar air system in summer, air–water heat transfer units are used (see Figure 8.9). These can transfer the excess heat to the domestic water and hence make an additional contribution to the reduction of the conventional energy requirements.

Figure 8.9. Air–water heat transfer units. Source: Grammer, Amberg

8.2.5 Control

Standard temperature difference controllers can be used for the control of solar air systems. The controller compares the room air temperature and the collector temperature. If a given temperature difference (normally 3 K) is reached, the fan is started up. For dual-source heating systems the set temperature of the rooms can be controlled independently of the solar controller by a room temperature regulator installed in the room. Therefore the solar system can raise the room temperature within certain limits above the set temperature of the classical heating system purely with solar energy.

Systems that transfer excess heat to the domestic water require a priority controller. Hot water preparation is then controlled as a low priority. If the solar air system has reached the desired room temperature, the air stream is used for hot water heating (see section 8.3). At times when room heating is not required, the hot water is heated exclusively by solar energy.

For large systems, in buildings with complex ventilation and air-conditioning systems, the solar air system can be linked into the controller for the ventilation/air-conditioning system. Figure 8.10 shows an example of a façade-mounted collector.

8.3 Systems

8.3.1 Air collector systems in housing construction

Air collector systems can be operated in different system configurations. The following can be used in housing construction.

Figure 8.10.
Solar air collector on the façade.
Source: Grammer, Amberg

8.3.1.1 SOLAR FRESH AIR SYSTEM (AIR HEATING WITH FRESH AIR)

This is the simplest sort of solar air system. The collectors have fresh air flowing through them, which is then blown into the building in heated form. There is no exhaust air system; the exhaust air leaves the building by leakage or through exhaust air flaps (Figure 8.11). The required installation for blowing solar-heated feed air into the building is minimal. All that is necessary is to conduct the air into the required rooms. This system is used mostly for retrofitting into existing buildings. If the hygienically required air change is brought into the building via the solar air collector system, every degree of temperature increase brings with it an energy saving. Thus, for example, with an outside temperature of –10°C and a desired room temperature of +20°C, a temperature increase in the solar air collector of only 15 K reduces the ventilation-heat requirement by 50%.

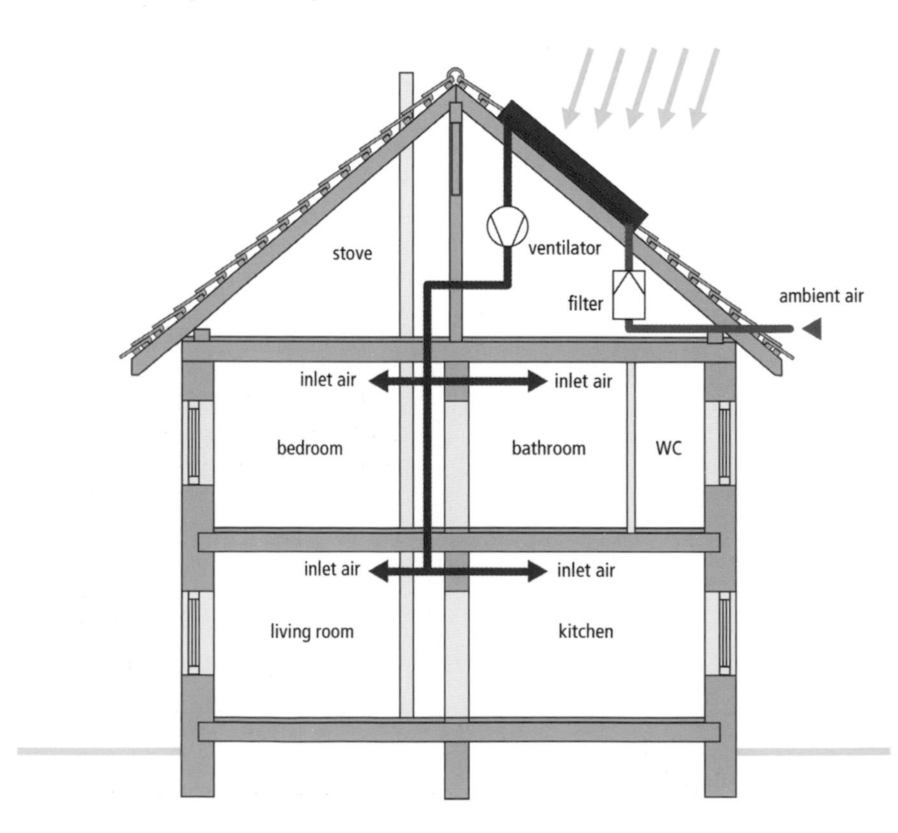

Figure 8.11.
Elementary diagram of a solar air system with fresh air operation

8.3.1.2 SOLAR-SUPPORTED HOUSE VENTILATION

Because of the increasing improvements in the thermal insulation standards of new buildings, the sealing of the walls of buildings is constantly improving. This requires the use of a system for controlling the feed and exhaust airflows, into which the solar air system can be easily integrated. With corresponding solar radiation the fresh air is led through the collectors and heats the building (see Figure 8.12). The heat recovery system provided in many ventilation systems provides additional heating for the fresh air. In more complex systems a portion of the circulated air can be returned to the collectors.

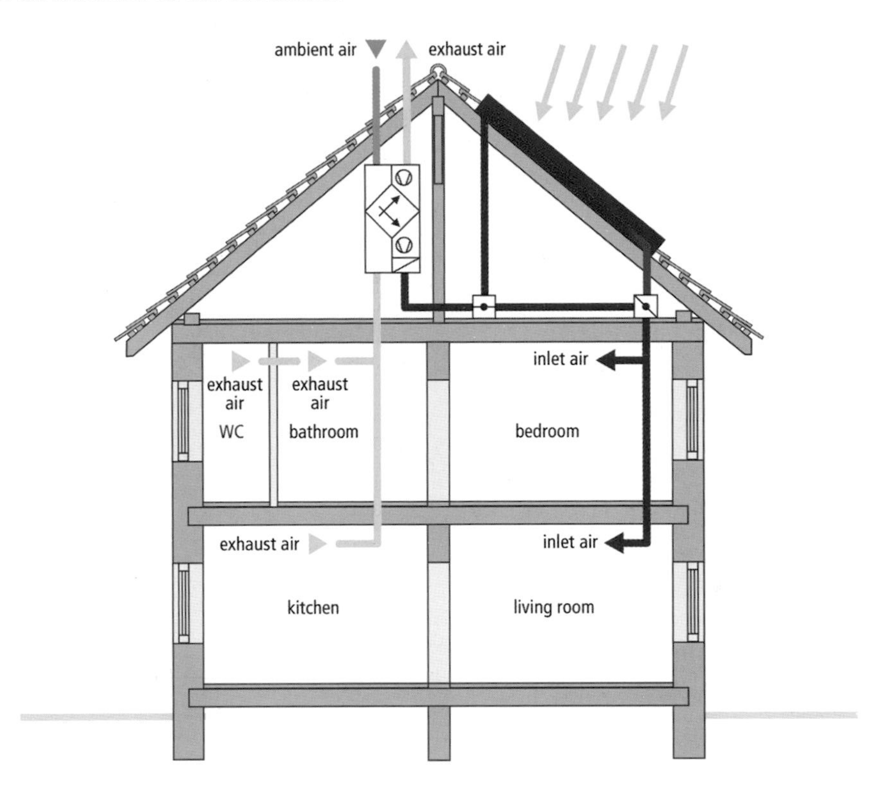

Figure 8.12.
Elementary diagram of a solar air system with heat recovery. Shown here is a feed air operation without inclusion of the collectors

8.3.1.3 SOLAR AIR HEATING WITH STORAGE

In order to use solar air heating of a building at times that do not coincide with the solar radiation, the heat must then be stored. This plays a very important role, particularly in dwellings, as heat is required in the evening and at night.

To store the generated heat it must be transferred to a suitable medium. For this purpose gravel or stone storage vessels can be used, for example (see Figure 8.13), but these are usually linked to high costs.

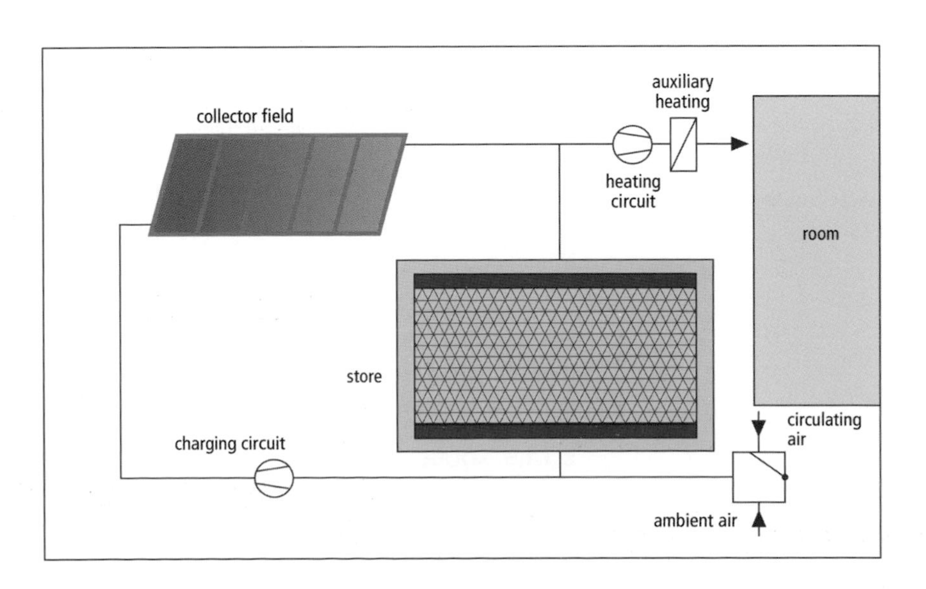

Figure 8.13.
Elementary diagram of a solar air system with (stone) storage vessel

A comparatively cheap method of intermediately storing the solar heat is possible with *hypocaust systems* (see Figure 8.14), as long as a solar air system is planned for a building project from the very beginning. Here the warm air stream is led through parts of the building, such as walls or floors. The heat is transferred to the building component, and the component transfers this heat to the adjacent rooms with a time delay. Conventional heating can also be operated via this system in order to use it optimally.

In houses equipped with such a system, the collector surface area is often subdivided. One part feeds the hypocaust system; another feeds a controlled direct room heating system. In this case it is also sensible to incorporate optional switching of the hypocaust system to controlled ventilation, as the corresponding building component can only absorb a limited amount of heat.

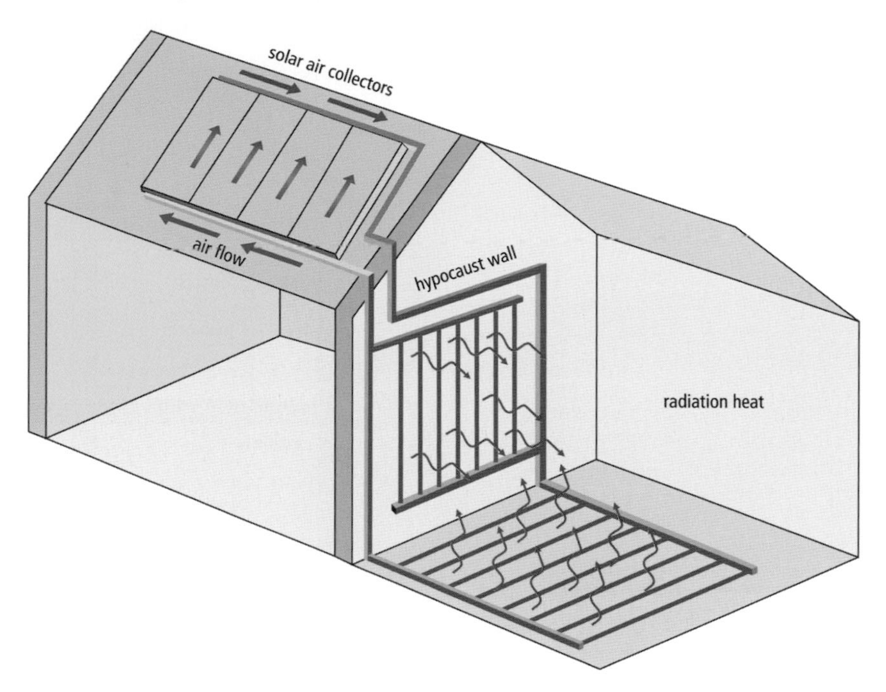

Figure 8.14. Example of a building with hypocaust wall and floor

8.3.1.4 SOLAR AIR HEATING AND DOMESTIC WATER HEATING

A sensible extension of the solar air system is the transfer of excess heat in the sunniest months to the domestic water. In the summer months room heating is normally not necessary, which means there is excess heat available. Between May and September (in temperate climates; November to March in the southern hemisphere) the solar-heated air can thus be used almost exclusively for domestic water heating, and the existing system can thus also be used sensibly within this period.

During the other months the domestic water heating is operated as a low priority, as the air collectors operate more effectively for room air heating and achieve greater solar yields. Once the set room air temperature is reached, a bypass valve is triggered. The system is now operated as a circuit in the same way as for water-based solar systems. The air–water heat transfer unit, which transfers the heat from the air to a separate liquid circuit with safety module and pump, is installed in the bypass circuit. This transfers the heat to the domestic water storage tank via an internal heat transfer unit (Figure 8.15). The liquid-based circuit is operated with frost protection as, otherwise, frost damage to the heat transfer unit can occur at air temperatures below 0°C.

A two-storage control system is necessary to control the system. The cost–benefit ratio for this additional investment for the heat transfer unit, domestic water circuit and safety module must be critically checked, but for most systems can be seen as being worthwhile.

8.3.1.5 SOLAR AIR SYSTEMS IN LOW-ENERGY HOUSES

In new, very well-insulated buildings in temperate climates, it is almost possible to dispense with conventional room heating (combustion of fossil fuels) where a solar air system is installed. Because of the low heat requirement, the solar system can often

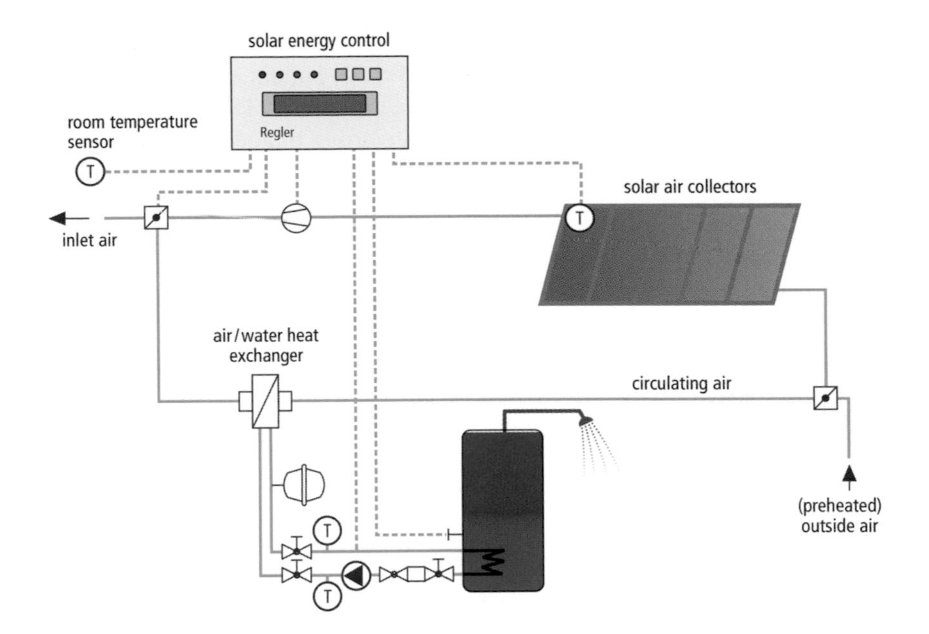

Figure 8.15. Basic diagram of a solar air system integrated into the domestic hot water system

supply the necessary heating energy in combination with a heat pump as a heat recovery system. However, it is necessary to consider here the primary energy balance in connection with the number of hours of operation of the heat pump over the year. For safety, low-temperature electrical convectors can be installed to cover the residual heat requirement on very cold days.

The system set-up is similar to that shown in Figure 8.16, but instead of the recuperator or rotary heat transfer units, the heat pump is used for heat recovery.

Because of the very low energy requirement of low-energy houses, it is still necessary to weigh up whether a solar air system should be installed. Because of the significantly higher proportion of hot water treatment in the total energy requirement, a solar thermal system for domestic water heating may well be more sensible under these circumstances.

For so-called passive houses (heat requirement $\leq$ 15 kWh/m^{2}a) the use of solar air systems is often found not to be worthwhile, as their energy requirement is very low even during the transitional periods. The solar air system would only provide good yields in three or four months of the year.

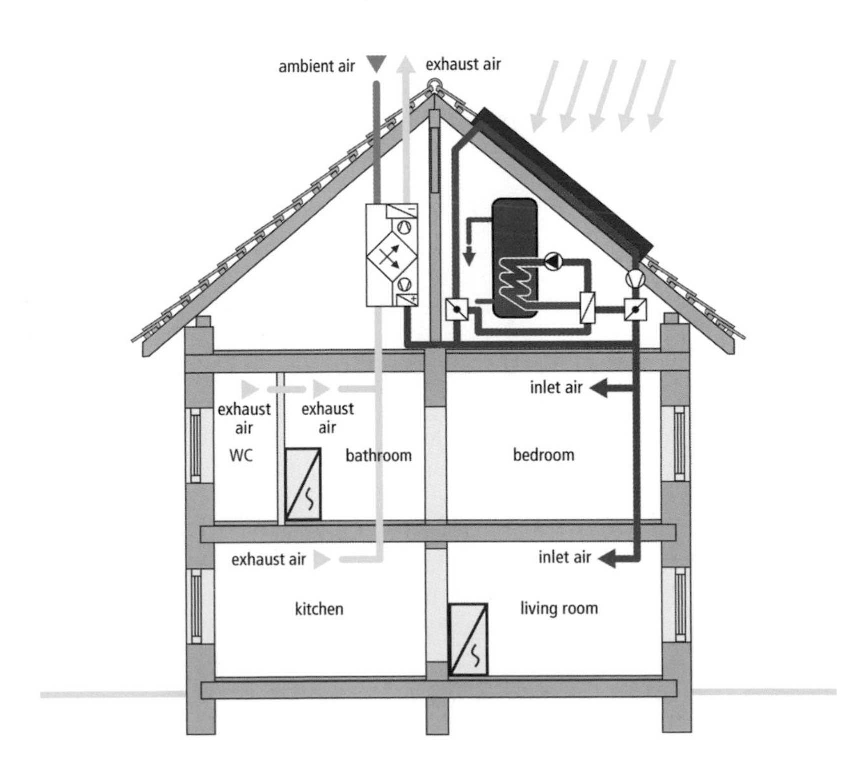

Figure 8.16. Basic diagram of a solar air system with heat recovery (heat pump) and domestic water heating

8.3.2 Factories, halls and office buildings

For factory and office buildings the integration of solar air systems is an interesting and inexpensive application. Here controlled air feed and return lines are already available in many cases. In addition, the heat is required at the same time as the solar radiation is available, as the buildings are generally only used during the day. As a result the solar air system can be easily installed. This is particularly truc for new buildings, in which the large collector surface areas required can replace part of the building shell if integrated into the roof or façades. The additional costs for the solar system can be reduced by setting them against the savings.

Such systems can be used in numerous applications such as sports halls, indoor pools, warehouse and factory buildings, supermarkets, office buildings and schools; an example is shown in Figure 8.17. A particularly promising application for solar air collector technology is that of indoor swimming pools. Here there is an almost full-year requirement for heating energy, with a simultaneous need for large air exchange rates for hygiene reasons.

Figure 8.17. Solar air system in the 'Märkischer Polstermarkt' (furniture store) with partial façade substitution (south and west façades). Source: Grammer, Amberg

The system construction is similar in principle to the small systems in dwelling houses. Because of the size of the collector surface area it is essential to ensure an adequate circuit arrangement for the collector and collector field circuits to maintain the optimum airflow through the collectors. Rows of collectors connected in parallel have been found to be advantageous. The shortest possible pipe routes should be aimed for between the rooms and the collector field.

The airflow and the size of the collector surface depend primarily on the requirements of the building (see section 8.4). Surface- or personnel-related air change rates are specified according to the type of use. With respect to the surface area (surface area = useful area) this can be between 3 and 20 m^3/m^2h.

8.4 Planning and dimensioning

8.4.1 Fundamental considerations

The design of the solar air system is heavily dependent on the climate and the type of application. A precondition is knowledge of the air change rate within the building. This prescribes how often the air volume in a room has to be exchanged within one hour. The air change rate depends on the type and utilization of the building, and on the number of people in the building.

Furthermore, limits are defined in local regulations for the hourly ventilation requirement, depending on the type of activity. These may pose minimum flows: for example a minimum of 40 m^3/h per person for seated work and at least 70–90 m^3/h per person for physical activities. The relation between air change rate and ventilation requirement can be calculated from the building volume:

$$\text{Ventilation requirement } (m^3/h) = \text{Air change rate } (h^{-1}) \times \text{building volume } (m^3)$$

For the design of the system the specific volumetric flow (m^3/h for each m^2 of collector) that is necessary for a useful temperature increase in the air, as recommended by the manufacturer, should be observed.

The following is an example of the design of an air collector system for a warehouse.

8.4.2 Calculation of required flow rate, required collector surface area and connections of collectors

The warehouse under consideration has a floor area of 25 m × 38 m = 950 m^2 and a height of 8 m. The recommended ventilation rate (resulting from local regulations) for warehouses is for example two to six changes per hour. In our example we wish to exchange the air three times per hour. From this information the ventilation requirement is calculated as

$$\text{Ventilation requirement} = 3\ (h^{-1}) \times 7600\ m^3 = 22{,}800\ m^3/h$$

Furthermore, the manufacturer recommends a specific volumetric flow per square metre of collector surface of between 40 and 80 m^3/h. This results in a maximum temperature increase of about 35 K. A specific volumetric flow of 60 m^3/m^2h is selected for our example.

The required air collector surface area we wish to calculate can now be obtained from the ventilation requirement and the specific volumetric flow:

$$\text{Collector surface area } (m^2) = \frac{\text{Ventilation requirement } (m^3/h)}{\text{Specific volumetric flow of collector } (m^3/m^2h)}$$

From this we get a collector surface area of

$$A = \frac{22{,}800\ m^3/h}{60\ m^3/m^2h} = 380\ m^2$$

The flow speed across the free-flow cross-section of the absorber should be between at least 2 m/s (for thermal reasons) and a maximum of 7 m/s (for dynamic reasons). The free-flow cross-section can be found in the product literature of the collector manufacturer.

The minimum and maximum number of collectors connected in series can be calculated from these conditions.

The minimum area of collector surfaces to be connected in series is

$$A_{min} = \frac{A_Q v_{min}}{V_{tot}}$$

where A_Q is the free-flow cross-section of the collector (m^2), v_{min} is the minimum flow speed (m/s), and V_{tot} is the specific volumetric airflow (m^3/m^2h).

The free-flow cross-section is the product of the height and width of a flow channel in the absorber: in our example $A_Q = 0.96 \text{ m} \times 0.095 \text{ m} = 0.0912 \text{ m}^2$. Thus the minimum area of collector surface connected in series can be calculated as follows:

$$A_{min} = \frac{0.0912 \text{ m}^2 \times 2 \text{ m/s} \times 3600 \text{ s/h}}{60 \text{m}^3/\text{m}^2\text{h}} \approx 11 \text{ m}^2$$

For the highest permitted collector surface area connected in series, it is necessary to use the maximum flow speed v_{max}:

$$A_{max} = \frac{0.0912 \text{ m}^2 \times 7 \text{ m/s} \times 3600 \text{ s/h}}{60 \text{ m}^3/\text{m}^2\text{h}} \approx 38 \text{ m}^2$$

These results establish the arrangement of the collector modules within a collector field. The principle of uniform flow through individual legs by means of the circuit arrangement according to the Tichelmann principle must be observed (that is, the pressure loss of each collector or each row of collectors is the same).

In our example, for the required collector surface area of 380 m^2 at a module size of 2.4 m^2 about 160 collectors must be interconnected in rows of at least 5 and a maximum of 16 collectors. With the available roof surface area of 570 m^2 and orientation of the long side to the south, a possible solution would be the installation of 16 parallel rows of 10 collectors, whereby possible mutual shading of the rows of collectors must be taken into account (see 'Mutual shading' in Chapter 4, section 4.3.3.2).

8.4.3 Calculation of fan power

In order to overcome the frictional losses that occur throughout the system, a particular *fan power* P_F (W) is required. Determination of the fan power requires knowledge of the total pressure losses of the system. These are made up of the pipe friction losses in the absorber channels, the inlets and outlets to the rows of collectors, the pipe bends and the pipelines themselves. Values for the pressure losses, depending on the volumetric flow and the flow speed for the individual components, can be found from the manufacturer's nomographs.

In our example this results in a total pressure loss of 254 Pa (1 Pa = 1 Ws/m^3). See Table 8.2. The fan power P_F (W) can then be calculated as follows:

$$P_F = \frac{v_V \Delta p_{tot}}{\eta_{fan}}$$

where v_V is the hourly ventilation requirement (m^3/h), Δp_{tot} is the total pressure loss (Pa), and η_{fan} is the efficiency of the fan.

By assuming a fan efficiency of 0.7 the fan power can then be calculated as

$$p_F = \frac{22{,}800 \text{ m}^3/\text{h} \times 254 \text{ Pa}}{0.7 \times 3600 \text{ s/h}} = 2298 \text{ W}$$

Table 8.2 Example of calculation of pressure loss

Component	Individual pressure loss	Qty/piece quantity	Pressure loss (Pa)
Pipeline	1 Pa/m	49 m	49
Inlets and outlets	9.5 Pa	4	38 Pa
Pipe bends	13 Pa	4	52 Pa
Collector	4.6 Pa/m	10 × 2.5 = 25 m	115 Pa
			Total 254 Pa

8.5 Installation

8.5.1 Collector installation

In describing the different collectors and system variants we have already discussed the possible methods of installation. As for liquid-based solar systems, on-roof and in-roof installation is possible, as well as mounting on stands and integration into façades.

8.5.1.1 ON-ROOF INSTALLATION

PITCHED ROOF

Fixing of the collectors takes place in the same way as for other flat collectors. The collectors are mounted on the roof with roof hooks (rafter anchors), which are led for example through a tiled roof and fixed to the rafters. The collector is then screwed down using rails and fixing straps. Normally two fixing straps are required per collector, with two rafter anchors each. As the collectors are connected to one another directly by means of butt joints, a rigid unit is produced: therefore only one strap is required for a middle collector. Pitched roof installation is used mainly for domestic buildings.

FLAT ROOF

When using solar air systems on halls or factory buildings they are mounted either on the façade or on stands on the flat roof (Figure 8.18). Here too the installation is similar to that used for normal stand-mounted collectors. Some manufacturers offer a complete installation set of aluminium profiles. This consists of one bearing rail and two further supporting rails, which are screwed together to form a triangle on the bearing rail. If several rows are installed on a roof a grid structure is obtained, as the bearing rails of the different rows are connected together. The underside of the rails are coated with a closed-cell foam material, which prevents damage to the roof skin and simultaneously ensures uniform load distribution. The collectors are mounted on the aluminium rails with screws, and hence take up the inclination angle of (normally) 45°. Between the collectors in the rows a silicon bead is applied for sealing purposes and to prevent the ingress of moisture. According to the manufacturer's information, an additional roof load from the collectors of about 13–15 kg/m² can be expected with this system, which should not lead to any structural problems on most roofs.

Figure 8.18. Stand-mounted solar air system on the roof of a hall. Source: Grammer, Amberg

8.5.1.2 IN-ROOF INSTALLATION

Optimum roof integration can be achieved by means of the modular design of the collectors.

VARIANT 1

The collectors can be mounted on the roof boarding or lathing (Figure 8.19). The air collectors are screwed to the roof boards through the sheet metal trough. The collectors are connected to each other with sheet metal collars. A wooden frame is erected around

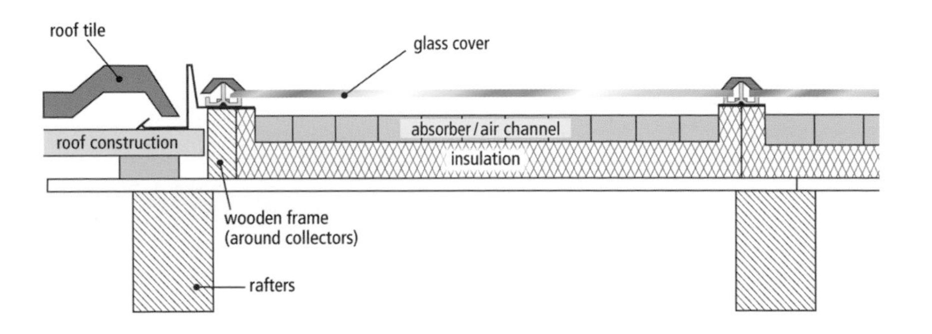

Figure 8.19. In-roof collector installation on the roof rafters

the collector field. A sealing strip and a lead sheet apron create the sealed transition between collector and roof surface. The glazing bars are screwed to the surrounding wooden frame or to the butt joints between two collectors. Additional retaining brackets are installed at the bottom side, which should prevent slipping of the glass cover. The glazing bars are fitted with a rubber profile strip after the glass has been fitted.

VARIANT 2

The collector troughs can also be installed between the rafters (Figure 8.20). The wooden frame around the collector field can be dispensed with, and the glazing bars are mounted directly on the rafters. Around the edge of the collector field a rain-proof transition to the roof surface is still necessary. This variant has a much lower height.

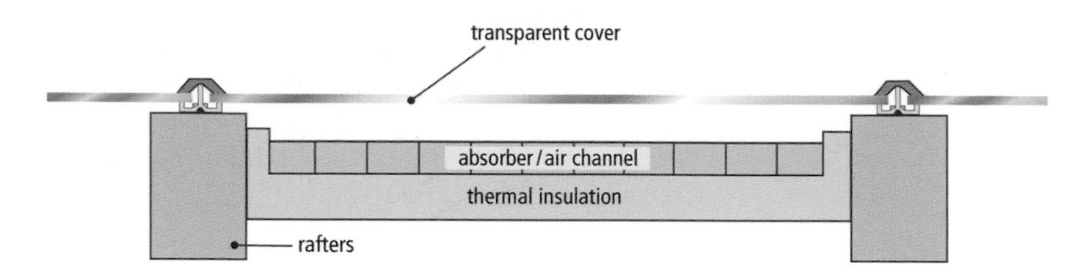

Figure 8.20. In-roof collector installation between the rafters

8.5.1.3 FAÇADE INTEGRATION

The mounting of collectors within a façade-integrated solar system depends mainly on the solar air system being used, as well as on the degree of building shell substitution.

One example is the *Grammer system*, in which the collectors replace only part of the building shell and are placed, for example, on a concrete façade. In another possibility the collectors are integrated into the post-beam or steel frame construction of the façade, and the whole surface area is covered by glass. The collectors used here are not standard collectors, but are made according to the building requirements. The butt joints of the collectors are designed so that they can be optimally bedded into the post-beam construction. Owing to the good thermal properties of the collectors they simultaneously serve as thermal insulators for the building. In a market building a *k*-value of 0.5 W/m^2K was achieved for the collector.

The *façade absorber system* made by Solarwall is installed in mass-produced sizes onto existing building walls. In the case of larger systems the trapezoidal absorber sheeting is mounted on, for example, double T-sections. A back-ventilated absorber façade is thus created. The absorber sheet can be made and mounted in lengths of up to 16 m without joints. On the upper side the collectors have an intake nozzle with a diameter of 100 mm. The corresponding collector pipes or boxes, which distribute the solar-heated air into the building, are fitted to these nozzles. In the summer the absorber can be bypassed through an additional opening to avoid overheating in the ventilated rooms, and to maintain proper ventilation.

The Solarwall column collector can be used in dwellings and office buildings. It is a 1 m wide element with defined closures on the sides, which can be mounted in lengths of up to 16 m in façades between rows of windows, for example. It can also be integrated into a roof if it has a pitch of at least 20° (Figure 8.21).

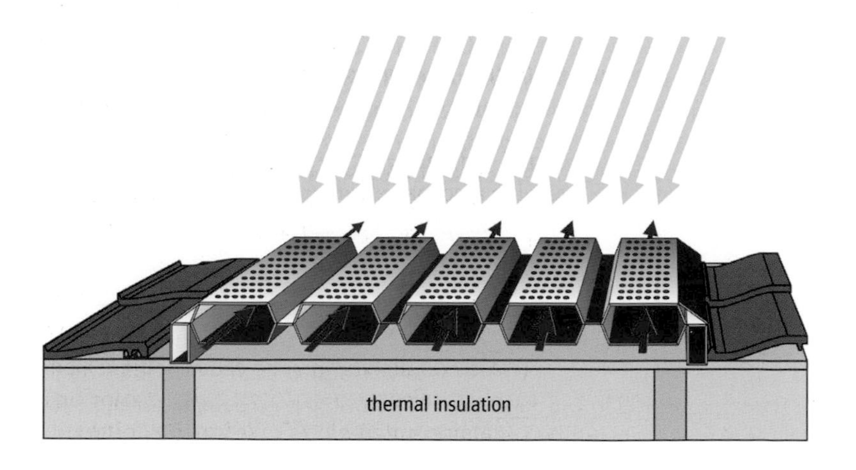

Figure 8.21. Diagram of a roof-integrated Solarwall column collector

8.5.2 Installation of pipelines

The main principle for the installation of the piping is to choose the shortest possible routes between the rooms to be heated and the collector field. Because of the large dimensions of the lines, careful planning is thus essential to achieve this aim. The wrapped spiral-seam tube and the necessary shaped parts are fixed together with nipples and sleeve connectors. Wall and ceiling breakthroughs are made with a plastic collar piece. Where necessary the lines are attached to the wall or ceiling with anchor clamps. In principle, for aesthetic reasons, it is preferable to integrate visible installation parts into walls or ceilings, though this is possible only in certain cases because of the large pipe cross-sections.

8.5.3 Positioning of the fan

When installing the fan, a suitable position should be chosen. To prevent energy losses from unavoidable leakage at the pipe connections, the solar air system should be operated at low pressure. At the same time the fan should be at a sufficient distance from the rooms to avoid noise disturbance.

If a heat recovery unit is used, this must be installed as centrally as possible and without vibration. Short connecting routes should also be used here.

8.6 Costs and yields

The average cost of an air collector without assembly, installation and VAT, with respect to the absorber surface area, is approximately €250–300/$250–300/£175–210 per m^2. The average costs for façade air collectors or Solarwall façade collectors, depending on the wall/roof construction and the desired airflow rate, are about the same level as for comparable façade structures. Of course the complete installation may be much more expensive, depending on the complexity, heat exchangers etc.

Information on system efficiency is not meaningful collectively for air collector systems, as there are no average values for the thermal load that can be compared with the values for hot water consumption. The system efficiency and hence the specific yield of an air collector system are strongly dependent on the type of application:

- system efficiency up to about 30%: air collectors to supplement a ventilation system with heat recovery (priority switching for heat recovery)
- system efficiency > 30%: similar, but with use of excess energy for domestic water heating
- system efficiency up to 65%: fresh air preheating for an indoor swimming pool; drying of wood chippings.

8.7 Examples

8.7.1 Domestic building, Potsdam, Germany

This slim 16-storey block of flats in the centre of Potsdam was subjected to basic restoration in 2000. Thermal insulation was improved, and the sanitary installation was renewed. Controlled feed and return air ventilation with heat recovery was installed in the bathrooms of each of the 93 flats. The old windows were also replaced by modern well-insulated windows. The slim south façade, in which only the windows for the landings are installed, was partially equipped with air collectors (Figure 8.22).

To the left and right of the windows, two rows of 2 × 15 series-connected collectors of type G LK 3M from Grammer, Amberg have now been installed. The 60 collectors provide a total collector surface area of 150 m^2. The air collectors transfer their heat to the ventilation system. In the summer months, or if for other reasons heating of the bathrooms is no longer necessary, the solar air system is switched over to domestic water heating. Here the heat take-up is designed for an average yield of 300 W/m^2, so that the heat exchangers have a maximum transfer capacity of 2 × 22.5 kW.

The solar air system preheats the room air in the bathrooms of the building. This is achieved by two central ventilation systems installed in the plinth. The collector

Figure 8.22. Solar air system at Breite Strasse, Potsdam. Source: DGS

system is also divided into two pieces; two parallel rows operate respectively on a central ventilation system. Two storage tanks of 1000 l each are installed for domestic water support in the summer months. Conventional domestic water heating takes place in a through-flow system via the domestic substation, which supplies all the other rooms with heat.

Control of each of the two parallel collector fields takes place by means of a temperature difference controller for twin storage tank systems. Bathroom heating has the priority. When the maximum temperature is reached, the system switches over and charges the domestic water storage tank. A maximum temperature limit (80°C) is also set here, at which the fans switch off.

Some system data:

- Location: Potsdam
- Collector surface area: 150 m^2
- Alignment: Inclination 90°, azimuth 0°
- Commissioned: 2000
- Volumetric flow: 3400 m^3/h
- Power: 100 kW_{peak}
- Solar planning: PAi-PLAN GmbH, Potsdam; Grammer KG, Amberg
- Installation: Frohberg
- Yield: 65,076 kW/a
- Solar fraction: heating 56%; hot water: 37%
- CO_2 savings: 6749 kg/a.

8.7.2 Solarwall on General Motors building in Ontario

Solarwall cladding was installed in 1991 at a General Motors lead–acid battery plant in Oshawa, Ontario (Figure 8.23). The product, trademarked Solarwall[R], performs like a solar panel, but looks like a conventional metal wall. The system provides heated ventilation air for the plant. It operates by drawing air through a large number of small holes that penetrate the cladding. Heat that is absorbed from the solar radiation falling on the surface is transferred to the air, which then moves upwards behind the wall to an overhanging, enclosed canopy. The system is driven by two ventilation fans installed inside the building, which use the canopy as a manifold, and which distribute the warmed air to the building via ducting.

Figure 8.23
Solar wall on General Motors building in Ontario. Source: Conserval

The GM Solarwall installation incorporates modifications and improvements from several years of Solarwall testing and development. Total energy savings result from four different mechanisms:

- active solar heating of ventilation air
- recapture of heat flowing out through the south building wall
- reduction of conduction through the south wall due to increased local air temperature
- destratification of building interior air.

The solar collector consists of black aluminium wall cladding 6.25 m high. Holes 1.6 mm in diameter cover 2% of the surface. The collector is fastened vertically to the south building wall, leaving a 150 mm airspace behind. The enclosed canopy overhangs the top of the cladding by 0.9 m. The effective collector area is 420 m^2.

The system uses two high-efficiency fans, designed to supply a total airflow of 20,750 l/s. The solar-heated air is mixed with recirculated air, and distributed through 122 m of fabric duct. The mix of fresh and recirculated air is controlled automatically to provide a preset air temperature.

Based on 1993–94 monitoring results, the GM Solarwall delivers 455 kWh/m^2 of solar energy per year. Additional savings provided by the Solarwall amount to 300 kWh/m^2 annually, for a total annual contribution of 755 kWh/m^2. For the full 420 m^2 collector area, this represents a total contribution of 317 MWh annually.

The installed cost of this system is estimated at C$92,400 or C$220/m^2, including the cost of design modifications. Compared with alternative ventilation air systems, this represents a cost increment of between C$15,500 and C$23,200. Depending on the fuel displaced, energy cost savings would be between C$4700 and C$12,200 per year, and payback between 1.3 and 4.9 years.

The GM Solarwall installation is providing increased ventilation in the battery plant, resulting in improved indoor air quality. Additional environmental benefits derive from displacement of the conventional ventilation (make-up air) unit that would otherwise be required. This results in reduced emissions from combustion. In the case of a conventional natural gas unit, for instance, the reduction in carbon dioxide releases is estimated at 81 t annually, based on a seasonal efficiency figure of 70% for a gas-fired make up air unit.

9 Solar cooling

9.1 Introduction

The growing desire for comfort has led in recent years to a considerable worldwide increase in the number of buildings with air-conditioning. This trend can also be observed in Europe, and particularly in southern Europe. A similar development can be seen in the automobile industry, where a steadily growing number of cars now have air-conditioning. This is leading to a vast increase in the number of people who are used to living in a climatized environment, and such an environment is also desired and expected at work or at home. Moreover, the results of investigations show that the working capacity of human beings significantly decreases with room temperatures above 24°C.

As the energy consumption of air-conditioning systems is relatively high – and standard refrigeration-based compression systems use electricity – future-oriented solutions for a sustainable energy supply demand renewable energy systems.

Producing cooled air by making use of solar power may seem paradoxical at first sight. Generally, the sun tends to be viewed as a source of heat. However, there exist thermal processes to produce coldness, in which water is cooled or air-conditioning is driven directly by a heat input. These processes are generally suitable for using heat provided by solar thermal collectors as the principal source of energy. Of course, solar radiation can also be converted to solar electricity by photovoltaic systems to drive conventional refrigeration compression systems. However, the latter approach will not be dealt with in the following section, for several reasons:

- In the short term, generating solar electricity in northern and central Europe will be more costly than producing solar thermal heat (exceptions are very small-scale applications, in the range of a few watts up to several hundred watts).
- Although the commonly used refrigerants for compression refrigeration systems generally no longer harm the ozone layer of the atmosphere, they still intensify the greenhouse effect significantly. Compared with CO_2, today's refrigerants could

Figure 9.1.
100 m² collector system for a solar-powered air-conditioning system in Freiburg, Germany. Source: Fraunhofer ISE

potentially boost the greenhouse effect several thousand times over. Thus these refrigerants contribute considerably to the anthropogenic global warming of the atmosphere.

- The refrigerants used in commercially available thermal refrigeration systems contain only substances that do not influence the greenhouse effect. Adsorption- and sorption-based systems just use water as the only refrigerant.

To date, the traditional design of solar thermal collector systems for providing hot water in moderate climate zones such as central Europe has generally been based on the idea that excess heat in summer should be avoided, or at least kept to a minimum. With the systems increasingly deployed in recent years to support heating, a summer excess in principle cannot be avoided. For reasons of economy – the specific collector yield falls with the increasing degree of solar fraction – a larger collector surface is often also rejected, even if this would ultimately be associated with greater environmental relief.

The use of the summer excess heat for solar thermal cooling therefore offers the opportunity to improve the efficiency of solar thermal systems for providing hot water or heating support. Moreover, the new application largely eliminates system shutdowns with the high shutdown temperatures that are a stress on materials. In addition, a promising new application field results for the solar thermal industry. There are also corresponding export opportunities to countries where room air-conditioning is already standard today on account of the climatic conditions.

In this chapter, we repeatedly make a distinction between cooling and air-conditioning. By cooling we mean reducing the temperature, for example in a room or even the temperature of a machine in industrial processes. By air-conditioning, on the other hand, we mean the 'conditioning' of rooms in respect of their temperature and humidity properties, so that people in the room feel comfortable. Hence the term 'air-conditioning' is drawn wider and comprises, in addition to lowering or raising the room temperature in summer and winter, a reduction or increase in air humidity in the room to comfortable values. Despite this distinction, in industry circles the term *solar cooling* has become entrenched as the umbrella term for talking about solar-driven cooling and air-conditioning. Hence if we say 'solar cooling' in this chapter, we mean this general category.

Another argument for solar cooling – diametrically opposed to solar space heating – is the chronological coincidence, in principle, between demand (dissipating the cooling load) and energy supply in the form of solar irradiance. Figure 9.2 illustrates this relationship for central European conditions. It shows the seasonal correspondence between irradiance and cooling load to be dissipated for a seminar room in Perpignan (southern France). It is clear that there is a very good match with cooling requirements. The result is that no large seasonal heat storage facilities are required.

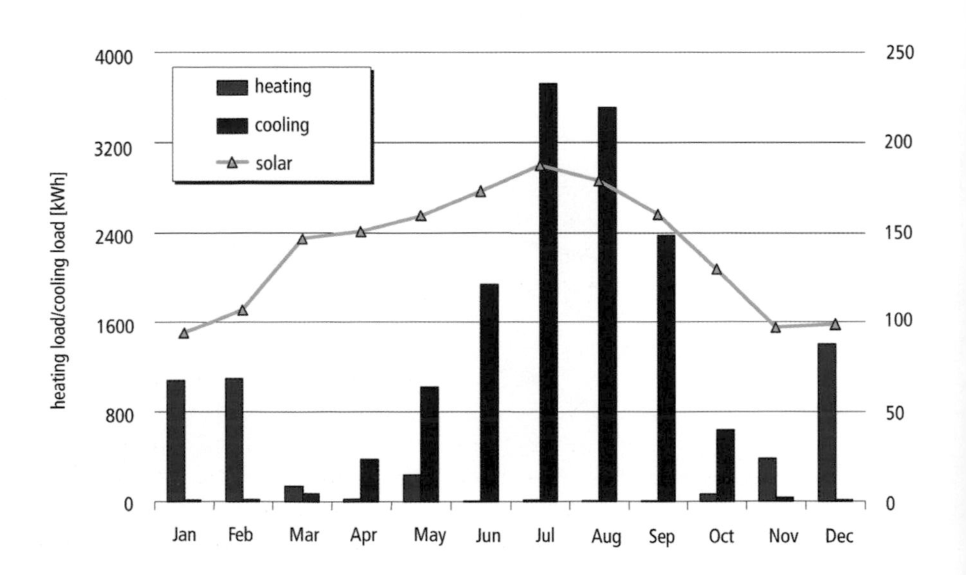

Figure 9.2. Correspondence between solar irradiance and cooling load/heating load for a seminar room in Perpignan (southern France). Source: Fraunhofer ISE

9.2 Theoretical bases

This chapter is intended to give an overview of the processes for solar thermal cooling. It will become apparent that the greatest challenges lie in the system integration of collector installation and cooling technology. As a result, so far there has been a real shortage of planners who are able to plan a solar cooling system from scratch. However, compared with designing a solar thermal system for domestic water heating, the planning task is also much more complicated. The task requires knowledge in very different areas, from building systems to collector technology to cooling systems. In addition, so far there has been very little practical experience with the technology. Some of these systems are presented in this chapter.

9.2.1 Overview of thermally driven cooling processes

Table 9.1 shows the various thermally driven cooling processes. Among the processes available on the market, it is possible to distinguish between the closed absorption and adsorption processes and the open process of desiccant cooling.

Table 9.1. Overview of thermally driven cooling technologies

Process	Absorption	Adsorption	Desiccant cooling system
Type of air-Conditioning	Chilled water (e.g. chilled ceilings)	Chilled water (e.g. chilled ceilings)	Air-conditioning (cooling, dehumidification)

In the closed processes, the cooling medium is not in direct contact with the environment. First of all, cold water is produced. This cold water can then be used in chilled ceilings, in concrete core conditioning, or also in the classical way in the air cooler of an air-conditioning system to reduce temperature and/or humidity.

By contrast, in the open process of desiccant cooling the cooling medium (water) comes into direct contact with the air being conditioned. The cooling and dehumidification functions are directly integrated into the air-conditioning system. This is why one frequently also encounters the term *air-conditioning without refrigeration*.

9.2.2 Absorption cooling

Absorption chillers (AbCh) differ from compression chillers in that they use a thermal compressor instead of a mechanical one. Figure 9.3 shows a schematic diagram of a system of this type.

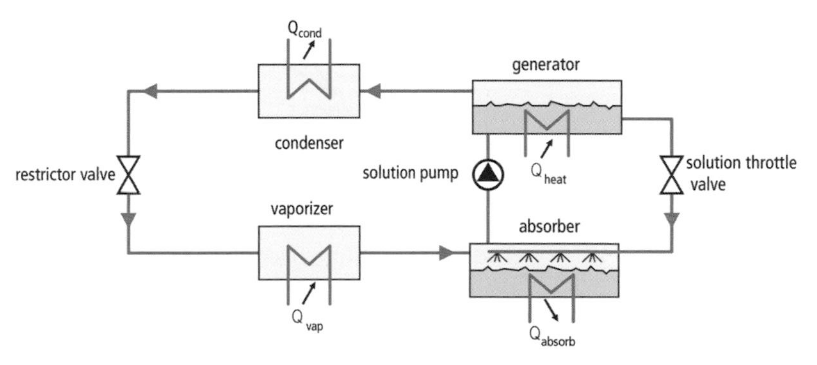

Figure 9.3. Diagram of an absorption chiller[46]

The condenser, cooling medium restrictor valve and vaporizer form the cooling part of the system, through which only the cooling medium flows. The thermal compressor comprises absorber, solution pump, generator and solution throttle valve, constituting the driving part of the system.

The cooling part of the AbCh is no different from a conventional compression chiller. The necessary compression of the cooling medium to the condenser pressure is performed by the thermal compressor. The vaporized cooling medium flows into the absorber, where it is absorbed by the solvent. The released absorption enthalpy must be dissipated, as the absorption capacity of the solvent decreases as the temperature rises. The absorption process enriches the absorption medium with cooling medium. The rich solution is pumped using the solution pump to the generator (also known as a de-aerator or boiler). Here, by supplying thermal heat, the cooling medium is separated from the solvent and the two-substance mixture becomes depleted of cooling medium. The depleted solvent is depressurized in the solution throttle valve to the absorber pressure, where it is once again atomized in order to absorb the cooling medium.

Apart from the electrical power requirements of the solution pump, the AbCh is driven only by thermal energy. However, the energy requirement of the solution pump is very low, with approximately 0.5–2% of the refrigerating capacity achieved in the vaporizer. The efficiency of the AbCh shown in Figure 9.3 is usually improved by installing a solvent heat exchanger. This is arranged so that the rich, cold solution after the absorber and the warm, depleted solution after the generator flow in opposite directions through the heat exchanger. This makes possible savings on thermal heat in the generator and on cooling water in the absorber.

9.2.3 Adsorption cooling

Currently only two Japanese manufacturers of adsorption chillers (AdCh) are known on the market. Their systems are very similar in design. The physical process is identical in both chillers. The description of the function and the design is taken from the technical documentation of GBU[47]. Figure 9.4 shows the schematic structure of a low-temperature AdCh.

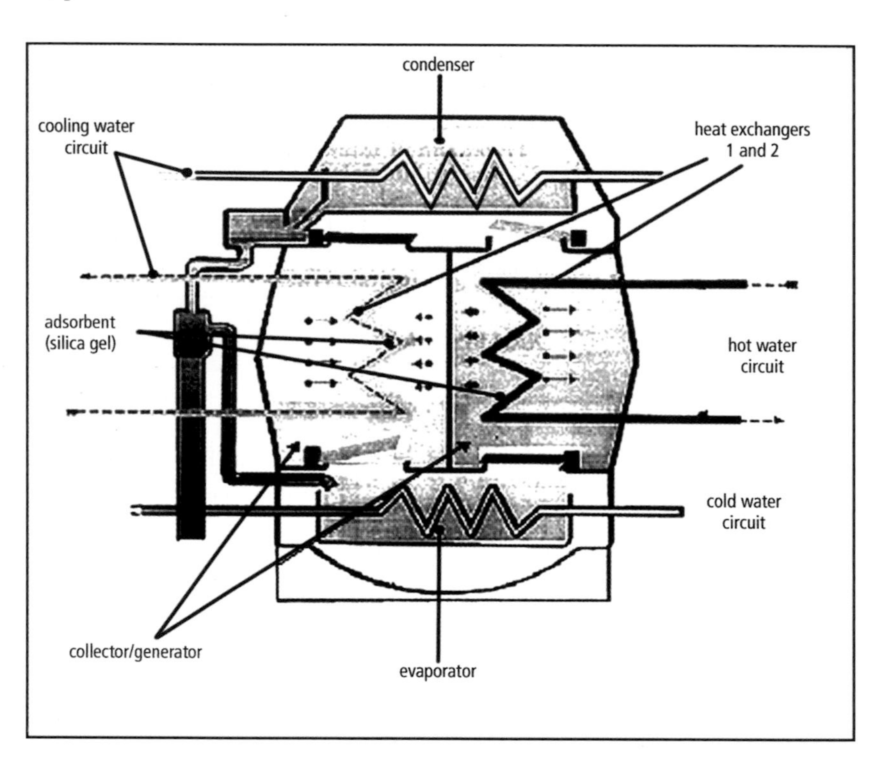

Figure 9.4. Design for a low-temperature adsorption chiller[47]

The AdCh essentially consists of a vacuum tank divided into four chambers. These are the evaporator (lower chamber), the generator and collector (middle chambers), and the condenser (top chamber). The generator and collector are each linked by flap valves, which open and close fully automatically as a result of the prevailing pressure differences in the chiller, to the condenser above them and to the evaporator below them. The AdCh uses water as the cooling medium and silica gel as the adsorbent. The physical properties of water and silica gel are used to produce refrigeration. At low pressures, water vaporizes at low temperatures and silica gel can bond large amounts of water without loss, reversibly and without increasing in volume, and release the water again when heat is applied.

The chiller is fully automatic in a working cycle of 5–7 min that essentially comprises the following four steps, which take place simultaneously. In the first step, water is injected into the evaporator, where it vaporizes. Heat is taken away from the cold water circuit. In the second step, the vaporized water is adsorbed in the collector. This process lasts until the silica gel is saturated. Then it is switched to the second adsorber chamber. In the following third step, the adsorbed water is desorbed after the application of thermal energy. The collector becomes the generator. In the fourth and final step, the desorbed water is condensed in the condenser. The condensation heat is dissipated via a cooling water circuit (recooling system). The circuit is completed as the condensed water is fed back into the evaporator through a valve.

A central characteristic of the AdCh is that the collector and generator are alternately heated and cooled, with the chiller therefore working discontinuously. Thus, alternating periodically, one side is cooled by the flow of cooling water in order to dissipate the heat arising from the adsorption, while the generator is heated for desorption. The periodic change is controlled via pneumatic valves. The chiller is controlled by measuring the cold water exit temperature. An advantage of the AdCh process compared with the AbCh is that it is not limited by a crystallization limit of the solvent.

Adsorption chillers are a new technology, currently with only a few demonstration projects in operation, and in the first market introduction phase. In respect of the use of AdCh for solar cooling, it is of particular note that they can be used to produce cold with temperatures starting from 55°C. Hence both evacuated tube collectors and high-end flat-plate collectors are suitable as solar thermal energy converters.

9.2.4 Desiccant cooling system

In contrast to the absorption and adsorption processes, the desiccant cooling system is termed an open process, as here the air is conditioned by coming into direct contact with the cooling medium. Water is used as the cooling medium, which gives this technology excellent environmental characteristics. In addition, the sorbent – either solid or liquid – also comes into direct contact with the conditioned air. This achieves the required dehumidification of the air.

9.2.4.1 DESICCANT COOLING SYSTEM USING SOLID SORBENTS

Figure 9.5 shows a diagram of a desiccant cooling system (DCS). Compared with a conventional air-conditioning system, the additional components – the desiccant wheel, the regeneration air heater, and the humidifiers for the extracted air and the incoming air – are integrated. On the other hand, the supply air cooler is no longer needed if the necessary scope of dehumidification can be completely implemented by the sorption wheel. A chief characteristic of desiccant cooling systems is that the dehumidification and cooling stages, which in a conventional air-conditioning system take place in one step at the air cooler, are separated.

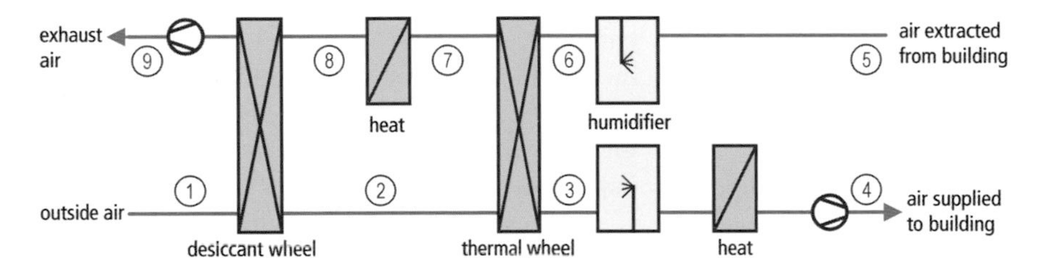

Figure 9.5. Diagram of a DCS system with desiccant wheel[46]

Figure 9.6 shows the process (summer usage) of a desiccant cooling system in a temperature/humidity diagram. The isopleths (lines of the same relative air humidity) are shown as a set of curves.

First of all, outside air is sucked in and dried as it passes through the dehumidification wheel (1–2). The stream of air is warmed at the same time by the adsorption heat that is released. In the thermal wheel (2–3) that comes next, the heat is transferred to the exhaust air or cooling energy is recovered from the exhaust air, and hence the dried incoming air is pre-cooled. In the supply air humidifier (3–4), the air is humidified and cooled adiabatically. The additional heat exchanger before the supply air humidifier is provided only for the heating scenario. The dried, cooled air is supplied to the building and undergoes slight warming in the ventilator (usually 0.5–1 K). As a result of sensitive and latent cooling loads in the room, the room air becomes warmer and is extracted from the room as warm, moist air (4–5). The extracted air is then humidified in the exhaust air humidifier until it is close to the dew point, in order to utilize the cooling potential of the extracted air (5–6). This

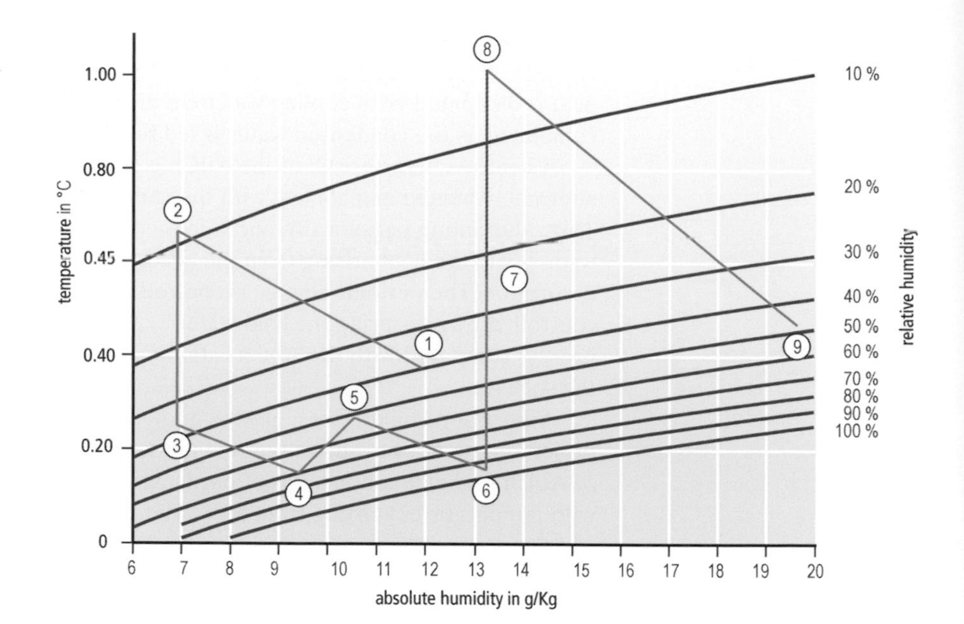

Figure 9.6.
Process diagram (summer) of a DCS system with sorption rotor. Source: Fraunhofer ISE

cooling energy is then transferred to the supply air via the thermal wheel (6–7). In order to regenerate the desiccant wheel, the exhaust air has to be heated sufficiently so that it is able to strip the moisture from the wheel. An external heat source is required at this point in the process (7–8). The heated exhaust air flows through the regeneration side of the desiccant wheel as regeneration air, where it strips the moisture taken up from the outside air (8–9).

9.2.4.2 DESICCANT COOLING SYSTEM USING LIQUID SORBENTS

As well as the desiccant cooling system using solid sorbents, which has been established in the market for several years and for which there are already several hundred systems in operation in Europe, there is also a desiccant cooling system that uses liquid sorbents. This process is still under development, and the first pilot systems are currently being tested in Germany and in Israel[48,49]. It is unlikely that a fully developed product will appear on the market before 2005.

As the name implies, the essential difference is in the physical state of the sorption material. In systems that use liquid sorbents, there is a preference for using water-based salt solutions (for example LiCl or $CaCl_2$). The results of various independent investigations have shown lithium chloride to have the greatest potential in respect of practicality. When selecting the optimum salt solution, as well as the purely thermodynamic properties, aspects such as corrosiveness and cost are of course also important.

Figure 9.7 shows a highly simplified diagram of the principle of a liquid sorption system. Only the two core components – the absorber and the regenerator – are shown in the figure. Looking at the basic principle, the open process of desiccant cooling using liquid sorbents has many parallels with the closed absorption refrigeration method.

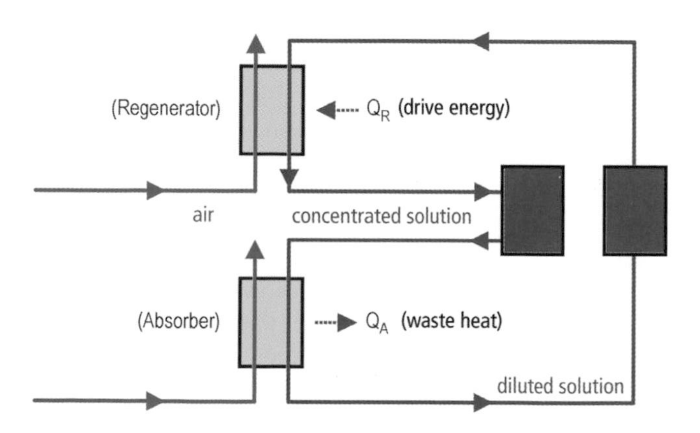

Figure 9.7.
Operating principle of a DCS system with liquid sorbents

The conditioned air is dehumidified in the absorber. Here the air is brought into contact with the liquid sorbent. Because of its highly hygroscopic properties, the sorption medium absorbs the water vapour that the air contains, thus becoming diluted. In the absorber, the heat – referred to as *absorption enthalpy* – is released, and this has to be removed. If this heat were not carried off, there would be a sharp rise in the temperature of the solution, and the absorption process would gradually be disrupted. In practice, the released absorption heat can be removed either directly in the absorber or in a heat exchanger downstream of the absorber. Thus, in the *absorber*, with the removal of heat there is a *reduction in the concentration of the water-based salt solution*.

In the regenerator, the concentration of the diluted salt solution is increased. For this, the diluted salt solution is brought into contact with the air again. Outside air is generally used for this process. In the regenerator, heat must be supplied to the process. As the required driving heat needs to be available at temperatures of 'only' 60–80°C, the integration of solar thermal energy suggests itself strongly here. Thus, in the *regenerator*, with the application of heat there is an *increase in the concentration of the water-based salt solution*.

The water-based salt solution therefore travels round a circuit, in principle, but comes into contact with air in the absorber and regenerator. In order to prevent the loss of salt – but also the associated potential corrosion problems in air ducts – one of the most pressing tasks in the development of this technology is to prevent the discharge of salt in the absorber or regenerator (known as *carryover*). This has a marked influence on the design of the absorber and regenerator. Both are essentially very similar in structure. The main task consists of bringing the air and the salt solution into contact with each other, either in counter-flow or in cross-flow, in such a way as to ensure good mass and heat transfer properties and at the same time prevent droplet formation.

Apart from these development challenges, solar air-conditioning (SAC) technology using liquid sorbents does, however, have some significant advantages over SAC using solid sorbents:

- Isothermal dehumidification of outside air is possible. Because it is possible to remove released absorption heat directly in the absorber, it is therefore possible to dehumidify the air being conditioned without causing an increase in temperature. This is impossible, even in principle, with sorption wheels. As a result, significantly higher dehumidification ranges can be achieved with the same air temperatures, or, with the same dehumidification ranges, significantly lower supply air temperatures are possible.
- Heat recovery between the released absorption heat in the absorber and the required driving heat in the regenerator is possible in principle. This enables the thermodynamic efficiency of the process to be increased.
- Because the sorption material is liquid, it can be used at a later time in the absorber or regenerator if solution reservoirs are installed. This first allows dehumidification even if there is no solar energy available at the time. Second, solar energy can be stored in the form of concentrated salt solution (see Figure 9.7). This is particularly promising as *this type of solar energy storage is in principle free of losses*. Hence this technology offers the potential to significantly reduce the required collector surface while keeping the same high solar fraction and hence helping to improve the cost-effectiveness of the solar air-conditioning system.

9.3 Integrated planning of solar cooling/air-conditioning systems

In contrast to solar thermal domestic hot water supply, system planning and system design for solar air-conditioning systems is significantly more complicated. Figure 9.8 shows various subsystems, which are linked to each other in a concept for a facility's solar air-conditioning system. The four subsystems are:

- building
- air-conditioning system
- heat supply
- cold supply.

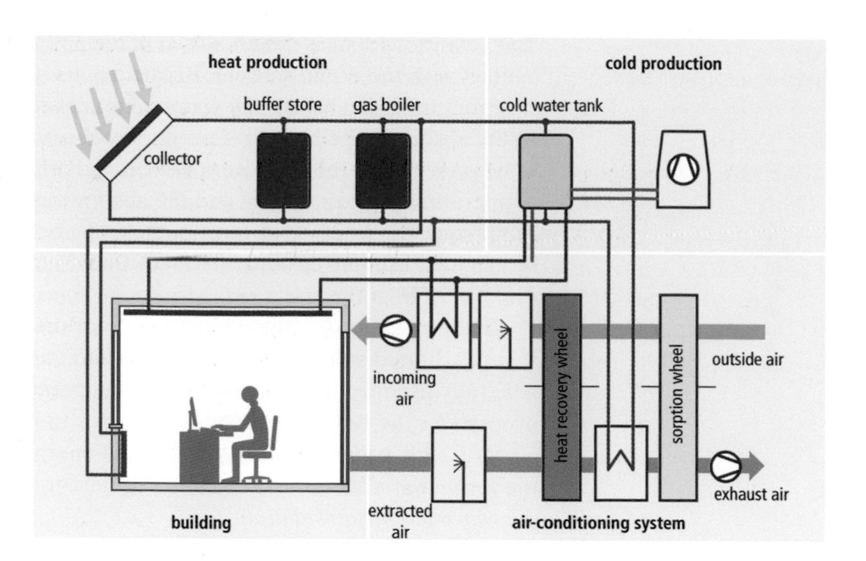

Figure 9.8. Subsystems of solar air-conditioning. Source: Fraunhofer ISE

Depending on the requirements of the air-conditioning task and on the climate zone, all four subsystems or only some of them will be in use. There is a distinction to be made between full air-conditioning with all four thermodynamic conditioning functions (heating, cooling, humidification and dehumidification) and partial air-conditioning (for example only heating and cooling).

The greatest challenge is to connect these systems together in an intelligent way. Hence, to achieve an overall concept optimized in terms of energy and economics, integrated planning with good communication between the various disciplines from the beginning is very important.

In air-conditioning for buildings there is no silver bullet – instead there is only the particular customized and optimized solution for each site. At the same time, climatic conditions are also very important, and there is a demand for set-up variants adapted to the climate. Designing a system using rules of thumb, as for solar thermal domestic hot water supply, entails a number of difficulties and at best only allows the basic variables to be determined.

For an economically rational solution it is always important to check first whether anything can be done to the building to reduce the cooling and heating loads. The main strategies are:

- thermal insulation of the building shell
- integration of the exterior sunshade systems
- reduction of the internal loads by using energy-saving appliances.

In addition, the possibility of night ventilation should be examined early on in the planning phase, as should the possibility of activating the thermal building mass (for example using concrete construction elements as active thermal components).

The most important steps in a good integrated planning concept are set out below. As planning should always be based around an actual project, we do not claim that this list is complete, nor that the order must necessarily be followed. This recommended method is merely intended as an aid for future planning.

(a) Calculate the hygienically necessary rates of air change for the air-conditioned rooms. In this way, for example, it may be possible to do without mechanical ventilation completely for air-conditioning in an IT room in which people spend only short periods of time. This makes big reductions in investment costs, as in this case the air-conditioning subsystem can be completely eliminated. For rooms where only very low rates of air change are required, check whether these could be implemented via appropriate cross-flow openings from other rooms or possibly by window ventilation. In this context, check whether window ventilation, if it were possible in principle, would nevertheless result in unacceptable working conditions because of noise pollution (for example close to busy streets). Generally workplace guidelines should also be followed.

(b) If mechanical ventilation as at (a) is deemed to be necessary, then see whether this can be implemented with a pure supply air system or with a pure air extraction system. In this case, include appropriate cross-flow openings in the façade and/or partition walls to corridors/adjacent rooms in plans. From the point of view of investment costs, these types of simplified system should be examined in every case and, if need be, should be compared with the resulting operating costs. This can lead to different system decisions, depending on the climate zone.
(c) Check whether splitting the air-conditioning system into different zones would be a good idea. In some circumstances this can lead to large cost savings. In this context, examine in particular the possibility of bringing together locally those IT units that produce large quantities of waste heat.
(d) Examine measures on and in the building to reduce the cooling and heating loads, and incorporate these in the planning. Generally, for buildings that have large areas of glazing, the possibility of an external sunshade system should always be investigated. This measure brings a very considerable reduction in cooling loads in the building, and it is cost-effective. Calculate the resulting cooling and heating loads for the building. It is highly recommended as well to take into account the partial load characteristics in respect of the heating and cooling load from the beginning, as the system operates with a partial load for the greater part of the year and hence this has the greatest influence on the energy consumption of the system as a whole. For more complex building structures, building simulation calculations are always a good idea.
(e) Check what requirements exist for the building in respect of flexibility of the type of use in the expected lifetime of the air-conditioning system. This question can have a very strong influence on deciding the type of air-conditioning technology.
(f) Check what temperature and humidity limit values will be desired and accepted in the building. Moreover, check whether it is acceptable to users and the technical process to exceed or fail to meet these temperature and humidity limit values for a particular number of hours per year. Checking this point can in some circumstances dramatically reduce the investment costs.
(g) Select the air-conditioning technology that is right for the particular use of the building (here questions such as the presence of toxic exhaust air, for example, should be checked as well) and the climatic zone (here it is quite possible that, for large buildings, different solutions will be found for different zones).
(h) Check whether, and which, thermally driven cooling systems come into consideration in principle. Check possible (partial) shading of the collector field.
(i) Check whether cost-efficient waste heat is available at the required temperature level, and whether the heating energy available from the waste heat is sufficient.
(j) Check to see which solar technology comes into consideration in principle for the appropriate cooling technology.
(k) Produce a rough design of the solar system with annual simulation calculations. At this stage the various collector technologies and their specific system costs can and should be investigated. When estimating the solar system costs, always take the effect of scale into account.
(l) Check whether the existing roof and/or façade surfaces are sufficient to implement a large part of the required driving heat with the solar thermal system. If it is possible and architectonically acceptable to integrate the collector system into the façade or on the façade, the collector system can serve a dual purpose – that is, for use to supply heat and as exterior shading. In this case, go back to point (d) and investigate its influence on the heating and cooling load.
(m) Produce a first cost estimate. For this, take the following information into account:
- Prices for electrical power (kilowatt-hour and capacity price).
- Prices for providing heat (kilowatt-hour and capacity price).
- Prices for the relevant collector technology. (Here the influence of the collector field size must be taken into account, as this can in some cases lead to very big price reductions. Thus, by enlarging the collector system from 10 m^2 to 100 m^2, in some circumstances it may be possible to reduce the specific system costs by around 30–50%.)
- Possible subsidies for the solar technology.

- Prices for the air-conditioning technology and cooling technology being considered. (Here bear in mind that, because of the intense competition in the air-conditioning industry, there may well be considerable price differences between the cost estimate and the results of the tendering process. This should be taken into account in the planning process.)

In general, the influence of the electrical capacity charge has a considerable influence on the cost-effectiveness of a thermally driven cooling system compared with an electrically powered compression chiller. However, in this respect it is critical to check whether the capacity peaks for the electrical power consumption are actually brought about by the air-conditioning task. If this is not the case, then this potential saving cannot be included in the economic efficiency evaluation. For future economic efficiency evaluations it may also be of interest to cash in on the possible carbon dioxide saving through emission trading.

After steps (a) to (m), iteration steps may still be necessary if the basic conditions in the planning process change. At first sight the process appears to require a lot of work, but the planning effort will pay off for the customer by providing a solution that is optimized in terms of energy and economics.

9.4 System technology

This section provides an overview of the system technology for solar air-conditioning systems. It will explain the fundamentally different concepts of autonomous solar-powered systems and solar-assisted systems. It will consider which collector technologies can be best implemented for the various thermally driven cooling technologies. In addition, it will describe various switching variants for the solar connection.

9.4.1 Autonomous solar-powered systems versus solar-assisted systems

In central European climates, autonomous solar thermal systems (related to the thermal power) are not economically viable as solar heating systems, because of the seasonally very different solar irradiance conditions. For solar cooling systems, on the other hand, two concepts are possible that are different not just technically but also in economic terms (Figure 9.9):

- solar-assisted systems
- autonomous solar-powered systems.

In *solar-assisted systems*, the solar energy supplies only part of the driving heat required for the summer air-conditioning. Whenever the directly available and/or

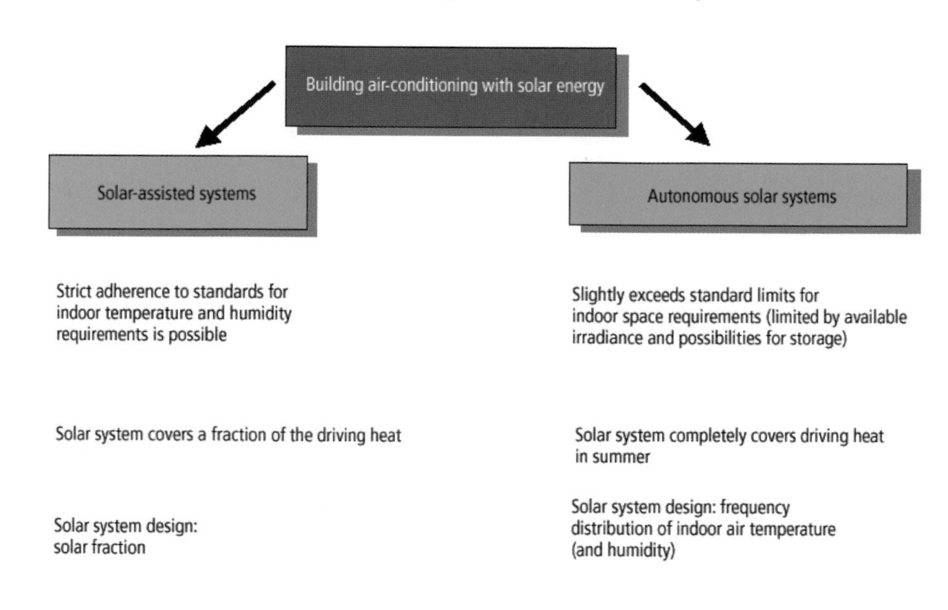

Figure 9.9. System concepts for solar cooling systems. Source: Fraunhofer ISE

stored solar energy is insufficient, enough driving heat can be provided via the conventional auxiliary heating system. This enables the design requirements for indoor space always to be attained (if the cooling and air-conditioning systems are correctly designed). In these systems the attainable solar fraction is utilized for sizing the design. In this context, the solar fraction for cooling is used. This is always less than 100%. The solar fraction for heating can also be calculated based on the fraction of heat energy used for the heat supply and/or averaged for the heating and cooling season based on the entire energy used for the heat supply. The conventional auxiliary heating system is always necessary and is generally designed so that the complete driving heat can be provided via the auxiliary heating system. With detailed prior planning (simulation calculations of the entire plant are necessary), and depending on the building, climatic region and size of the selected heat storage tank, under certain circumstances it is possible to dispense with part of the thermal output from the conventional heating system. In this case, use of the solar system enables costs to be saved in terms of the auxiliary heating system. This can have a positive impact on the economic efficiency of the solar air-conditioning.

With *autonomous solar-powered systems*, the complete driving heat for the summer cooling and air-conditioning is provided by the solar collector field. In these systems, as much air-conditioning is achieved as possible with the available solar heat. An intelligent control strategy is very important in these systems. By their very definition, these systems dispense with an auxiliary heating system. The solar fraction for cooling is therefore always 100%. Thus it does not make any sense to design these systems using this parameter. As there is no auxiliary heating system in autonomous solar-powered systems, depending on the system design there may be hours with extreme external conditions when limit values for indoor temperatures and/or humidity are exceeded in the building. The number of these excess hours as well as the size of the deviations from the limit values can be determined through simulation calculations and utilized as design criteria. This is why, when designing autonomous solar-powered systems, it makes sense to use simulation calculations that consider the solar system *and* the building in *one* simulation. For buildings with a very large temporal correlation between the solar irradiance and the cooling load to be removed, self-contained solar systems are interesting not only technically but also economically. A typical example of such a building could be an office building used predominantly during the day with extensive glazed areas in the façades. The advantage of these systems lies in the simplified system technology. Costs can be saved by dispensing with the auxiliary heating system and the back-up heat exchangers. For autonomous solar-powered systems, integral planning is even more important, as changes in the building during the planning without a corresponding adaptation of the solar heating system can have an immediate impact on the later comfort conditions in the building.

9.4.2 Which collector technology for which cooling technology?

One of the most important questions concerning the system technology of solar air-conditioning systems is the choice of the correct solar collector technology.

To answer this question it is important to compare the driving temperatures of the various thermally driven cooling technologies. These are depicted in Table 9.2. The driving temperatures for single-stage absorption chillers are between 85°C and 110°C. For two-stage models they are even higher, at around 150°C. For adsorption technology the driving temperatures are somewhat lower at 55–90°C. The driving temperatures for desiccant cooling systems (DCS) are the lowest. This applies for desiccant cooling systems with both solid and liquid sorbents. With the latter, according to the current state of knowledge, the minimum driving temperatures are slightly higher (approximately 55°C); however, there is potential for improvement here.

Table 9.2. Overview of thermally driven cooling technologies and driving temperatures

Process	Absorption	Adsorption	Desiccant cooling system
Type of air-conditioning	Chilled water (e.g. chilled ceilings)	Chilled water (e.g. chilled ceilings)	Air-conditioning (cooling, dehumidification)
Driving temperature (°C)	85–110	55–90	45–90

When deciding which collector technology is best suited for a particular cooling technology, the efficiency curves play an important role for the various collectors. In Figure 9.10 the efficiency curves for the most widely used solar collectors are displayed as a function of the mean liquid temperature. If it is assumed that the collector efficiency for the respective application should amount to at least 50–60%, then this provides the following picture with regard to Table 9.2.

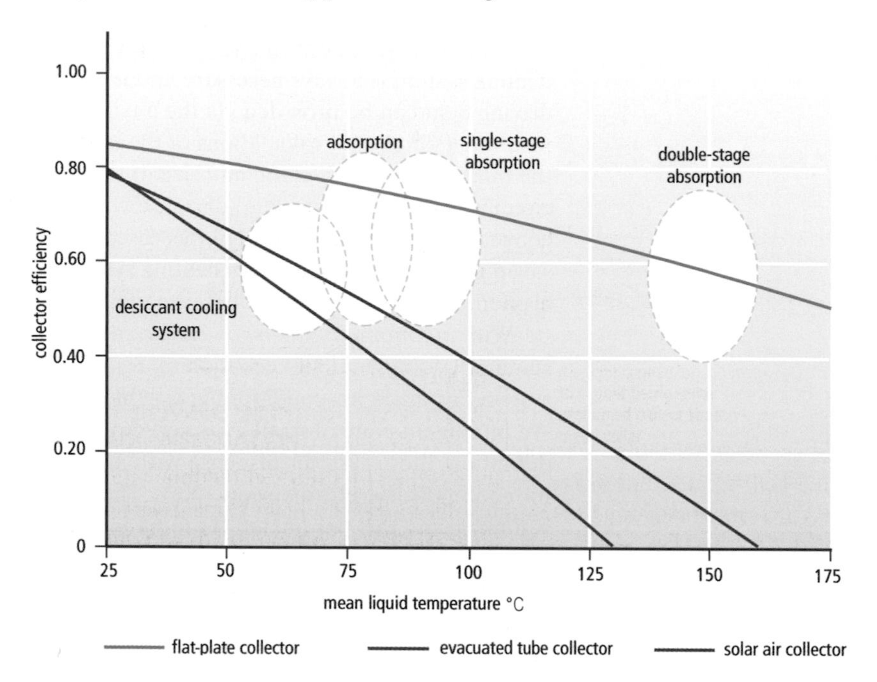

Figure 9.10. Collector efficiency for typical solar collectors. Source: Fraunhofer ISE

For desiccant cooling systems, both flat-plate collectors and solar air collectors can be considered. In principle, evacuated tube collectors are also highly suitable; however, their higher price counts against them. For adsorption chillers, very good flat-plate collectors or evacuated tube collectors can be considered. For single-stage absorption, evacuated tubes are more suitable. From the economic point of view, flat-plate collectors would probably not be considered for absorption applications, even if technically feasible. If they were used, then this would be only with very large collector areas. For two-stage absorption, only evacuated tubes come into consideration.

Apart from the efficiency curve aspect, the following factors play an important role when choosing suitable collectors:

- *Specific collector price.* Here you can compare both the price per installed square metre of collector field and the price per kW of thermal output at the desired target temperature. The latter figure enables better comparability; the resulting collector field size should also be considered, however, as this has an impact on the costs of the substructure.
- *Local shading situation.* It is important to analyse the local shading situation. This can also considerably influence the choice of the collector type if, for example, there is insufficient area for the flat collectors because the intended roof area is partly shaded.
- The possible *integration of solar collectors in the façade* can be tested in relation to the project. In this case structural aspects and, in particular, the appearance can be decisive in deciding for or against a specific collector technology. In such cases the possible substitution of façade components with the implementation of flat-plate or solar air collectors is an important factor. With many types of evacuated tube this is not possible.

9.4.3 Circuit variants

In terms of the system technology, the possibilities to connect the solar system to the entire solar cooling system are also very important.

In general we can differentiate between the following systems:

- systems with solar liquid collectors
- systems with solar air collectors
- systems with solar heat input only via the heat storage tank
- systems with direct input of the solar heat into the air-conditioning technology
- systems with auxiliary heating in the heat storage tank
- systems without auxiliary heating in the heat storage tank (only with solar heat storage tanks).

In turn, there are also hybrid versions based on these systems, and of course other systems are also feasible. The following illustrations are simply schematic diagrams to clarify the integration of the solar technology; they should not be understood as hydraulically complete circuit diagrams.

In terms of the cooling technology, all the relevant possibilities are illustrated; for investment reasons, however, it is very seldom that they are implemented at the same time. When using adsorption chillers, a cooling storage tank should definitely be integrated into the system in order to control the stability of the system. With absorption chillers, a cooling storage tank can also be advantageous. Which system is used in the end should always be decided in terms of the particular project.

Figure 9.11 shows a solar cooling system with solar liquid collectors whereby the solar heat input is only via the heat storage tank. There is also auxiliary heating in the heat storage tank. The system has the following features:

- It is suitable for absorption and adsorption chillers as well as desiccant cooling systems (solid and liquid).
- It is possible to integrate flat-plate collectors and evacuated tube collectors.
- There is dynamic decoupling of the solar system and cooling technology/air-conditioning through the heat storage tank.
- There is increased heat storage loss through auxiliary heating in the buffer storage tank.
- 'Traditional' control of the solar thermal system is generally possible.
- There is good control of the overall system through dynamic decoupling.

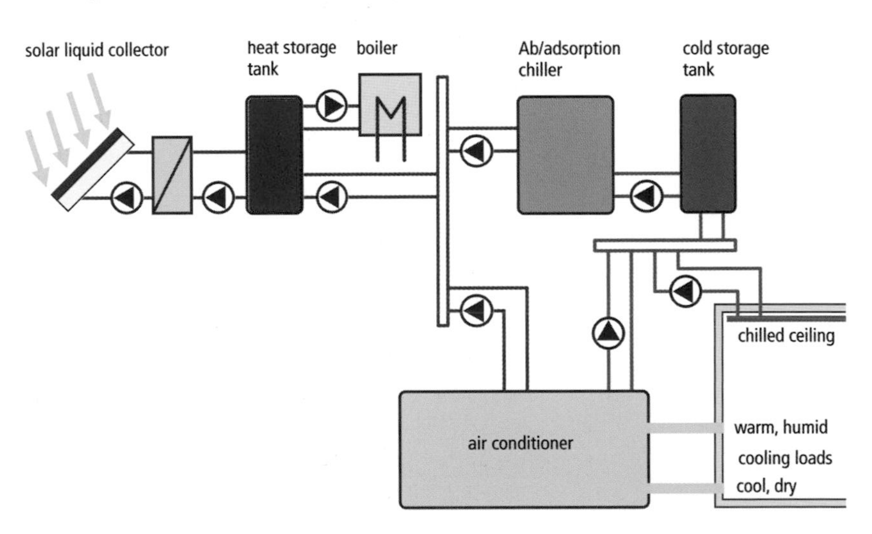

Figure 9.11. Solar cooling system with solar liquid collectors (the solar heat input is only via the heat storage tank) and auxiliary heating in the heat storage tank

Figure 9.12 shows a solar cooling system with solar liquid collectors whereby the solar heat input is only via the heat storage tank but with no auxiliary heating in the heat storage tank. The system has the following features:

- It is suitable for absorption and adsorption chillers and desiccant cooling systems (solid and liquid).
- It is possible to integrate flat-plate collectors and evacuated tube collectors.
- There is dynamic decoupling of the solar system and cooling technology/air-conditioning through the heat storage tank.

- There is reduced heat storage loss through auxiliary heating outside the heat storage tank (it is best to connect the auxiliary heating in series).
- The storage tank contains only solar heat.
- 'Traditional' control of the solar thermal system is generally possible.
- There is good control of the overall system through dynamic decoupling.

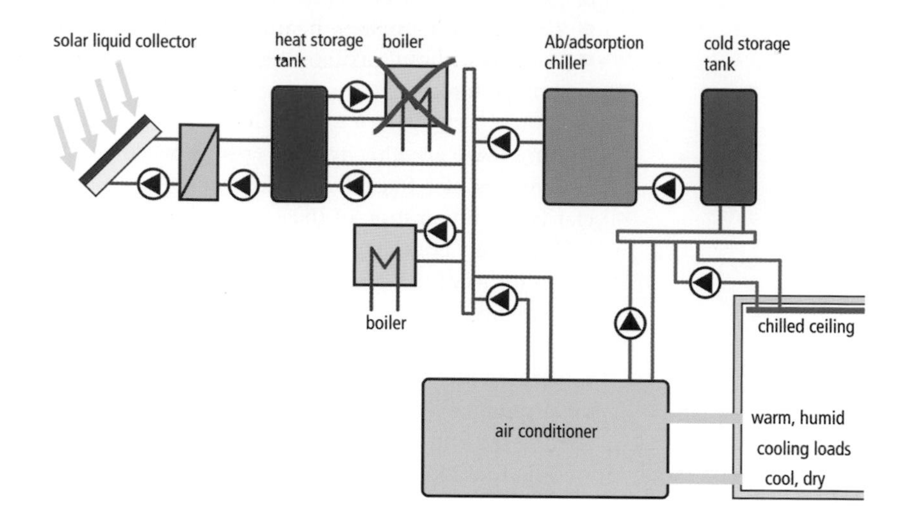

Figure 9.12.
Solar cooling system with solar liquid collectors whereby the solar heat input is only via the heat storage tank but with no auxiliary heating in the heat storage tank

Figure 9.13 shows a solar cooling system with solar liquid collectors whereby there is direct solar heat input and solar heat input via the heat storage tank, but without auxiliary heating in the heat storage tank. The system has the following features:

- It is suitable for absorption and adsorption chillers and desiccant cooling systems (solid and liquid).
- It is possible to integrate flat-plate collectors and evacuated tube collectors.
- There is dynamic decoupling of solar system and cooling technology/air-conditioning through the heat storage tank.
- There is reduced heat storage loss through auxiliary heating outside the heat storage tank (it is best to connect the auxiliary heating in series).
- The storage tank contains only solar heat.
- The collector yield is potentially higher, as the direct connection of solar heat is possible.
- The hydraulics and control are more complex owing to dynamic coupling.

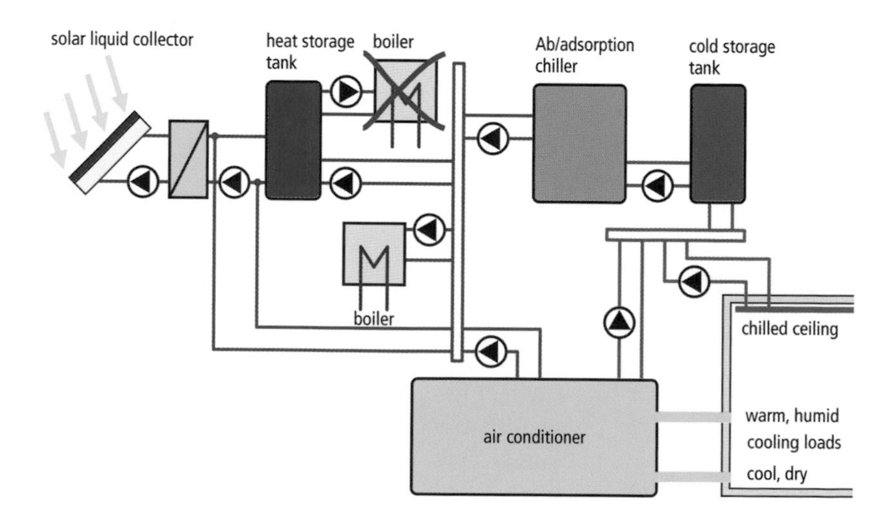

Figure 9.13.
Solar cooling system with solar liquid collectors whereby there is direct solar heat input and solar input via the heat storage tank, but without auxiliary heating in the heat storage tank

Figure 9. 14 shows a solar cooling system with solar air collectors equipped with an ambient air intake. It has direct solar heat input but no heat storage tank. The system has the following features:

- It is only suitable for desiccant cooling systems (solid and liquid).
- It is possible to integrate solar air collectors.
- There is direct dynamic coupling of the solar thermal system and the air-conditioning system.
- It is a storage-free system.
- It is a solar air system with ambient air intake.
- The hydraulics and control are easier, thanks to the simplified system technology, but control must be well adapted to fluctuating irradiance.
- Very low investment costs are possible.
- It only makes sense for buildings with a high temporal correlation between the solar irradiance and the cooling load.
- When heating, the direct solar heating of the ambient air means that diffuse irradiance is also usable in the systems.

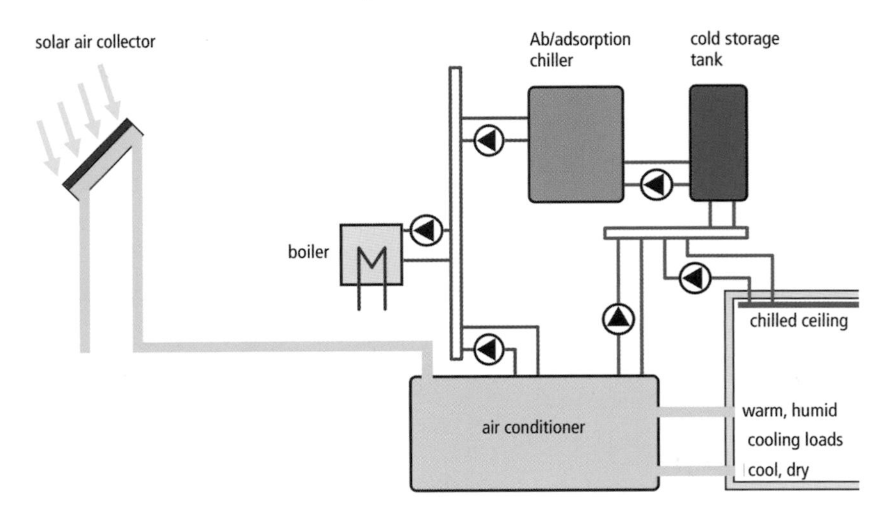

Figure 9.14. Solar cooling system with solar air collectors equipped with a ambient air intake. It has direct solar heat input but no heat storage tank

Figure 9.15 shows a solar cooling system with solar air collectors that have a regeneration air intake. It has direct solar heat input but no heat storage tank. The system has the same features as the system in Figure 9.14 with the difference that, when cooling, the regeneration air is drawn into the solar air collector after leaving the heat recovery wheel. This enables higher collector outlet temperatures to be attained for the same amount of irradiance. However, the higher absolute humidity after the heat recovery wheel means that, in central European climates, the regeneration of the desiccant wheel is roughly just as good (even with slight advantages compared with the system shown in Figure 9.14). This switching can be advantageous in warm, humid climates. However, the investment costs for this system are higher because of the additional air ducts.

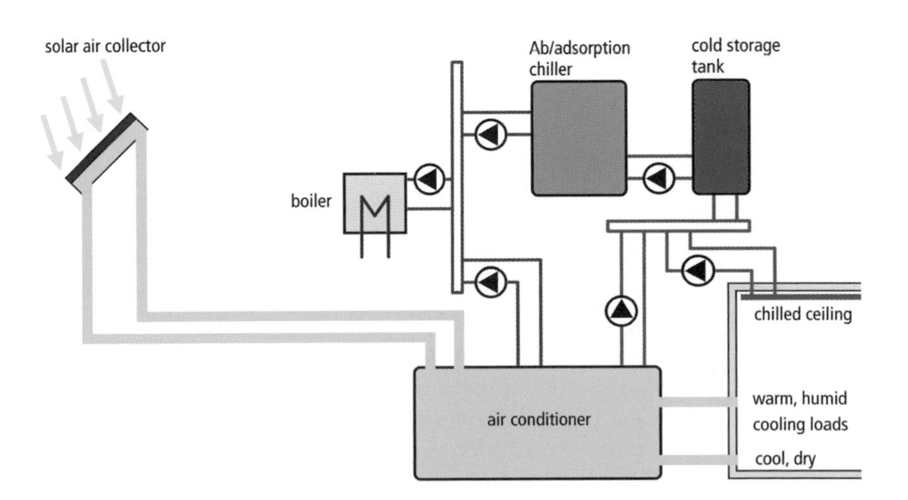

Figure 9.15. Solar cooling system with solar air collectors that have a regeneration air intake. It has direct solar heat input but no heat storage tank

Figure 9.16 shows a solar cooling system with solar air collectors that have an ambient air intake. It has direct solar heat input and integration of a heat storage tank but no auxiliary heating in the storage tank. The system has the following features:

- It is only suitable for desiccant cooling systems (solid and liquid).
- It is possible to integrate solar air collectors.
- It is a solar air system with ambient air intake.
- There is direct dynamic coupling of the solar thermal system and air-conditioning system.
- It is possible to decouple excess solar heat to the buffer storage tank.
- The hydraulics and control are easier, thanks to the simplified system technology, but control must be well adapted to fluctuating irradiance.
- Low investment costs are possible.
- It also makes sense for buildings with an average temporal correlation between the irradiance and the cooling load.
- When heating, the direct solar heating of the ambient air means that diffuse irradiance is also usable in the system.

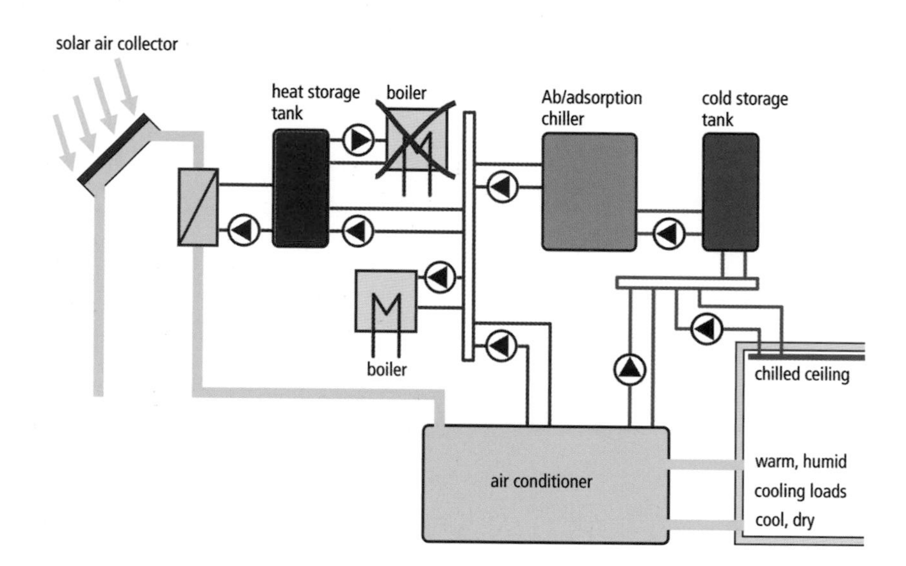

Figure 9.16. Solar cooling system with solar air collectors that have an ambient air intake. It has direct solar heat input and an integrated heat storage tank but no auxiliary heating

9.5 System design

For air-conditioning systems, the question of the system design cannot be answered as easily as it can be, for instance, with solar domestic hot water heating. It depends substantially on two influencing factors:

- The required cooling and/or heating loads of a building or part of a building always depend on the project, and can *only* be precisely determined *in relation to the project.*

This is the largest and most significant difference from designing systems for solar water heating, where the number of persons and the resulting hot water consumption almost exclusively determine the heating load to be met by the system. Although in air-conditioning technology there are also empirical values for the specific internal cooling load (in watts per square metre) for typical uses (for example offices, seminar rooms), it is the *architecture* and *design of the building* and the *climatic zone* that substantially influence the resulting cooling and/or heat load to be met by the system. For example, the existence or non-existence of a good external shading device on an extensively glazed building can easily halve or double the required cooling load. The use of rules of thumb for the schematic design of solar air-conditioning systems is, as a matter of principle, more difficult, and these themselves can only be used as rough reference values. *Moreover, a substantially larger number of influencing factors must be determined first in order to determine the reference values.* It is precisely this last point that obliges the planner to plan carefully in advance.

- Until now there has been relatively little experience with operating systems.

Most installations currently in operation are part of demonstration projects. The systems were mostly conceived and planned by research institutes and then implemented in collaboration with a designer on site. Therefore there are still very few designers in planning offices who have experience of implementing several solar cooling systems. This is shown by the specific collector surfaces per kilowatt cooling capacity found in the pilot systems, which range widely from 0.8 $m^2_{col}/kW_{cooling}$ to 8 $m^2_{col}/kW_{cooling}$.[50] Although these differences are due partly to project-specific circumstances (for instance, the collector areas are not always determined by the solar cooling), it is also clear that in some cases the planners were uncertain as to the design.

9.5.1 General procedure when designing systems

Despite the complexity in individual cases, it is still possible to provide general guidance on the system design. The considerable advantages to be gained by integral planning also apply here. This particularly concerns the effect of the system design on the economic viability of the solar cooling. Here iterative steps and even compromises between energy-optimized and economically realistic system designs are often necessary.

The following steps for designing a solar cooling system should be carried out in every case:

(1) Calculate the cooling/heating load of the building.
(2) Optional: Calculate the cooling/heating load time series for every hour of the year based on a building simulation.
(3) Design the air-conditioning and/or cooling technology. At this point check which air-conditioning and/or cooling technology can be utilized or is suitable for the particular building.
(4) Optional: If step (2) has been carried out, decide whether the design of the air-conditioning and/or the cooling technology should be based on the peak load or whether a deviation during individual hours/days should be accepted. Quantify the potential effect of reducing the air-conditioning and/or cooling technology on the ventilation conditions, using a combined building and system simulation.
(5) Calculate the driving heat capacity of the thermally driven cooling technology.
(6) Optional: Calculate the driving heat time series for each hour of the year based on a system simulation.
(7) Design the solar thermal system and, if intended, the heating storage tank to cover the driving heat capacity completely or partly. Design the auxiliary heating system if envisaged. In this step compare various system configurations using simulation calculations; this makes sense if (and only if) step (6) was carried out. If a façade-integrated solar system is planned or one that partly shades the façade, then at this point it can make sense to recalculate the heating/cooling load of the building according to step (1).
(8) Calculate all the energy and water consumption in the entire system.
(9) Calculate the costs of the solar system and the auxiliary heating system according to step (7). Take into consideration both the consumer price and the price per unit for the electricity and heat.
(10) Calculate the costs of the entire system.
(11) If the costs are not acceptable, iterative steps are recommended from step (3) or from step (7).

As well as the technical and economic design, both the realizable and the primary energy savings are also of interest when designing the system. This applies particularly for systems for solar cooling. This aspect will be considered in detail in subsection 9.5.5.

For a successful system design, it is always sensible to work with partners who already have practical experience in this area. The best results can be achieved through the use of simulation software that can depict both the building and the system technology, including the solar thermal system.

9.5.2 System design using empirical values

Based on experience from previously implemented pilot systems, the following provides a rough estimate in sizing the solar collector field in advance:

- For desiccant cooling systems with solid sorbents: approximately 5–10 m^2 collector area per $1000 m^3/h$.
- For ad/absorption chillers: approximately 1–3 m^2 per kW cooling capacity.

The use of these values should mean that any excessive over- or under-dimensioning can be avoided. These reference values enable the general cost of the solar system to be estimated in the early stages of the project during the rough planning. Deviations from these reference values are nevertheless possible when there are special project conditions.

For adsorption and, in particular, absorption chillers that use flat-plate collectors the specific areas can lie above this in individual cases. The information for the desiccant cooling systems applies for both solar air collectors and liquid collectors. For desiccant cooling systems with liquid sorbents, it is expected that smaller specific collector areas will be possible, owing to the possibility of having loss-free storage. This still has to be confirmed by pilot systems monitored with measurement equipment.

It should be emphasized, however, that the reference values should not replace a detailed system design.

9.5.3 System design according to the peak load

Another design method consists of designing the solar system in a way that the nominal cooling load of the cooling and air-conditioning system, $P_{cooling}$, with maximum solar irradiance on the collector field (1000 W/m^2) can only be provided by the driving heat of the solar system. The collector area, A_{col}, is calculated for this according to the following formula:

$$A_{col} = \frac{P_{cooling}}{COP_{cooling,nom}} \frac{1}{\eta_{col,nom}}$$

where $P_{cooling}$ is the nominal cooling capacity of the cooling/air-conditioning system (kW), $COP_{cooling,nom}$ is the nominal COP (coefficient of performance) of the cooling/air-conditioning system, and $\eta_{col,nom}$ is the nominal efficiency of the solar collector at the necessary temperature for driving the cooling/air-conditioning system. In the case of heat transfer in heat exchangers, the temperature differences should be taken into consideration, too. This leads to increased temperatures in the solar collectors compared with the temperatures given as driving temperatures in the technical specifications of the cooling/air-conditioning systems. Table 9.3 shows the respective specific collector areas per kW of cooling capacity derived from the above equation.

Thus for a single-stage absorption chiller with a typical nominal COP of 0.7 and for an evacuated tube collector with a nominal efficiency of 71% (typical for mean collector temperatures of 100°C), the resulting specific collector area is 2.01 $m^2_{col}/kW_{cooling}$.

As can be seen in Table 9.3, the specific collector areas for adsorption and absorption chillers calculated according to the peak load method are also within the area covered by the reference values. Since with the desiccant cooling systems the rated cooling capacity of the air-conditioning system depends directly on the external conditions, the specific collector areas here are given with respect to a nominal flow rate of 1000 m^3/h. It can be seen in Table 9.3 that the values calculated according to the peak load method lie clearly above the empirical values of 5–10 m^2_{col} for a nominal flow rate of 1000 m^3/h. Thus use of the peak load method with desiccant cooling systems will generally lead to over-dimensioned systems. Therefore for desiccant cooling systems it would appear sensible to reduce sizes calculated according to the peak load method by at least 30%.

The advantage of integrating solar air collectors is that no temperature differences occur at the heat exchangers. As the efficiency of 63% relates to the mean collector

Table 9.3
Collector areas per kW cooling capacity (see text for explanation)

Cooling	COP	Collector	Temperature of condensor (°C)	Average temperature of collectors (°C)	Irradiance (kW)	η_{col}	m^2_{col}/$kW_{cooling}$	$kW_{cooling}$/$1000m^3$	m^2_{col}/$1000m^3$
AbsCh	0.7	ETC	90	100	1.0	0.71	2.01	–	–
AbsCh	0.7	FPC	90	100	1.0	0.52	2.75	–	–
AdsCh	0.6	ETC	80	90	1.0	0.73	2.28	–	–
AdsCh	0.6	FPC	80	90	1.0	0.57	2.92	–	–
DCS: solid	0.8	FPC	70	85	1.0	0.59	2.12	6.00[a]	12.71
DCS: solid	0.8	SAC	70	60	1.0	0.63	1.98	6.00[a]	11.90

[a]This value depends substantially on the ambient climatic conditions.

temperature, and because there are temperature differences of approximately 35 K in solar air collectors, the necessary collector output temperature of 75°C is achieved with a mean collector temperature of approximately 60°C. This approach already takes into consideration the temperature losses of approximately 5 K that occur between the collector output and the air-conditioning system.

As when designing according to empirical values, designing according to peak load methods only leads to points of departure – even if these values are more precise. Moreover, a precise analysis of the frequency with which the peak cooling load occurs is highly recommended to avoid over-dimensioning the system. This analysis can also contribute to the determination of a sensible storage size. It should also be considered, however, that the maximum irradiance of 1000 W/m^2_{col} occurs only for a very short period each day. This particularly applies with moderate climates such as those found in central Europe. Building simulations with subsequent or coupled system simulations, in which design values according to the peak load method are used as starting values, generally lead to substantially more precise results and are therefore highly recommended.

If a solar air-conditioning system is to be used both in summer *and* in winter, a design produced according to the peak load should only ever relate to the summer peak *cooling* load and the simultaneously occurring maximum *summer* irradiance; otherwise the system will be too large and will also be definitely uneconomic.

9.5.4 Design of autonomous solar-powered systems

The detailed design of autonomous solar systems always requires a combined building and system simulation. The peak load method is well suited for providing an initial guidance value for the size of the collector field.

As has already been described in section 9.4.1, the quality of the design of autonomous solar systems can be assessed by the number of hours for which the limit values for the indoor temperature and humidity are exceeded. In a combined building and system simulation, the entire system is calculated and the dynamic building response is determined for the changes in the cooling/air-conditioning system's cooling capacity as the irradiance fluctuates. In this process, possible changes to the architectural design can be examined at the same time. This is a clear benefit of this procedure.

Based on the starting values, system variations can be made in terms of the collector and, if necessary, the heat storage tank. The results relating to the excess hours provide information on the possible acceptance by subsequent users of the building. Here it is particularly important that a possible later change in use of the building is already taken into consideration in advance.

Because of the generally good temporal conformance between the irradiance and the cooling load to be removed, autonomous solar systems – particularly with buildings used mainly during the day – can provide good solutions not just in terms of the comfort criteria indoors but also in terms of the energy performance and economic aspects.[51]

9.5.5 System design according to primary energy saving

The main motivation for deploying solar air-conditioning systems is to save energy. This in turn means a lowering of the running costs for the operator. Therefore solar cooling systems will only have a chance in the industrial sector in the long term if the *energy saving* and *cost reduction* aspects can be reconciled with one other.

In addition to the cost reduction aspect, it is also critical to examine whether primary energy can also be saved with the solar thermal cooling system. The concept of *primary energy* will therefore be briefly explained here. Primary energy actually means the energy content of the original energy source before it is converted. It can be divided into *non-renewable energy sources* – frequently described as conventional sources (coal, oil, gas, uranium) – and *renewable energy sources* (sun, wind, water). The concept of primary energy consumption is, however, frequently used only for energy consumption from *non-renewable* energy sources. Thus, when primary energy saving is referred to in the context of using solar technology, it is assumed that the converted solar energy is *not* a primary energy use: that is, primary energy is *saved* when gas consumption is replaced by solar energy use. This is not strictly speaking correct, but it is the normal usage, and will be so used in the remaining text.

In order to compare the various energy sources such as gas, oil and electricity with one another, they can be converted into primary energy units. For instance, in Europe 3 kWh of primary energy are used on average to produce 1 kWh of electricity. As primary energy consumption (with the exception of uranium) is directly associated with the emission of carbon dioxide, the assessment of energy systems in terms of primary energy is of considerable importance as far as the impact on global climate change is concerned.

It can be assumed at this point that a good, thermally driven cooling system uses less electricity than a compression-based cooling system. If this is *not* the case, then it should not be deployed. For this reason, in order to verify this question of electricity use, all users (including the small parasite users such as pumps or auxiliary motors) should be recorded and assessed during the planning phase.

The decisive question, however, is: Does a good, thermally driven cooling system need less primary energy than a good reference system with an electrically powered compression chiller? This is an important question for designing a sustainable solar cooling system.

Because of the clearly higher COP of an electrical compression chiller (typically 3) relative to a thermal chiller (typically 0.7 for a single-stage absorption chiller), a thermal chiller supplied purely with gas uses more primary energy! Therefore from the ecological point of view it is very important to supply a thermal chiller at least partly with 'neutral' energy forms such as waste heat or solar energy. This relation is clarified in Figure 9.17.

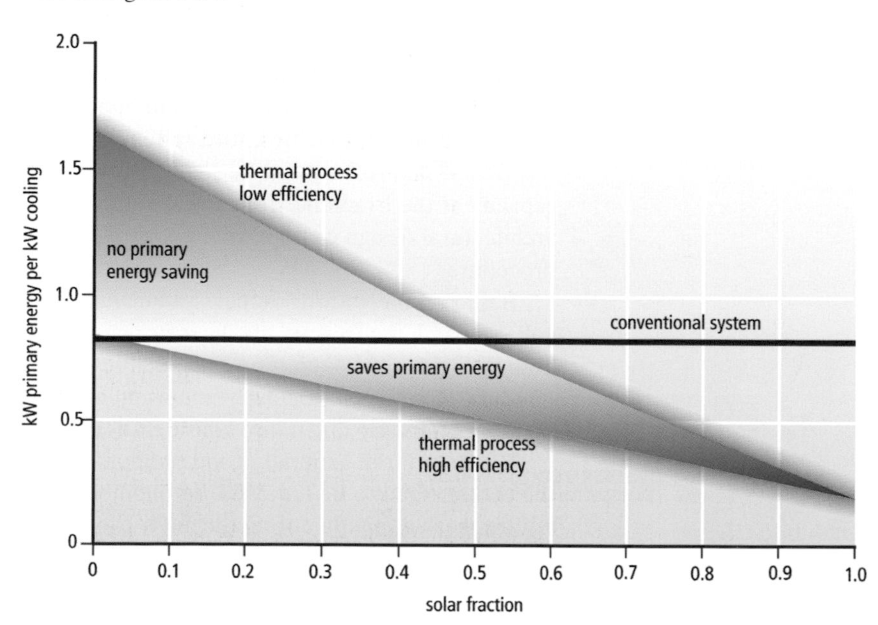

Figure 9.17.
Primary energy saving as a function of the solar fraction. Source: Fraunhofer ISE

The higher the solar fraction used for cooling relative to the overall primary energy use, the more primary energy is saved. The minimum amount that has to be replaced by waste heat or solar energy to be equal to the primary energy is smaller or larger depending on the efficiency of the thermal process. Therefore, in order to be able to save significant primary energy with a solar cooling system, the solar fraction relative to the driving energy when cooling should always be considerably greater than 50%. It is therefore recommended to design solar cooling systems for a solar fraction of at least 70–80%.

EXAMPLE: A SOLAR ABSORPTION SYSTEM IN SOUTHERN FRANCE FOR COOLING OF A WINE CELLAR

The solar cooling system was started up in August 1991 and has been working as planned ever since. The installation (Figure 9.18) consists of:

- 130 m^2 (solar aperture area) of Cortec vacuum tube solar collectors manufactured by Giordano. They were installed on the roof, oriented south-south-west with a 15° tilt angle.
- Technical premises on the second basement level containing:
  - flat plate stainless steel heat exchanger VICARB with a heat exchange area of 17 m^2
  - 1000 l buffer tank
  - liquid absorption cooling machine Yazaki type WFC 15, with a nominal cooling output of 52 kW
  - the pumps needed for the different circuits
  - an electricity meter and control box.

Figure 9.18. Solar absorption system for cooling of wine cellar in France

- An open circuit cooling tower with an output of 180 kW, installed on the northern façade.
- Three central air-conditioning units (one on each floor) equipped with filters, a chilled water storage tank (plus a hot water storage tank for the ground floor), and a 25,000 m^3/h centrifugal ventilator.

9.6 More information

The International Energy Agency Solar Heating and Cooling Programme has a special task on development of solar cooling technology. More information and examples can be found at http://www.iea-shc-task25.org/

10 Simulation programs for solar thermal systems

10.1 Introduction

Before a solar thermal system is built these days, a reputable planning office or installation company will first carry out a simulation of the system. Apart from the presentation of the results to the customer (marketing impact), this process is also increasingly being used to support the planning work.

Potential investors in or operators of solar systems ask for the optimum system solution, the expected solar yield and the level of energy saving. Simulation programs are essential to answer these questions. System planning should involve the optimization of the various system variants and system components on the basis of energy, economics and ecology. Whereas previously this optimization was possible only to a limited effect on the basis of empirical values, it is now made much easier – or even possible – with the aid of commercially available simulation programs. Nevertheless simulation does not replace draft design, estimation of the yield and reliable determination of the data that are necessary for planning. In order to obtain realistic results in the simulation and optimization of large or complex systems it is not only a powerful simulation program that is required but also the technical engineering knowledge of the planner. The results of a simulation are only as good as the realistic selection of the entry values and the simulation method. Many program makers have therefore begun to integrate plausibility controls at the data entry stage. This is helpful when gross dimensioning errors are made. Optimization can take place only if the planner knows which parameters or dimensions can improve the system to be planned, and if the program offers the optimization functions or variant comparisons. However, a critical evaluation of the simulation results is recommended for every planner because, in the end, the simulation program always assumes the optimum artificial conditions, which in reality cannot exist.

The time and cost savings in using simulation programs for dimensioning and planning have led to the increased use of these programs in planning offices. Installation companies are also increasingly using them for presentations to their customers. Some programs are particularly useful here because of the attractive graphics of the user interface. Determination of the yield, economic viability calculations and details of the saving in emissions show the advantages of solar systems, and supply sales arguments. Some programs combine the system layout, system data and results into a report that is ready for printing.

Some simulation programs are suitable for checking the calculations of existing systems if they permit the import of, for example, measured solar radiation and consumption data. These programs are indispensable during the daily handling of solar energy contracts or quotations. Which program is the most suitable for which application can generally be established according to the program classification, but it is finally up to the users to find this out for themselves by thoroughly examining the demonstration versions of the programs, readily found on the Internet.

Traditionally, in the areas of research and development or among the component manufacturers the various simulation programs have become indispensable. Increasingly, solar system component manufacturers are also offering programs that are tailor-made for their products.

10.2 Evaluation of simulation results

Solar thermal systems are used mainly in single-family houses as domestic water heating systems. Faulty forecasts due to simulation calculations are very rare in this area. However, if the systems are more complicated – for example a solar system that supports room heating, or a large solar system with several collector fields and buffer storage tanks – faulty forecasts are more serious.

For solar thermal systems it is essential to determine the planning bases, such as hot water consumption and heat requirements. If different consumptions or heat demands are found in practice, then the operating results will inevitably differ from the simulation results. Selection of the most suitable systems and components is also important, but it does not have such a significant influence on, for example, the solar yield.

The more complex the solar system, the more extensive are the input screens and the amount of input data required in the programs. Inexperienced users should make sure that they are clear about a parameter's significance before entering it. Most programs offer sensible guidelines for this purpose when the screens are opened. In spite of this, incorrect entries can never be fully excluded and this can lead to incorrect simulation results. The help functions also offer support here, although in some programs they could well be improved. A solar system can only be appreciably optimized by means of variation simulations if the user knows what effect any 'tweaking' will have.

The results should always be checked on the basis of empirical values. A solar system with a specific yield of below 250 kWh/m^{2}a is in general incorrectly designed. Even with good results, for example a specific yield of over 600 kWh/m^{2}a, there should be critical consideration.

10.3 Simulation with shading

The number of simulation programs with an integrated shading editor is constantly increasing. Programs that did not have this facility in older versions have integrated an additional calculation option for shaded systems during version updates.

Unlike photovoltaic systems, partial shading of the receiving surface at certain times of the day or year is less serious for the yield in the case of solar thermal systems. Collectors are much more able to convert diffused radiation into useful energy too. In addition, shading presents no risk for the safety of the system operation.

If the shading losses are known, the resulting losses in yield can mostly be compensated for by increasing the area of the collector surface. The degrees of utilization are naturally somewhat lower, but the desired system yield can be achieved at relatively little additional financial cost.

With the aid of comparative calculations the user or customer can be shown the effect of, say, a 20-year-old birch tree on the system yield, and this can help to eliminate the misconception that a solar system cannot be installed without felling the trees in the front garden.

10.4 Market survey, classification and selection of simulation programs

There are multiple applications for solar energy, and so there are also as many simulation programs. No matter whether it is used for domestic water heating, room heating support, swimming pool heating or solar air systems, any solar thermal system can be simulated on a computer. However, most programs are concerned with solar domestic water heating and room heating support. To summarize the market, the simulation programs can be classified according to their programming process into

- calculation programs
- time step programs
- simulation systems
- tools or auxiliary programs.

The simulation approach used in the program determines the accuracy, operating effort, flexibility, scope of application and calculating time, although the latter is now less significant because of the greatly enhanced performance of modern computers. These properties increase as we move from simple calculation programs to dynamic simulation systems: the more flexibly a program can be used, the greater the demands it places on the user.

The final category contains programs with sensible additions, specifications for detailed problems or for the design of the individual components of solar systems. For the selection of a simulation program, apart from the simulation process itself, the application is naturally also important. This brings in the question of the performance and the application options for the program. Therefore the system type or system configuration for which a simulation is required is important.

10.5 Brief description of simulation programs

In the following a selection of the most widespread simulation programs for solar thermal systems is introduced. For many programs a free demonstration version can be requested or downloaded via the Internet. Also, several manufacturers supply simulation software for their products: some of this has worldwide application.

10.5.1 Calculation programs

Calculation programs are simple programs based on static calculation processes. Usually it is only the average monthly values that are included for the individual locations. On the basis of the alignment, the collector type, the size of the collector surface area and the hot water consumption, they determine the yield for the whole system. They deliver quick results, mainly in the area of standard systems for domestic water heating. The behaviour of a system under specific conditions and at smaller time intervals cannot be considered. Also, the various system configurations available on the market can only be reproduced in a very restricted way.

10.5.1.1 F-CHART

The well-known and widespread calculation program f-Chart offers a simple-to-operate user interface for dimensioning a solar thermal system for domestic water heating with a solar storage tank. The entries are made via a screen, which is arranged for meaningful entry values from the start of the simulation. The help function also includes advice on the limits and size of the parameters to be entered. About 100 European weather data records are integrated into the program (monthly averages) together with 98 collector data records.

The accuracy of the f-Chart process with respect to the result of an annual simulation is sufficient even in comparison with time step programs. The degree of solar coverage, and the energy and emission savings, are output as monthly averages over the year in the form of graphics or tables in a three-page report.

Because of its clear layout and user friendliness it is recommended for installation firms, energy consultants and craftsmen, as such people often handle orders for standard systems for domestic water heating in one- and two-family homes.

More information and a demo version can be found at www.fchart.com

10.5.2 Time step analysis programs

Whereas calculation programs only carry out static simulations with monthly values, time step simulation programs permit a more dynamic evaluation in a particular time cycle. Simulation takes place on the basis of weather data and consumption values in an hourly or shorter resolution. Some programs permit instantaneous displays and for example the determination of storage temperature at a given time. The user selects the appropriate system type from preset types and the corresponding collector from the collector library, and enters the parameters for the system location and further system components.

The program user interfaces are arranged to be user friendly. Experienced Windows users will not require a lengthy familiarization period.

10.5.2.1 T*SOL

Since the beginning of 2001 the popular program T*SOL has been available in a fully revised 32-bit version. This fundamental revision of the extremely successful program took account of the increasing demands that are placed on time step analysis programs. T*SOL 4.0 is available in one variant (professional), which is described below. In the near future an expert's version should also be available.

The professional version replaces the previous version 3.2, and essentially retains all the previous useful functions. T*SOL program is popular because it can perform simulations over any time period. The temperatures can be observed during the simulation by means of a coloured display. The results (temperatures, energies, degree of coverage and utilization) can be output at a resolution as short as one hour as tables or graphics. An easy-to-use viability and emission calculation is integrated. A helpful feature has been found to be the simple copying function and the export/import of consumption and results data for accurate analysis in other programs (for example in table calculation programs).

Furthermore additional system variations are possible in version 4.0, which result from the integration of further libraries. Apart from the well-known collector and location libraries, a storage tank and heating boiler library is also integrated for the first time into a simulation program. As well as domestic water and buffer storage tanks, combined storage tanks (tank-in-tank systems) are now also included. Therefore it is now possible to simulate such systems with accurate product data as measured at the manufacturers' premises. Free configuration as in simulation environments is not possible, however, as otherwise the user would have to first compile a newly created system, but it is possible to reproduce individual company-specific systems that differ from the preset program options (see also Figure 10.1).

The standard version of T*SOL comes with the system configurations illustrated below:

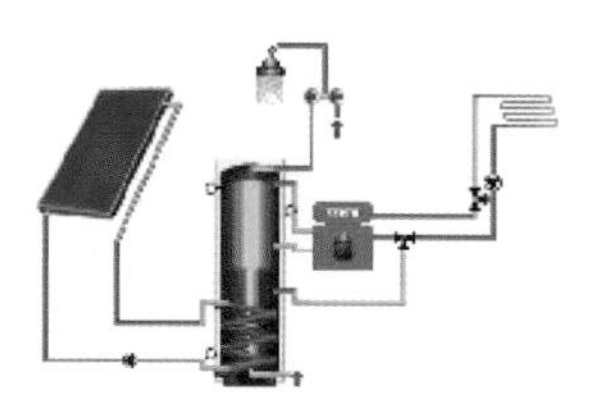

System with a combination tank for the provision of hot water and space heating

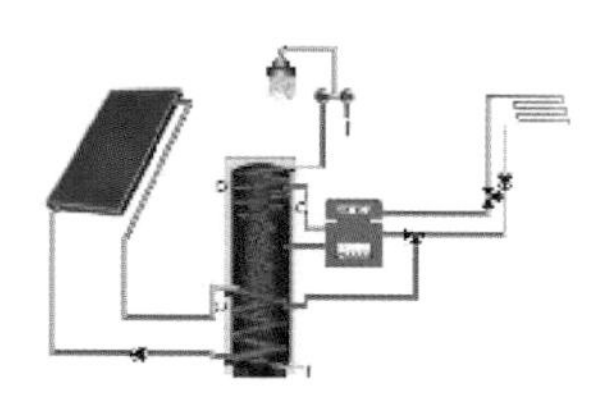

System with a combination tank and internal heat exchanger for the provision of hot water and space heating

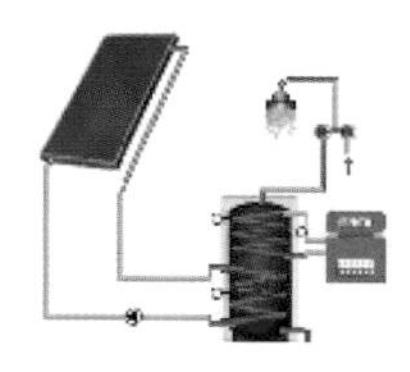

System with a bivalent hot water tank

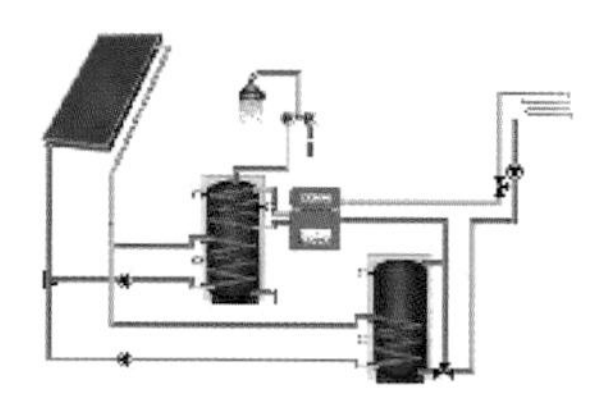

System with a bivalent hot water tank and space heating

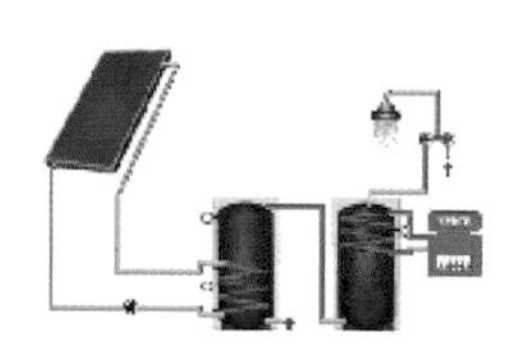

System with two hot water tanks

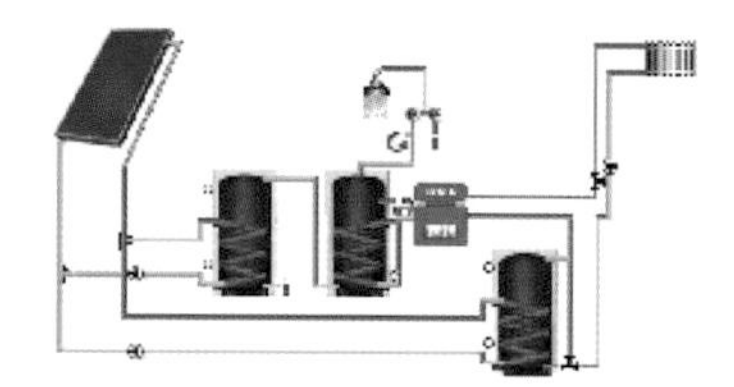

System with two hot water tanks and a space heating tank

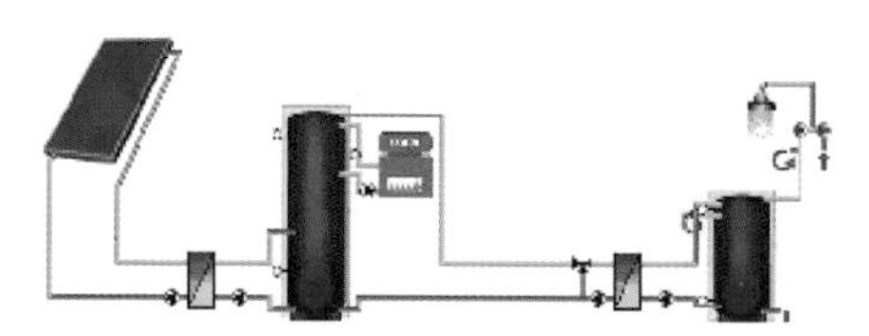

Large-scale system with solar buffer tank

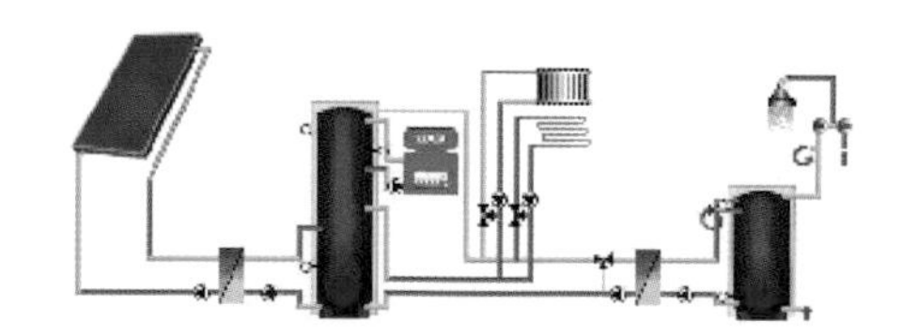

Large-scale system with solar buffer tank and space heating

*Figure 10.1. System configurations supported in the standard versions of T*Sol (source: www.tsol.de).*

The suppliers of such systems can then distribute this as desired (for example ready to download from the Internet) and hence make them available to the planners or installers. The program supplier Dr Valentin GbR has even taken the step of carrying out product support and updates, for example also for collector and storage types, mainly via the Internet: http://www.tsol.de/.

As in the previous version, numerous collectors' and weather data are included in the program. It is also possible to continue to procure additional weather data for outside Europe from the manufacturer or to import it using the METEONORM program (see section 10.5.4.2).

For better calculation of systems that support room heating, a single-zone model for thermal building simulation is also now integrated into the program.

Additional new tools for the professional version of T*SOL 4.0 are a shading editor and a design assistant. This permits draft designs to be produced for different degrees of coverage, and also carries out rapid simulations for three possible variants. The results can be shown in two different project reports. Visualization of the results and the simulation itself have also been improved: for example, it is now possible to observe how the layers of a solar storage tank function during the simulation.

Simulation of swimming pools (both indoor and outdoor) can be carried out with a supplementary module in the same way as for version 3.2. A further supplementary module, SysCat, permits the calculation of large solar systems.

In the expert version there will be multiple variation and simulation options for the scientific user. Measured values for radiation and water consumption can be imported with an hourly resolution, and for experimental simulation all the characteristic data for collectors and storage tanks can be changed.

Further plans are being made to simulate different alignments of collector fields, to dimension the expansion vessel, and to optimize the design of the pump. Through the modular construction of the program these options should be capable of integration into the existing program with no problems.

T*SOL is available in different languages (English, German, Italian, Spanish and French); more information can be found at http://www.tsol.de/englisch/startseite-e.htm.

10.5.2.2 POLYSUN

The Swiss Windows program Polysun is comparable to T*SOL in its scope of performance.

The weather data from more than 300 European locations are integrated into the program with an hourly resolution (Figure 10.2). Eleven different system variants, including one with a combined storage tank, can be selected. Version 3.3, which has been available since December 2000, is able to simulate large tanks better. Here the problem often arises that, at the start of the simulation, an unrealistic situation takes place in the storage tank. If the conditions in the storage tank differ by more than 5% from the starting values after the year's sequence has been completed another annual simulation is automatically appended.

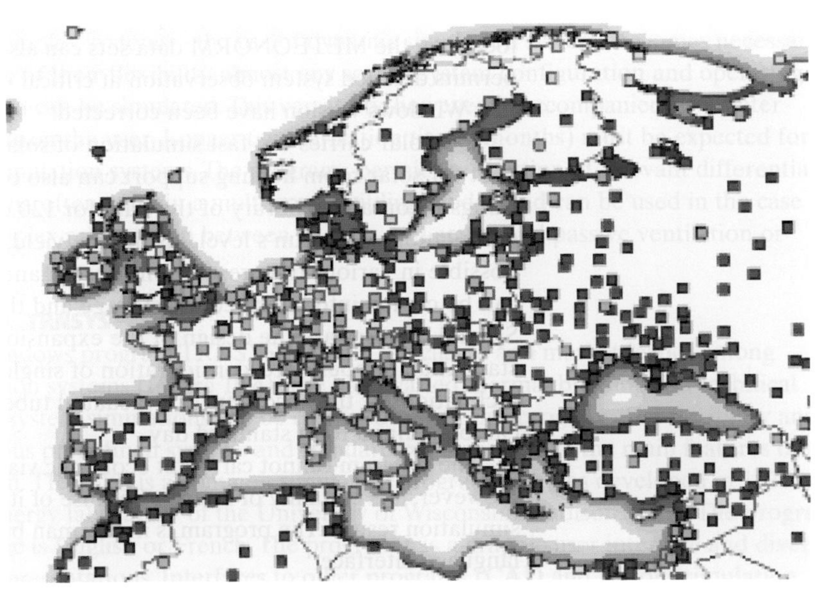

Figure 10.2. Meteorological sites available in Polysun

The integrated collector library from SPF Rapperswil now includes over 170 types of collector, whose measurements took place according to the new EN/ISO standard. Updates can be obtained from the respective sales offices or on the Internet (http://www.spf.ch/). The hot water consumption can be edited in the same way as in T*SOL, but in a somewhat more restricted fashion. Four types of typical daily profile can be selected. Holidays, which are periods in which no hot water is needed, can on the other hand be freely selected. An instantaneous value display of the simulation is

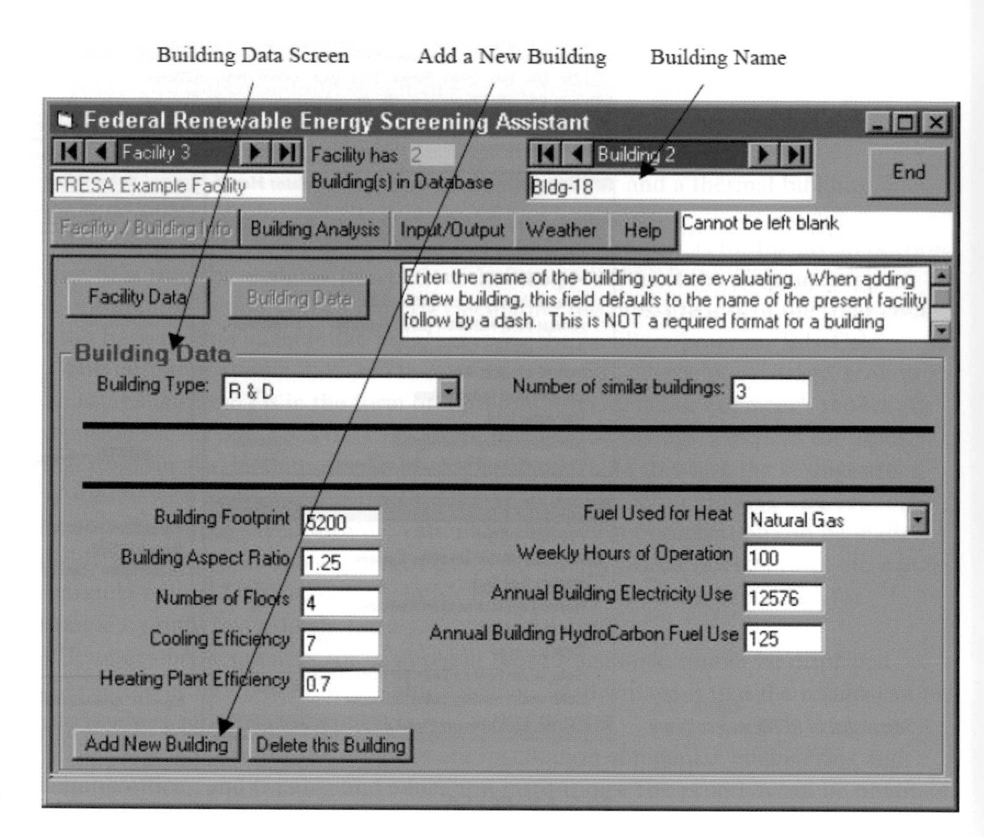

Figure 10.4.
Screenshot of FRESA software

and firms specializing in building and system simulation. TRNSYS is also used extensively in universities and colleges and for research and development. In the area of solar thermal systems the various types of collector are prefabricated and can then be parameterized. Certain solar system examples are also included in the standard package. Thus for example thermosyphon systems, various heat exchangers (counter, parallel flow and cross-flow etc.), concentrating collectors, stratified and stone-filled storage tanks can all be simulated.

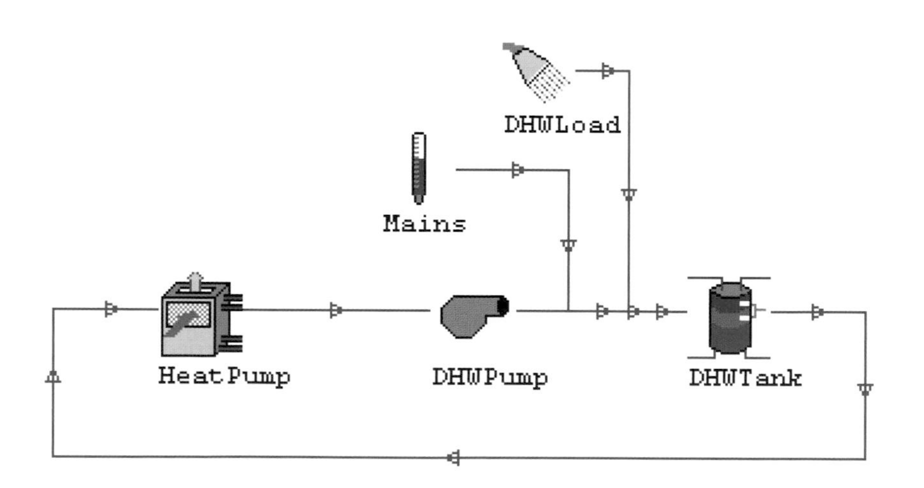

Figure 10.5.
Screenshot of TRNSYS software

10.5.3.2 SMILE

SMILE is a simulation environment developed by the Technical University of Berlin and the GMD (Society for Mathematics and Data Processing). It provides an object-oriented programming language for describing models and computer experiments as well as all the component parts required for all simulations. The applications are mainly in the energy area and extend from solar thermal systems, photovoltaic, heating and air-conditioning technology, hydraulic networks via building simulation to power station technology. The particular strength of SMILE is its ability to combine all these applications by means of integrated examination.

SMILE has proved its capabilities in the university area in various research projects over a number of years. This scientific origin is evident when SMILE is used: thus a graphical user surface that is also appropriate for occasional users is only now being developed). It is certainly an attraction that the program can be procured free for non-commercial use. To date the program has only run on various UNIX platforms such as LINUX.

For special applications, such as building simulations, graphical user surfaces and visualization tools already exist. In a current research project the linkage of CAD architecture software to thermal building simulation has been realized in SMILE in order to be able to carry out energy analyses for geometrically complex buildings within a reasonable time. Also visualization has been integrated in the area of pipe network simulations, such as local area solar heating networks, which permits the simulated pipe network to be closely examined. In the meantime the user has been offered through the support of the SMILE home page by the TU Berlin a reference manual, a tutorial and an expanded component library. Commercial licences can be obtained for SMILE from the firm Dezentral GbR (at www.dezentral.de). Here complete support can be obtained, the program can be optimally set up, and an introduction to the program is possible in user training courses. Dezentral also supplies SMILE solutions for special applications that conceal the complexity of the system from the user by appropriate user interfaces.

10.5.4 Tools and help programs

Numerous small help programs (tools), which as well as the simulation programs also provide supplementary help for daily work in the area of solar energy, are available. A few programs are described in more detail in the following.

10.5.4.1 SUNDI

The Windows program SUNDI is a simple-to-operate program for shade analysis. On the screen it is possible to represent sun-path diagrams throughout the world with the shading as entered and hence determine the times at which the shade occurs. Furthermore the irradiation losses as a result of shade can be calculated and issued. The SUNDI program can be procured free over the Internet. SUNDI is a sensible means of supplementing programs that do not have integrated shade editors, such as f-Chart.

More complex statements can also be made on shading, and finally the results can be easily copied and used in other programs, such as PV simulation programs or thermal building simulations.

10.5.4.2 METEONORM

This well-known Windows program contains a worldwide weather database (with 626 weather stations and 359 towns) and a weather generator. It calculates the hourly values of insolation and interpolates the horizontal global radiation and the temperature of every location in the world. In addition further climatic data such as air humidity, dew-point temperature, air pressure, wind direction and speed can be issued as hourly values. METEONORM contains worldwide sun-path diagrams and can convert the global solar irradiance for inclined surfaces. A shade editor permits horizontal shading to be taken into account. The results can be read into various programs. More info can be found on http://www.meteotest.ch/en/mn_home?w=ber

11 Marketing and promotion

11.1 The fundamentals of solar marketing

11.1.1 Customer orientation: the central theme

What actually *is* marketing? We like this definition best:

> *Marketing is the totality of the measures that make it easy for your customer to choose you and your product.*

We have divided the subject into three parts: the fundamentals, systematic marketing planning, and the sales discussion – the most direct form of contact with the customer.

Solar energy is an attractive energy. It represents the future and a respectful attitude to the natural and the human environment. In selling a collector you are partly selling this idea. Customer benefit should be the main theme of your marketing strategy.

As the reader you are our customer. What therefore is the benefit of this chapter to you? More enjoyment in sales and acquisition – and more success with less effort. The additional success you will have is shown in the following:

WHAT WILL YOU GET OUT OF SELLING SOLAR SYSTEMS?

You will be:

- **Promoting something good (of benefit to the environment and society and of benefit to you as well).**
- **Advising your customers (your customers' benefit is your benefit).**
- **Strengthening customer relationships (paving the way for further sales, increasing the level of familiarity).**
- **Gaining in competence (practice makes perfect, technically and consultatively).**
- **Improving your image (at the very least showing the customer that you have their welfare at heart).**
- **Gaining personal satisfaction (emotional self-interest).**
- **Safeguarding workplaces (safeguarding your own workplace, future opportunities).**
- **Adding value.**

11.1.2 The iceberg principle

Just as you only see one seventh of the volume of an iceberg, similarly only one seventh of our communication is conscious; by far the greater part is invisible (Figure 11.1). However, this part determines the behaviour and the decisions of the person to

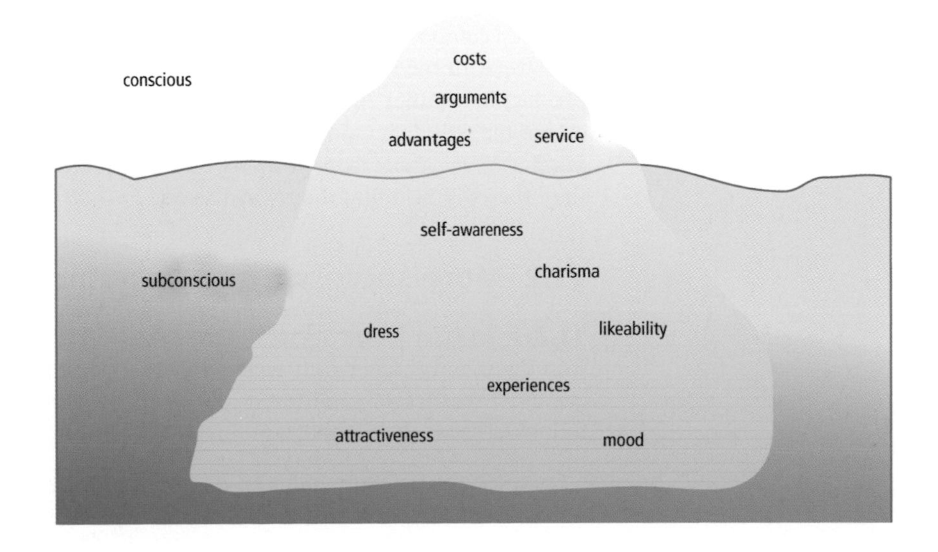

Figure 11.1. Only one seventh of communication is conscious

whom you are talking. This is true even when trust and a long-term customer relationship with mutual benefit is involved.

This means that, to obtain new business, you must deliberately plan using your common sense. For continuing success, however, your attitude, your feelings and the feelings of your customer are important. It is also important how you dress, and how you behave. You are not only selling a product, you are also selling what you 'represent' as a person. Are you credible? Are you good company? Do you have a good rapport with the customers?

The iceberg principle also means that a customer notices very quickly whether you are genuine – for example whether you are really interested in what *they* want. It is therefore worth considering fundamental questions such as: What is the aim of your work? What impression do you want to leave with your customers?

11.1.3 The 'pull' concept

He or she who wishes to sell more, first feels a pressure to be rid of something. According to the iceberg principle this pressure is transferred to the customers. They react in the same way that you react to pressure – with resistance! It is much better if you can see yourself as the answer to your customer's prayers. You read the desires of your customers and can provide all sorts of good products and services for their benefit. In this way you build up a 'pull', which attracts the person you are talking to.

Customers will buy something because they are convinced of its benefit. Attitudes and feelings play an especially important role in the pull concept. In addition to the factual information that you provide, they convey an image of your company and your product. You will acquire business if you can portray your product and company so attractively that your customers want to buy exclusively from you.

The most important differences between push and pull are listed in Table 11.1.

Table 11.1. The difference between push and pull (according to Binder-Kissel[52])

Push	Pull
Use every discussion to 'grab' potential customers Aggressive selling Sale at any price	First build up trust through general discussion, identifying the starting points for cooperation Sale only if the customer really wants something
One discussion = one completion	Honest, serious advice Acceptance of a 'no'
Company image, second priority	Image as significant factor → attractiveness for potential customer
Completion = success factor	Word-of-mouth recommendation A recommendation to somebody else is also a success
→ Critical customers, who were forced into buying High complaints rate Price sensitivity	→ Satisfied customers, who come by free will Fewer complaints Lower price sensitivity

Modern marketing operates according to the pull concept. As the differences between the products of competitors become smaller, so the significance of the relationship between company and customer becomes more important. Marketing therefore does not stop with the sale but sees it as the beginning of a new sale. That may be another product or another customer. What is most important is your behaviour when you create the pull. Is it enjoyable to work with you, to recommend you to others, to purchase something from you? If you make it easy for customers to see the benefits then they will also find their way to you.

11.2 More success through systematic marketing

11.2.1 In the beginning is the benefit

You know and respect your product from a salesman's point of view. However, each customer sees it from a different point of view. It all depends on perspective! The following shows how important this is:

> A novice said to an older monk: 'I have noticed that you smoke during your prayers. The abbot has forbidden me from smoking during prayers.' To this

the older man replied: 'It all depends how you approach the abbot: I asked if I could pray while I was smoking.'

The customer is not interested in the fact that a solar thermal system is 6.6 m^2 in area and has a selective coating with an emission coefficient of 5%. What he or she wants to know is whether it will supply hot water between April and October. Craftsmen and engineers often concentrate on the technical data as the significant features of their product. Customers, however, are mainly interested in what they will get from it – this is often different for men and women, as it is for private individuals and industrial customers.

Take a look at your products and services and gather together the customer benefits (Table 11.2 shows some examples). Note that there are several benefits to each feature. Identify all the possible benefits from the customer's point of view. The following list of potential benefits will help you in your detective work:

- time saving
- solution to a problem
- cost saving
- environmental protection
- security
- prestige
- enjoyment
- comfort
- health
- information
- entertainment
- public relations (for business customers)
- motivation of staff (for business customers).

Table 11.2. From features to benefits

Feature	Advantage	Benefit (for customer)
Tempered glass	Resistance to hail	Long service life, no troubles and no repair costs
In-roof installation of collector	Better thermal insulation	Fuel cost savings Visually attractive through harmonic integration into the roof skin
22 m^2 collector system	Higher thermal yield	Low fuel costs Noticeable use of the sun even in winter Secure supply in emergencies Reduction in cost risk when energy prices rise
800 l storage tank	Large storage capacity	Hot water reserves in times of high consumption Better utilization of high irradiation Longer bridging of poor weather periods
Storage tank with flow heater principle	Hygiene	Fresh, healthy domestic hot water

11.2.2 The four pillars of the marketing concept

Tie your new business development and sales efforts into a marketing concept. Instead of random calls, cultivate the market systematically. Even two days spent on concentrated planning will save you lots of time, money and stress in the future. You will be able to target your efforts and fill the gaps in your sales cycle; turnover will increase and will continue to do so.

Such a marketing concept is supported by four pillars: analysis of your company, your products, the market and your marketing turnover. In this chapter we are concerned mainly with the last two points. However, it is worth examining the other two areas, and some suggestions in the form of questions are given below.

11.2.2.1 THE COMPANY

WHAT ARE YOUR OWN OBJECTIVES IN YOUR WORK? WHAT IS IMPORTANT TO YOU?

In marketing, your aim should be to acquire orders that serve your own objectives. Your concept of yourself is also important here: for example, do you want to sell as cheaply as possible, or is quality the most important aspect? How important to you

Figure 11.2.
The bait must be tasty for the fish, not the fisherman!

are ecological and social concerns? Are you happy with the external image of your company?

WHAT CAN YOU OR YOUR COMPANY DO PARTICULARLY WELL (STRENGTHS)?

Are you a person who likes fiddly jobs and enjoys designing complicated systems? Or can you handle large-scale standard projects particularly well? The more you concentrate on your core competences the more readily you will be recognizable externally. Customers can then classify you more easily: 'the specialist for heating support from the sun'.

WHAT ARE YOUR OR YOUR COMPANY'S WEAKNESSES?

How does your after-sales service system operate? How long does it take to answer a customer enquiry? Are your most important members of staff – those who are customer-facing – properly trained? You should also consider resources with respect to your marketing. If your efforts were to be successful, could you deal with all the orders and still meet customer expectations?

11.2.2.2 THE PRODUCTS

WHAT ARE THE MOST IMPORTANT BENEFITS THAT YOU OFFER YOUR CUSTOMERS (STRENGTHS)?

This includes both your services (quality, planning services) and equipment (collectors, boilers etc.). Are you up to date? Have you got suitable products for your target group?

WHAT WEAKNESSES DOES YOUR PRODUCT HAVE IN COMPARISON WITH THE COMPETITION?

What reasons do interested parties give for deciding against your offer? What complaints do you get from past customers? Perhaps in the medium term you could include an alternative product in your range, or you could balance, for example, a higher price with extra services such as a free check after one year.

FROM WHICH PRODUCT DO YOU EARN THE MOST? WHAT GIVES YOU THE MOST PLEASURE?

These questions can have consequences for your marketing, your calculations, or the general aims of your company.

11.2.2.3 THE MARKET

WHO ARE YOUR COMPETITORS?

In which area do you work and what are your target groups? What is your reputation, your price structure? How big, how aggressive are you? It is a good idea if now and again you can have exchanges with your competitors, perhaps even carry out common campaigns. If you select your target groups carefully you will generally get on well alongside each other.

WHAT IS SPECIAL ABOUT YOUR OFFER WHICH MARKS IT OUT FROM THE COMPETITION (POSITIONING)?

Formulate this in one or two sentences:

> We are specialists for pellet heating with solar involvement. Our firm includes roofers and heating engineers.

You can use this both in printed leaflets and at exhibitions in order to introduce yourself to interested parties. In this way you will be better remembered. Pay particular attention to your positioning. Price, service, quality – all appear equal in one sense. However, customers may be persuaded to opt for your system if, for example, you have an attractive display that graphically shows the solar gain.

WHO ARE YOUR TARGET GROUPS?

Whose problems can you solve best? Who is most likely to be persuaded by the benefits you are offering? (See Figure 11.3.)

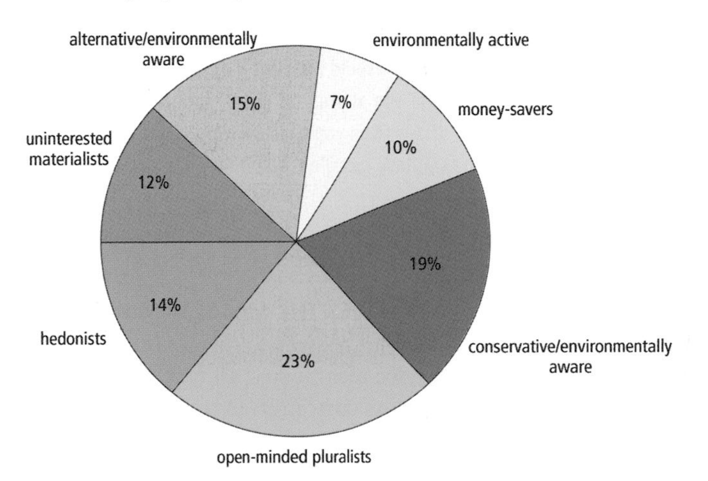

Figure 11.3. Values and consumption patterns of households asked about saving energy Source: Prose & Wortmann (1991)[53]

You can develop a separate acquisition strategy for each group. So, for example, alternative/environmentally aware, environmentally active, and conservative/ environmentally aware can be targeted with ecological benefits – however, each of the three groups has a different attitutude towards life. Correspondingly important are the peripheral elements of your marketing, such as your dress and appearance. This is particularly evident in personal discussions. It is sensible for each person to accurately seek out and talk to his or her target group(s).

MARKETING TIP

Working with schools, you could offer a day of orientation for school leavers. In this way you could ensure new 'solar recruits' and at the same time reach two multiplier groups: pupils and teachers.

With the watering can principle (Figure 11.4) you will never be able to penetrate through the general mass of advertising. It is only with a 'targeted stream' that you can achieve the necessary intensity. And with a targeted stream you can maintain this intensity, as you are only trying to reach a restricted group.

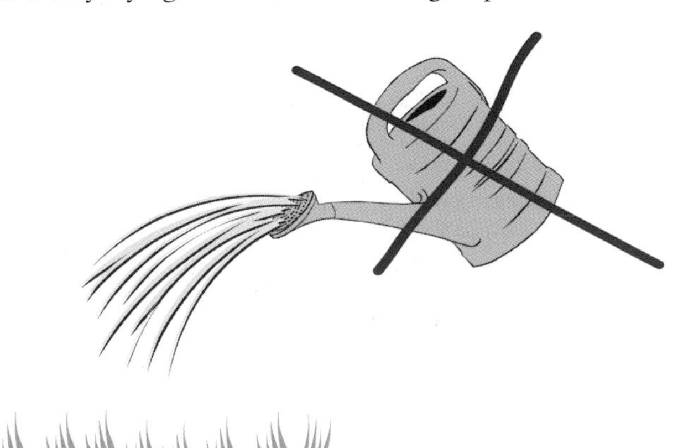

Figure 11.4. Target group derives from target. You won't get rich with the watering can

In your examination of target groups, start with your own experiences. What moves people? Which groups do you feel will be the ones that respond? They may expect that you too should 'take a risk'. Seek opinions from uninvolved people, laypersons, acquaintances; exchange thoughts with your colleagues; survey your customers; read specialist magazines and your local newspaper.

In this way you can keep your ear to the ground and stay aware of what is going on. Perhaps a new building area is being planned, or a solar initiative is being formed in your community.

MARKETING TIP

In a new building development you could make yourself known by direct mail or a mailshot. Those who build do not usually have the money for additional things such as solar systems. You could play along with this and say:

> **We can offer you non-binding advice as to how you can prepare now for the installation of a solar system at a later date at very little cost.**

The threshold for such a customer to make contact with you is thus very low, as there is no need to invest anything at present. He or she will remember you, however, when the need for a solar system eventually arises. And you also offer services other than solar energy; perhaps the customer is unsatisfied with a sanitary installer, and if you are already known...

11.2.2.4 MARKETING TURNOVER

WHO DOES THE MARKETING IN YOUR COMPANY?

The best answer to this is – everybody! From the telephone operators to the customer service technicians, all employees should know how they can contribute to gaining and keeping customers. Are the external sales staff motivated? Is there a training need?

WHAT MARKETING RESOURCES DO YOU HAVE?

These resources extend from printed matter, brochures and sample letters up to standardized procedures. How do I carry out a complaints discussion? How do I behave on the telephone? How is an enquiry dealt with?

WHAT SORT OF MARKETING EXPERIENCES HAVE YOU HAD TO DATE?

Evaluate this in terms of time, money and enjoyment on the one hand and success on the other. Have there been unexpected effects that you can build on and use? What have you not tried yet?

Don't be satisfied with easy answers. If, for example, you were not satisfied with the results from an exhibition, ask yourself: 'Was this due to the exhibition itself or was this my fault?' Are there other exhibitions through which you could reach your target audience better? How can you prepare better? Did you do everything possible to interest the target group?

STRATEGIC CONSIDERATIONS: THE 7C STRATEGY

Before you study the range of marketing tools in more detail, we would like to introduce you to the 7C strategy (C = contacts). This is the basis for the targeted selection of marketing tools. The selection creates a marketing concept out of random marketing.

Figure 11.5 shows the phases involved in winning customers. Without regular reminders, customers will forget who you are – conduct regular campaigns to create interest and refresh their memories! This 'pre-sales' phase requires mental staying power; as soon as you detect some interest it becomes easier.

As a rule of thumb you need seven contacts to gain each new customer. This can extend over years. Do not be put off; the first contact hardly ever leads to success.

In the next section important marketing tools will be introduced. On the one hand the selection of an advertising medium depends on the target group and the message, but it also depends on the advertising budget. Be aware of the difference between fixed costs (for advertising) and variable costs (for example postage costs for mailshots), but also be aware that production costs can vary enormously. For example,

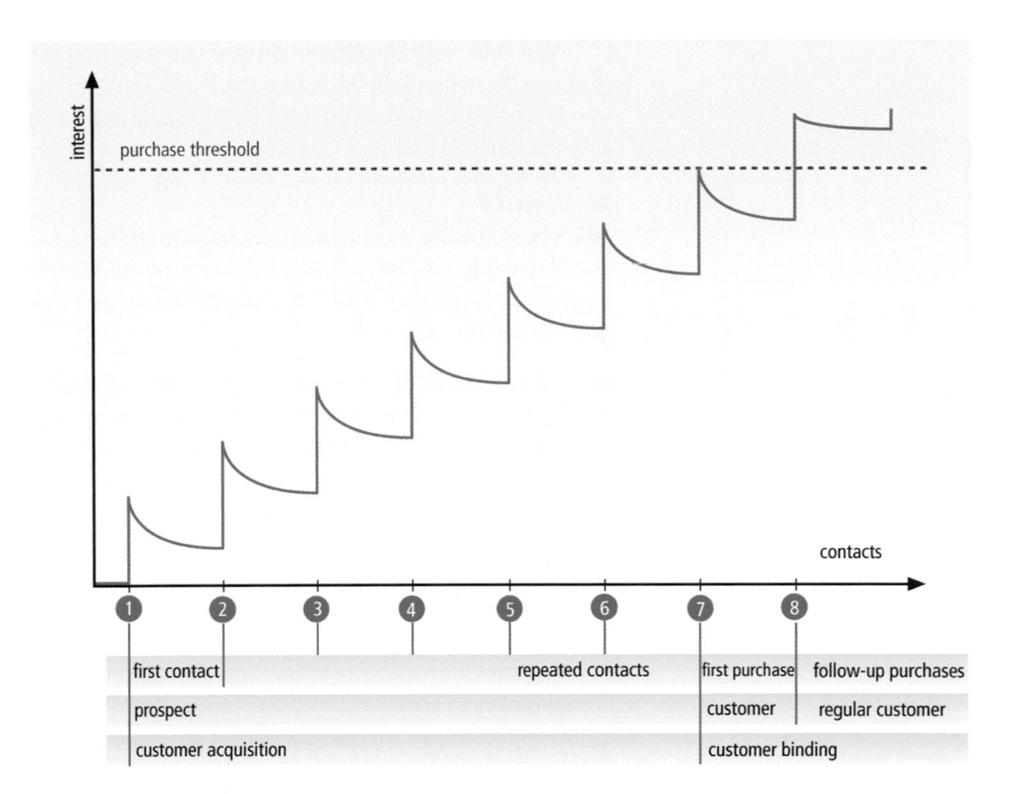

Figure 11.5. The phases of winning customers according to Wittbrodt[54]

the production of printed advertising material is only a fraction of the total production cost for television advertising.

THE COVERAGE OF MARKETING

This list is ordered according to coverage achieved. At the same time, as you move down the list, the mode of address changes from the direct (personal) to the indirect (for example displays), and the absolute costs (not the cost per contact) increase.

- **personal contact**
- **telephone**
- **event**
- **mailshot**
- **newspaper, TV, radio.**

Rule of thumb: The smaller the target group the more direct the approach.

11.2.3 The range of marketing options

11.2.3.1 DIRECT MARKETING

PERSONAL CONTACT

The iceberg principle (section 11.1.2) becomes evident with personal contact. Here you can use your personality to build confidence and create a personal bond. Personal contact is mostly used to convey comprehensive advice to interested parties, and is the preparation for final agreement in a sales discussion. It is dealt with in detail in section 11.4.

TELEPHONE

The telephone is an intensive form of personal contact in which all components of the iceberg play a part. With the telephone you create a personal closeness to the customer. You can hear the 'undertones'. How genuine is the interest? Is it an anxious customer who needs security? Or is the customer well informed, and just in need of clear facts? The telephone is ideal for helping hesitant interested parties over the contact threshold to the next stage.

As a rule, with the telephone you can pick up early expressions of interest which you can then supplement with a mailshot. Telephoning is an art that you and your staff can learn. Section 11.4 provides some suggestions; it is worthwhile pursuing this theme more intensively

MAILSHOTS

Mailshots are the most commonly used marketing tool, which is why they are given more space here. They are a good compromise between the direct and indirect approaches, and stay within reasonable financial boundaries. You can use mailshots for various purposes:

- to create interest – for example to create an initial contact
- to update wide target groups (offers, events, new products), for example within the scope of the 7C strategy (see section 11.2.2.4)
- as regular reminders in order to build customer relationships.

In general mailshots should trigger a reaction and the customer should then make contact with you.

THREE FINDINGS ABOUT MAILSHOTS

1. **It becomes more and more difficult to get the required attention. Even response rates of 1–2% are comparatively good.**
2. **A mailshot without a follow-up telephone call makes little sense.**
3. **From 1. and 2. it follows that you should go to a lot of effort and use all your creative powers when creating and organizing a mailshot. It is better not to do a mailshot than to do a bad one!**

DESIGN OF THE MAILSHOT

Have you heard of **AIDA?** This is what your mailshot should trigger in the customer:

A Attention
I Interest
D Desire
A Action

Attention

The first objective of a mailshot is that it should not be thrown away! This is the hardest part in the composition of a mailshot. Concentrate all your creativity on this. Here are some ideas:

- *Exciting title.* Create a vision in the mind of the reader. You could introduce a product in the form of a headline such as 'Hot Winter 2003', or you could announce an energy-saving week with a 'Squirrel Campaign'.
- *Elegance.* The design will shape the reader's first impression. Take time to arrange your text and any pictures so that they are pleasing to the eye. (See Figure 11.6.)
- *Picture.* If your product is visually interesting, use a good picture to catch the reader's eye and to help them remember it.
- *Gimmick – something playful.* An unusual method of folding, punched envelopes, a gift, a folded paper sundial – these all increase the chance that the letter will not be immediately thrown away; however, the cost also increases. Product and gimmick must always be balanced. The content of your message must stand up to the expectations raised by the gimmick.

Interest

This means:

- understanding
- recognizing the benefits.

Put yourself into the shoes of your target group. Which benefits are they interested in? How can you put this across so that it is easily understood? Brevity is the soul of wit.

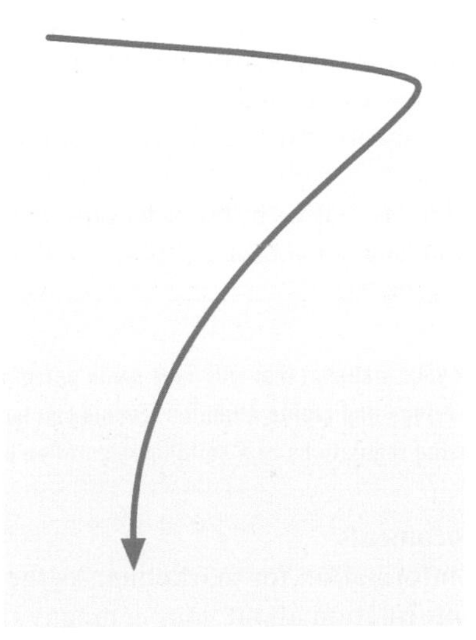

Figure 11.6. Typical eye progression when reading a text. In general it is true that: we look in the same way that we read – from left to right; central elements are perceived with particular intensity; pictures are registered before text. Therefore clarifying text should come after the picture or to the right of it; otherwise the reading process stalls

A mailshot should be no more than one page. And it is not just the complete text that benefits from being brief – keep sentences and words concise and powerful. The optimum amount of text for someone to view is 12 syllables. Text that is important for the readers' understanding of your message should not be any longer.

Desire

In order for recognized benefits to be translated into desire, the customer must believe you. The less pressure you put on the reader the better your chance. You should therefore be restrained in your design. In a time when everything flashes and glitters, and mega and giga bargains are to be had, a more solid, self-aware modesty goes down very well.

Action

Your reader should now do something. Make this as easy as possible for him or her, for example by:

- *Tempo*. Have even shorter sentences at the end than at the beginning.
- *Concrete appeal – what should be done now*. 'Reply to this e-mail'; 'Register by 3 July'.
- *Testimonials*. Let satisfied customers say good things about your product. In this way third parties confirm your statements, which lowers the resistance threshold.
- *Gift*. If the cost is reasonable you could offer a gift: 'Wide-awake people who respond before 10 August will receive a solar alarm clock from us.' An early-booker discount is also a gift.
- *Preprinted reply*. A fax/e-mail/web form is standard these days. Just fill in the sender's details. It is often sufficient just to send back your mailshot. You will know from whom it comes by the personalized address.

CHECKLIST

If you think your draft mailshot is good, give it to someone who is not involved and ask him or her:

- **Does the design appeal to you?**
- **Is the subject of the mailshot clear within five seconds?**
- **Have I formulated it clearly and positively?**
- **Does it have the desired effect on you?**
- **Is the message credible?**
- **How would you react to it?**

Organization of the mailshot

Your well-formulated and well-designed mailshot will be put to best use through exceptional organization. Here are some tips:

- Who will phone to follow up the mailshot and when? You should give the recipient from one week to 10 days.
- How do you ensure that each customer has to deal with only one employee?
- How is customer interest logged?
- What happens if they are interested? The customer should get a reply from you within two days.
- Would you be able to deal with the orders if interest is very high? You may decide to send off the mailshot in stages.

MARKETING TIP

Use a current event for your mailshot that will have some personal meaning for the recipient. This enables you to build a bridge and create attention. Events can be anything from the weather to new laws, for example energy-saving regulations or a building restoration programme.

Brochures, sales documents

You need printed information for marketing. As the basis for this we recommend a leaflet that gives information about your company – a flyer. This should be professionally designed and printed. It is like a particularly attractive business card; it is a memory aid that you can give, for example, to interested parties on an exhibition stand.

In each case it is necessary to decide how much information is required (see Figure 11.7). You don't need an expensive, four-colour brochure for every purpose. What is important is that you have current information to hand for the most frequently asked questions. This can be information that you have put together in house, for example reference systems with a photo. A uniform appearance is worthwhile for those who require more information.

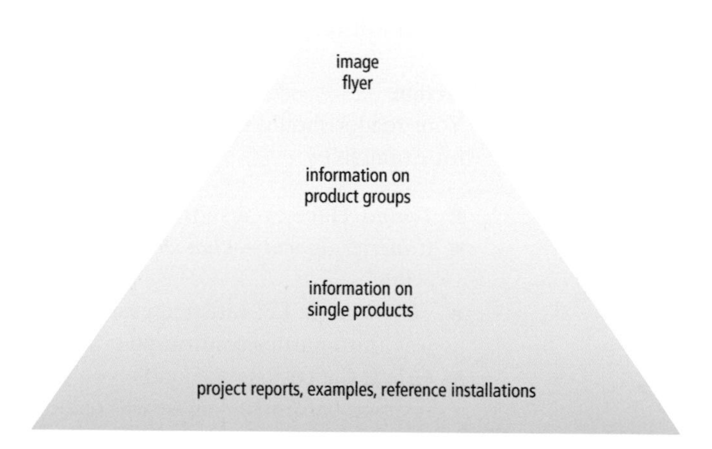

Figure 11.7. The publication pyramid. The further down the pyramid you go, the more simply the information can be configured

11.2.3.2 EVENTS

Events have a particular attraction. You make contact with lots of people over a short period of time. Preparation for and execution of an event is a motivating experience for the whole company. In addition, you may be able to get free coverage in the local press. However, events also take a great deal of preparation and require intensive personnel involvement. During an exhibition it is often only possible to continue with normal business in a limited way. In addition to the costs therefore, you may also have a lower revenue.

Some ideas for events:

- participation in exhibitions and trade fairs
- in-house exhibitions, open days
- participation in local campaigns or events
- product presentations, for example in a hotel
- customer seminars
- presentations.

Events are a tremendous opportunity to obtain addresses of interested parties. Don't take all your informational material with you – instead, offer to send information to anyone interested and in this way you will build up contact addresses. You will also prevent your expensive brochures from becoming yet more waste paper at exhibitions.

MARKETING TIP

Try thinking indirectly: for example, invite a local school class to a solar morning. You will not only enjoy passing on information, but will also gain 20–30 assistant sales staff – no one influences adults as well as children influence their parents!

11.2.3.3 PUBLIC MEDIA

If you want to reach a lot of people you have to use public media. Some examples are:

- the Internet
- press
- cinema
- radio
- television.

Today the Internet is more or less taken for granted. A complete section is devoted to it in this chapter (section 11.2.3.4).

With the press there are many options:

- advertisements
- inserts
- press information
- editorial coverage (a journalist writes about a subject that you have proposed).

Even smaller companies can make good use of the press, at least in the local area. Editors are always interested in new things: if you have installed a solar system on an eye-catching house or if a school class has created and installed a system with your help, then call the local press.

Sometimes a good picture is also sufficient to give you coverage: for example a mobile crane with a large system on its hook, or a system that is particularly architecturally attractive. Ask the journalists what interests them. Give them background information that is of general interest to them. Keep PR and advertising well apart. If you want pure advertising then go straight to the advertising department.

11.2.3.4 THE HOMEPAGE FOR SMALL BUSINESSES*

As a contact and presentation medium the Internet has great significance for marketing. Virtually all large and many small companies have an on-line presence. The following sections show why this is a worthwhile venture, and how resources can be optimally used to set up the Internet presence.

THE AIM OF A COMPANY WEBSITE

According to a corny joke on the Internet, company websites are created either for the boss, the designer or the customers. The aim of a commercial website must always be to gain customers and strengthen customer relationships. Internet sites that are created for pure vanity are therefore not a subject of this contribution. The practical use of a company website is to increase sales, either because the Internet site leads to and accompanies company development, or because it acts directly as a platform for the sale of goods. In the solar area the latter goal is still not easily realized and therefore requires special efforts, which we shall discuss at the end of this section.

CONTENTS OF A SMALL COMPANY WEBSITE

Websites should be of practical use to the customer. Content should be determined by the three W's rule: Who? What? Where? The following questions have to be answered:

- Who is being presented on the website?

---

*Author: Dr. Eduard Heindl, Heindl Internet AG, eduard@heindl.de

- What can I obtain from this company?
- Where do I find the company in the real world?

To enable the customer to easily relate to the Internet site, the complete company name and logo, and the location at which the company operates must be on every page. A search engine will not normally recognise the geographical location in which the company running the website is located. However, customers will be interested in finding a company in a particular area.

Detailed information about the company and the products should be presented on separate pages. A fundamental rule applies to the presentation of information, independently of the respective products; these pages may well be somewhat more extensive than the homepage as the visitor expects precise information on the Internet.

Text should be made more appealing by use of meaningful pictures and graphics. The interested visitor also expects to be able to print out these pages, so they should be configured in such a way that no important information is lost when they are printed.

Company sites should include an accurate description of how to reach the company. This should include a description of the route in text form, and also a map. It must always be easy to print this page.

CONTACT BY E-MAIL AND SPECIAL FORMS

The Internet is the only advertising medium that permits direct contact without a change of media – an opportunity not to be neglected. A contact form must therefore be included on every website, which the customer can easily fill in and then transmit directly to the company by e-mail.

Some simple rules apply to the design of a form. The user should have to fill in as few fields as possible, normally it is sufficient to just have the name and e-mail address or telephone number. It is also sensible to include a free text field, so that the customer can also place a special enquiry. A 'call-back' field is a nice gesture, and tells potential customers that you are prepared to cover the telephone costs. What is essential for the effectiveness of on-line forms is that enquiries are processed quickly. It is worth doing this as the user already has a real interest in the company's services. Users generally expect a reaction on the same day or at the latest on the next working day. To succeed in this the company must clearly establish who reads and who responds to e-mails. Arrangements to cover holidays are essential, as a reply after several weeks is useless. As an alternative, the e-mail address of the company should also be published so that a customer can email the company. This also means that the customer can send photos, for example, at the same time

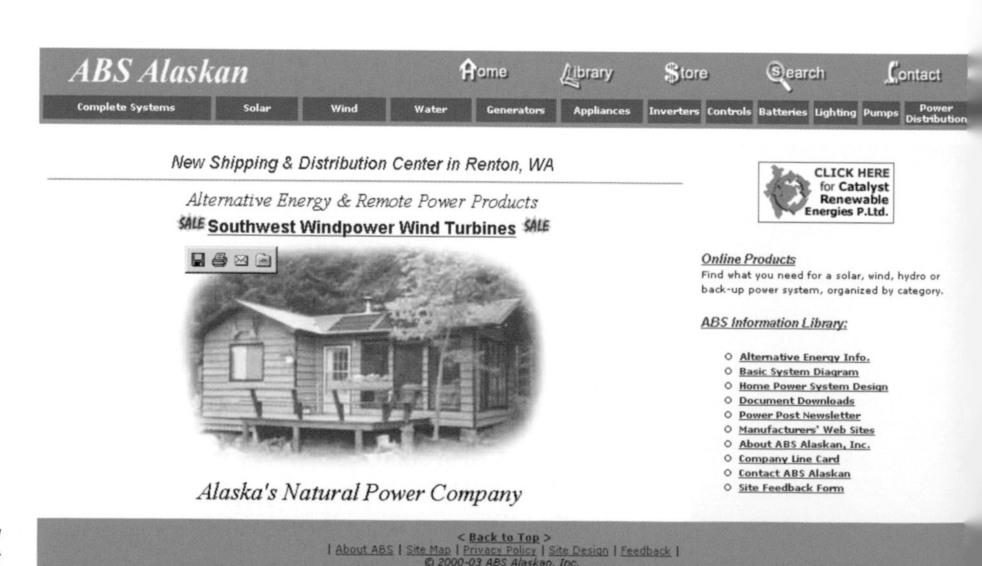

Figure 11.8. The homepage should look inviting and immediately impart the most important information
Source: www.absak.com

MEASURING SUCCESS

Another special aspect of the Internet is the possibility of determining the number of visitors to the website at little expense. Almost all providers offer a web statistics function for this purpose. The important characteristic variables here are the number of pages called up and the number of visitors. Other variables, such as hits and data transfer, are meaningful to computer professionals but are normally otherwise of little interest. A good website should have at least one visitor per day, otherwise the cost of appearing on the web is not justifiable.

The success of a website, however, is measured not only by the number of visitors, but also by the number of enquiries and finally the increase in sales. This variable is very difficult to establish, as an on-line sale rarely takes place; however, the purchasing support often takes place via the Internet.

A simple calculation should make this clear. For an economically justifiable website the extra revenue should be greater than the cost of setting up and running the site. The costs of an Internet website are made up as follows:

$$C = S + L \times O$$

where C is the cost of the website over the amortization period, S is the set-up cost (one-off), O is the operating costs per annum, and L is the length of time on the Internet until the next major revision (amortization period).

The Internet can lead to new customers, who bring additional net revenue and possibly additional profit:

$$P = L \times A \times T \times M$$

where P is the additional profit, A is the additional customers per annum gained via the Internet, T is the average turnover per new customer per year, and M is the profit margin.

For example, the following values could apply to a larger business:

New customers, $A = 5$

Average turnover, T = €5000/$5000/£3500 per annum per new customer

Profit margin, $M = 10\%$

Cost of setting up Internet site, S = €2500/$2500/£1750

Operating costs per annum, O = €800/$800/£560

Amortization period, L = 3 years

In euros:

$$C = S + L \times O = €2500 + 3a \times €800/a = €4900$$

$$P = L \times A \times T \times M = 3a \times 5 \times €5000/a \times 0.1 = €7500$$

In US dollars:

$$C = S + L \times O = \$2500 + 3a \times \$800/a = \$4900$$

$$P = L \times A \times T \times M = 3a \times 5 \times \$5000/a \times 0.1 = \$7500$$

In pounds sterling:

$$C = S + L \times O = £1750 + 3a \times £560/a = £3430$$

$$P = L \times A \times T \times M = 3a \times 5 \times £3500/a \times 0.1 = £5250$$

The additional profit is given by

$$P - C = €2600/\$2600/£1820$$

This means that an Internet appearance that costs about €2500/$2500/£1750 with annual running costs of about €800/$800/£560 is worthwhile in this case, as over three years it generates additional profit of €2600/$2600/£1820.

COST OF IMPLEMENTATION

The costs of an Internet website are made up of four cost factors:

- web design
- web server
- web directory
- Internet access.

Certain key values should therefore be used for the calculation, which naturally depend significantly on the individual wishes.

The web design should always be done by a professional agency, otherwise it cannot be guaranteed that operation of the pages will be technically trouble-free. A professional company can also advise on legal issues concerning the information displayed. Pages that do not function are effectively anti-advertising! A well-designed site with 10 pages and a contact form will cost in the order of €2500/$2500/£1750 (depending on country and level of sophistication).

The web server is not set up in the company but at a provider. A simple site does not require a high-performance server, and thus the costs are around €300/$300/£210 per annum. A website must also be promoted within the Internet: the best way is to enter it in websites with specific themes and high visitor numbers. The Internet agency must also ensure that the site is easily found by search engines. Another €500/$500/£350 must be allowed for this each year. Your own Internet access, necessary in order to be able to answer e-mails, does not have to be calculated separately, as it will be mostly used for other purposes, and the cost is negligible in comparison to other items.

SERVICING COST

Those who take Internet marketing seriously and wish to gain more than a handful of new customers each year must also service their Internet site. Servicing mainly consists of updating the contents without visual changes.

Updating means 'moderate but regular changes'. It cannot be said often enough that it is only continuous servicing of the content that leads to noticeable advertising success. No customer expects daily news from a workshop, and certainly not 'current' offers for Christmas or Easter. In most cases three to four hours per month are sufficient to add information such as completed projects to the reference list, to announce appointments, and to delete old information. This work can be done with inexpensive editing systems, which permit web pages to be changed without knowledge of HTML and without endangering the layout.

ACTIVE ADVERTISING ON THE INTERNET

Advertising on the Internet differs from classical print media because the potential customer first has to be made aware of your homepage. There are several ways of doing this, the best of which is offered by the Internet itself and which is described here.

Search engines

Anyone who has used the Internet to solve a problem goes first to a search engine such as Google and enters a keyword or keywords. Out of the first 10 results a suitable site is visited, and if the site is good this can already be the start of a business relationship. It is therefore very important that your company appears high up the list. Internet sites should be optimized for search engines. For this to be successful the businessman must know by which search words he wishes to be found. There should also always be a regional reference, in which the place name appears on the Internet site. The site should then be submitted once to the search engines; repeated submission is unnecessary.

Portals

In the traditional media market, specialist magazines provide a service by attracting and retaining a target public. On the Internet this is done by subject-orientated portals. These have very high visitor numbers, which an individual company could rarely reach. It therefore makes sense for a company to draw attention to their own homepage within a portal. This can be done by placing an entry in a classified directory or by means of banner advertising and links.

Classified directory

There are now a huge number of classified directories ('yellow pages') on the Internet (Google found 114,000), but only very few in which it is worthwhile being entered. A good reference number demonstrating the significance of a classified directory is the number of visitors to the directory per annum.

Banner advertising

The most established form of advertising on the Internet is banner advertising. With banners it is possible to attract the attention of a wide public to a company and to entice new visitors to your own website. A banner is a small picture integrated into another website (see Figure 11.9). This picture is linked to your own site by a hyperlink. The visual configuration of the banner should be restricted to what is most essential. A 468 × 60 pixel area offers little space for text. For a banner to work it is therefore sensible to commission an agency for the graphical work. Banner advertising has the greatest effect when the target public is searching for related products and services. Therefore solar companies should place their banner on websites to do with building, heating, solar energy and environment, as potential customers can be gained there.

The price for banner advertising is usually calculated according to the number of hits and the banner size. According to the situation, between €10 and €50 ($10–50, £7–35) is charged per 1000 PV (page views).

Figure 11.9.
Banner advertising should be placed where the potential customer surfs.
Source: www.mysolar.com

Links

A website should always be well linked within the Internet. Apart from the options of the classified directory and banner advertising already described, you should try to be mentioned on suppliers' pages with a link. Satisfied customers may also be prepared to have a link from their homepage to yours. In return, you should also be ready to include a link back from your own pages to the respective company.

DIRECT SALES THROUGH THE INTERNET

The Internet is not just a large shop window; it is also developing more and more into a marketplace. However, changes in purchasing behaviour are much slower than the acceptance of the new medium. The opportunities that are now on offer will be introduced here.

Suitable products

Not all products are equally suitable for selling over the Internet. On the one hand the product must be in a price class that represents a justifiable risk in dispatch and payment. On the other hand the product should be capable of being used without further support from the vendor. This is not the case for solar thermal and photovoltaic systems today; however, many other products in the photovoltaic area are well suited to the mail order business. In a sales analysis by Solarserverstores (www.solarserver.org) an optimum price of about €100/$100/£70 was established. Items that sell particularly well on the Internet are charging units and solar lamps, as well as solar-operated household articles and solar watches.

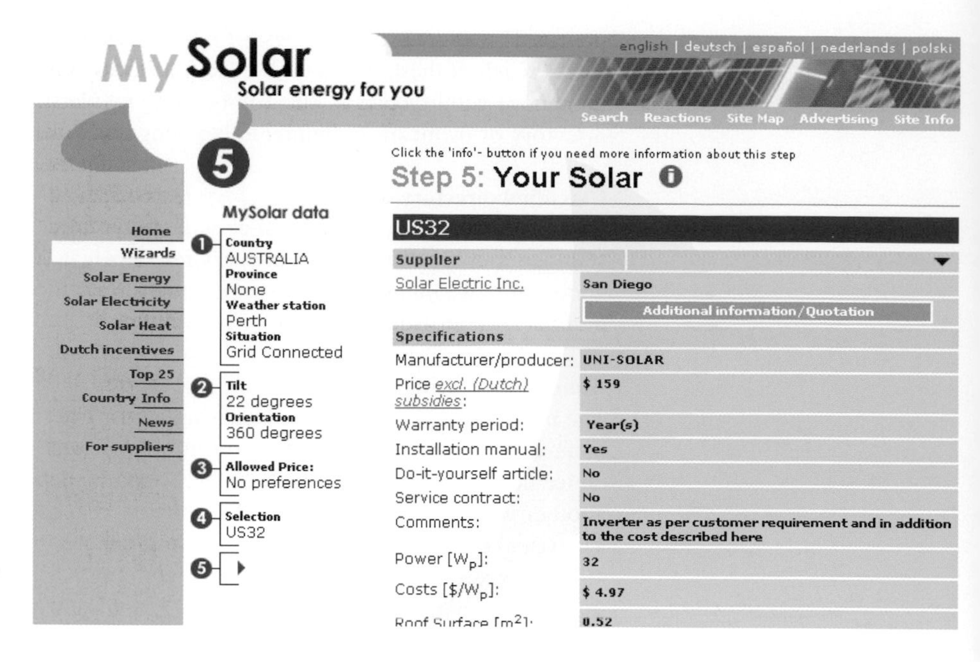

Figure 11.10. Direct sales on the Internet can be completed with shop systems. The product prices must be accurately calculated. Source: www.mysolar.com

Technical costs

An on-line shop requires very high investment, both for the software and the product entries as well as for the day-to-day operation. Today standardized program packages, the installation and operation of which require considerable experience, usually serve as shop software. This software must run on a powerful and very reliable computer, otherwise a high number of hits may cause delays or frequent server failures. It is therefore recommended that an experienced service provider, who can demonstrate comparable projects as reference, should carry out the implementation. The cost for technology and software in the first three years is at least €10,000/ $10,000/£7,000. On top of this there is a cost of about €50/$50/£35 for each product entered in the catalogue.

11.2.3.5 CUSTOMER RELATIONSHIPS

So far we have only discussed the winning of new customers. But what do you do with the customers that you have already gained? For most companies a large treasure chest remains buried in this area. It is seven times easier to reactivate an old customer than to gain a new one. Therefore customer focus as a theme is not only more enjoyable but is also the more successful approach.

Consistent customer focus in all customer relationships (Figure 11.11) is a culture in which the management must set an example and which they must also implement in all areas of the company. The installer should know exactly what was promised to the customer when the contract was awarded. Apart from neat craftsmanship he must also feel that he is responsible for cleanliness when working in the customer's house. He should also be aware how to professionally handle a special wish or a complaint by the customer. Some can do this naturally, but as a rule training is required.

MARKETING TIPS

- **Carry out a mailshot campaign to your customers. Use a questionnaire to find out about customer satisfaction and offer a small gift for returning it.**
- **Variant: Ask for details of acquaintances who may be interested in solar systems and offer a gift for every address.**
- **Alternative: Send a questionnaire with the invoice, which can be returned anonymously.**

Satisfied customers help you to sell. Ask whether they would be willing to give a reference. If they agree, show your appreciation. The opportunity to ask existing customers about their experiences is worth much more to a potential customer than anything you can promise. The offer alone strengthens credibility. At the same time you are displaying customer care: the existing customer feels honoured and will probably also enjoy passing on his or her knowledge.

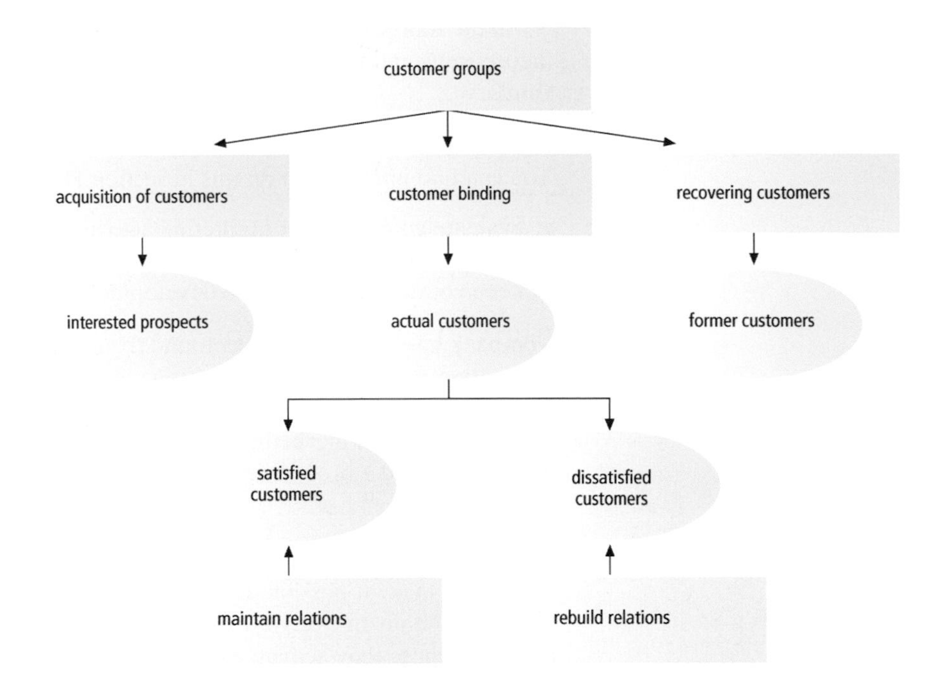

Figure 11.11. Customer relationships (according to Binder-Kissel[52])

11.2.4 To the goal in six steps

11.2.4.1 THE BASIC IDEA

To acquire customers for solar energy you need great staying power. So that you don't exhaust yourself, we recommend a systematic procedure in six steps: the *marketing cycle* (see Figure 11.12). For large campaigns it is best to work through this cycle twice. Start with a test run and just a few addressees. With the experience gained from this you can then improve the main campaign.

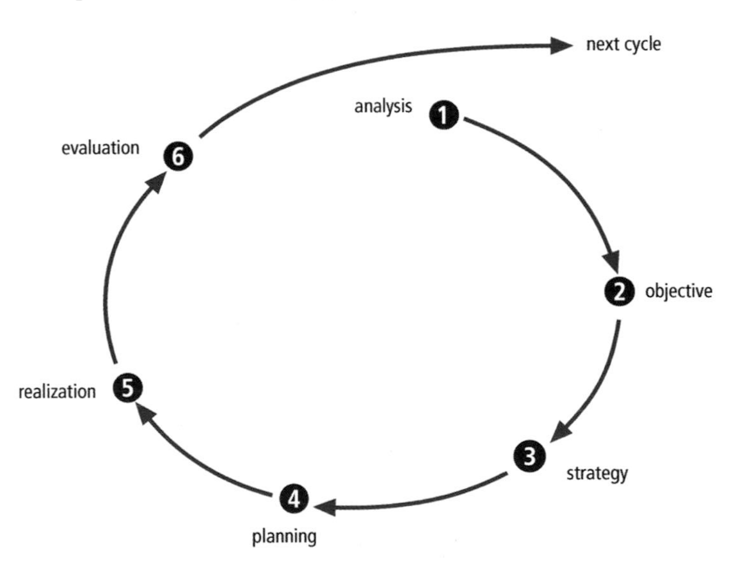

Figure 11.12. Marketing cycle in six steps: analysis; objective; strategy; planning; realization; evaluation

You may think you don't have time for all this. However, you need to find the time! Systematic marketing will ultimately take less time than random calls which don't produce the desired result.

For example, block out half a day each month for planning your marketing efforts. In addition, at a quiet time – such as between Christmas and New Year – take two days to consider the marketing concept for the next year. Of course, you will also need time for implementation, whether that is attendance at a trade fair or a mailshot. If you have planned well you can distribute the work for these campaigns over several shoulders. When everyone works on a joint campaign for a few days it strengthens the sense of belonging to the company. And the success of a well-planned campaign spurs everyone on!

In the following the six steps of the marketing cycle are discussed both theoretically and in an example – a marketing concept by the company Heating and Sanitary.

11.2.4.2 STEP 1: ANALYSIS

You can find information on this in section 11.2.2. In addition you should:

- evaluate your previous marketing activities in terms of cost and benefit
- evaluate your business situation
- keep your eye on the future development of the company.

Company boss Mr Heating has found from discussions with competitors and association officials that he is profiting less from the solar boom than his competitors. His wife and sons, as well as a glance at the balance sheet, show him that he has done little in the area of marketing. His customers previously came to him if they wanted a new bath or had a broken heater. Mr Heating Senior recognizes that this is no longer sufficient.

11.2.4.3 STEP 2: OBJECTIVES

Objectives make it possible to set and specifically select priorities. They make it easier for you to retain an overview, even in hectic situations. Objectives free up additional energy, because they activate your subconscious and support you in concentrating on the essentials.

So that this isn't merely a declaration of intent, you must really work through your objectives. Write down the results, and make them 'SMART':

- Specific: what exactly?
- Measurable: how do you check your success?
- Achievable: can it be done?
- Relevant: is the result important to the business?
- Time plan: by when do you want to have achieved it?

What do you think of the following objective: 'Next year I want to have more time for the family'? Exactly – too vague. It is better to say: 'From 1 March I am going home at 5 p.m. on Tuesdays and Thursdays.'

Among other things Mr. Heating Senior plans to do the following in the first days of the new year:

- By 9 January I shall prepare a marketing strategy for the current year.
- This year I shall actively gain 20 new customers for solar systems.

11.2.4.4 STEP 3: STRATEGY

Prepare a 7C strategy for your target group(s) (see section 11.2.2.4) and translate this into a rough time plan. In this you should mark the marketing activities month by month. Enter fixed dates such as trade fairs and holidays.

A few tips for this:

- As a general rule plan four to seven contacts per customer each year.
- Don't overload the contacts. Include some information but not a whole package.
- Change the medium. For example, use a Christmas greeting.
- The form of contact is not as relevant as its permanence.

Consider when the more favourable and less favourable times are. You must therefore consider the following dates in particular:

- mailshots: holiday dates, Christmas/New Year period
- telephone marketing: holiday dates, office hours
- daily rhythm: lunch breaks, meetings, Friday afternoon
- events: public holidays, trade fairs, parallel events, weather.

Reserve half a day in every month for updating your 7C strategy.

11.2.4.5 STEP 4: PLANNING

To obtain new business you are concerned mainly with operational planning. As a rule you prepare a marketing plan for one to two years. Do not give up if you don't have immediate success. Obtaining new customers requires persistence and long-term thinking.

The 7C strategy provides the rough timescale. For detailed planning you should set up a schedule of tasks for the individual campaigns (such as a mailshot). In this you must establish:

- Who? (responsibilities)
- What? (definition of tasks)
- By when? (time plan)

Plan the necessary initial and follow-up work, such as:

- Designing the advertising material, for example a circular.
- Obtaining the necessary details about the advertising medium, for example magazine deadlines.
- Gathering addresses (for example, from an address search in your own database).
- Allocate time for your own involvement. You control the campaigns. The success depends upon your availability. Do it in small morsels where possible. You don't have time to call 50 people on one day, but if you make two calls every day you can get through them easily in a month.
- Check availability of the necessary staff. Block out well in advance the time when you will need all the staff involved in the new business to get together. If you don't, no one will have time for it; it will also create a strong feeling of togetherness.
- Schedule follow-up work for those who are most interested. Good organization is worthwhile. You must know who gets what, who your A customers are, and when the next campaign is due. You must also know when to reduce your efforts, or when to reallocate addresses from category 'A' (very promising) to 'C' (less promising).

Mr Heating Senior has drafted the schedule of tasks shown in Figure 11.13 for the initial action in his 7C strategy (producing a flyer).

Who	What	When
Team	Brainstorm on content	10-1-04
Smith	Select marketing agency to take over text, graphics and production	10-1-04
Smith	Establish content and discuss with marketing agency ('briefing')	17-1-04
Jones	Clearance on rights of figures, deliver figures together with logo and other material to marketing agency	24-1-04
Marketing agency	Draft layout and text for proofreading	24-1-04
Smith, Marketing agency	First correction round	7-2-04
Smith, Marketing agency	Second correction round	14-2-04
Jones	Collect addresses for distribution of the flyer	19-2-04
Smith	Determine number of copies and give clearance to print	19-2-04
Smith	Check printed version	29-2-04
Jones	Distribute flyer to external addresses and internally to all colleagues; Secure documentation copies and spare copies; Reserve number of copies for trade fair and other events; Track distribution and signal timely (3 months in advance) when reprint is necessary	3-3-04
Smith	Collect suggestions for improvement in next version	From 29-2-04

Figure 11.13. Planning for the flyer for Heating and Sanitary

11.2.4.6 STEP 5: REALIZATION

There is not much to say here, except – do it!

The greatest danger of marketing campaigns is that, just when they start, something else much more urgent arises. Stay with your 7C concept! Marketing is important. You as the boss must provide the urgency that it requires.

Mr Heating Senior takes on board this recommendation. He has agreed all the dates well in advance with his staff, and has assigned one person for emergency customer help during the campaign. In this way the remainder of the team can concentrate fully on marketing.

Addresses are of central importance for marketing realization. How do you obtain them?

OBTAINING ADDRESSES

A common source is the telephone or address book. Member or company directories are often helpful. You can very easily collect the addresses of interested parties yourself on the exhibition stand or at an information centre in the town. Make it a rule to obtain the requirements and address of every interested party. Don't overdo the amount of advertising material you take with you; send the selected information to those interested – for this you need the addresses.

You can use the Internet to obtain addresses in two ways:

- *Homepage*. A concise, informative homepage is sufficient; it doesn't have to be graphically sophisticated. What is more important is that the homepage is registered with the search engines; always check that you can be found. On your homepage offer interested parties the option of requesting material.
- *Portals*. A register of the most important portals on your subject. A basic entry is often free, and it is often worthwhile to pay for a link to your own homepage. In the solar area: register with www.top50-solar.de, www.top50solar.nl, www.top50solar-esp.com, www.ises.org, the correct sublistings in www.yahoo.com and similar site listings. In this way you can also see how often you are accessed in comparison with the competition (see also section 11.2.3.4).

There are many service companies that have specialized in the sale of customer addresses. Many address suppliers have concentrated their developments on regions or fields. It is often worthwhile comparing suppliers. Addresses can be searched for according to various criteria.

ADDRESS MANAGEMENT

A well-managed customer database is a valuable resource for a company. Give it the necessary attention. Table 11.3 shows a good general structure of an address sheet.

Table 11.3. The three areas of an address sheet according to Binder-Kissel[52]

Address data	Name of the company and legal form Address, telephone, fax, e-mail, customer no., field, company size and turnover if possible
Profile data	Name, first name, function, position, extension and mobile telephone no., possible other information: date of birth, hobbies etc.
Action data	Date of initial contact, contact history (meeting minutes), action history, products ordered, ordering media, order value, any reminders and open positions

The following procedures have proved to be worthwhile:

- The addresses should be able to be viewed and changed by all participants. A notebook computer is recommended for external sales staff so that reports can immediately be made after a discussion.
- Notebook and server database must be automatically compared.
- Include a contact history in which you record important discussions.
- Keep a history of actions taken, in which you note all marketing activities. What does the customer already possess in the way of documents? Which invitations or greeting cards has he received?
- Don't just record all the contacts but also all additional information that has been obtained during the acquisition of new business, for example customers' preferences.

- Using 'search terms' mark which product your customer is particularly interested in. In this way it is easier to put together the address lists for mailshots etc.
- Assign your customers into two or three categories. 'A' customers are the ones you must concentrate on; for 'C' customers the interest in your products is weaker.

11.2.4.7 STEP 6: EVALUATION

Immediately after a campaign evaluate the cost and benefit, for example in your monthly marketing planning. At the end of a year it is worthwhile carrying out an annual review. It is only after fundamental evaluation that you can really enjoy the fruits of your work. In this way you can plan the next phase more effectively and efficiently.

At the company of Heating and Sanitary, marketing has been well worthwhile. After one year they have gained 23 new customers with an order volume of around €320,000/$320,000/£224,000. The cost was significant: €26,000/$26,000/£18,200, plus 40 person-days from the staff. However, much of the work that was done can still be used during the following year, and the customers were so satisfied that the first contract was surely not the last.

> *'Success is going from failure to failure without losing enthusiasm.'*
> (Sir Winston Churchill).

11.3 The Soltherm Europe Initiative: European network of solar thermal marketing campaigns

In Europe, the Soltherm Europe Initiative is a platform to bring together marketers and promoters in the field of solar thermal energy. The Soltherm Europe Initiative aims to create cooperation between sales and installation companies and information centres and information campaigns.

The Soltherm Europe Initiative has started its activities in the form of an EU Altener project that was initiated in 2002. The following key parties are included in this project:

- Europe: European Solar Thermal Industry Federation, Climate Alliance
- Austria: Arbeitsgemeinschaft Erneuerbare Energie, Ober-™sterreichische Energiesparverein, Grazer Energie Agentur
- Belgium: 3E
- Denmark: Danish Technology Institute, Ellehauge
- France: Ademe, Rhone-Alpes Energie Environnement
- Germany: Bundesverband Solarindustrie, Berliner Energie Agentur, Door2Energy
- Greece: Centre for Renewable Energy Studies
- Italy: Ambiente Italia
- Netherlands: Ecofys, Novem
- Spain: ICAEN, Trama
- UK: Building Research Establishment.

The project provides a suitable start for the Initiative, but Soltherm Europe has the vision to become a long-lasting and strong action network, supporting market campaign organizers throughout Europe in the coming years.

11.3.1 Publications

At present the structure of the platform is being built up to inform and support new campaigns. The first publications for market developers are already available:

- A European market report together with in-depth market analyses for the 10 countries and three regions represented in the project. What market stimulation actions worked in a specific case, why, why not, and under what conditions they would work again.
- Soltherm campaign guidelines that are based on the most striking campaigns throughout Europe and the combined experience within the Soltherm project team. The campaign guidelines contain a number of elaborated example campaign formats,

plus expert information on how and when these should best be used to design new campaigns. Success factors and common bottlenecks are analysed. The experience is condensed into a campaign checklist for organizers of market stimulation initiatives.

- A central Internet site, www.soltherm.org, serves as the central information vehicle, with entries for consumers, industry, installers, campaign organizers and various groups of consumers. The above reports as guidelines are available for download here.
- On the Internet site an expert campaign tools library is available, providing hundreds of tools and information sources, and thus serving as a library for the organizers of campaigns. The material includes campaign handbooks, promotion materials, model contracts, performance calculation tools, picture and photo collections, and much more. Each tool is presented with an expert comment sheet (in English and in the language of the tool) describing who should use the tool, when and how, and where and how to obtain it. The library has search and selection functions in a number of languages so as to be useful for local campaign organizers, installers and other parties.
- The Internet site is planned also to feature products and installers' databases, where both campaign organizers and consumers can find the available products and installers, together with basic information and objective quality indicators (such as local certificates, and compliance with the European Standards and the European quality certificate the Solar Keymark).
- On www.soltherm.org an on-line overview of quality standards and certificates for products is available. These provide campaign organizers with clear criteria for distinguishing quality products and installers.

11.3.2 Partnership

The Soltherm Europe Initiative has over 50 partners including government departments, ISES Europe, the IEA, industries, installers and energy agencies, and it is still growing steadily.

News and project opportunities are exchanged within this partnership; it also provides a network that enables Soltherm partners to access each others' experiences and expertise more easily.

11.3.3 Campaigns and market activities

The exchange of experience and the Soltherm Europe partnership are also used to develop new initiatives and enhance existing campaigns. Some examples of such campaigns (at the end of 2003) are as follows:

- In Austria, a Soltherm working group is being set up: two workshops were held, presentations were given at workshops/conferences, and newsletters have been disseminated.
- In Belgium, the Soltherm Walloon campaign is continuing, and a communication campaign is in preparation.
- In Denmark, a city campaign was started and a national workshop held. Another workshop is planned.
- In France, the Plan Soleil is continuing with publicity on TV, in national newspapers, brochures, a call centre and other measures. In the Rhône-Alpes region a campaign 'Coup de Soleil' is in preparation, a Soltherm workgroup is discussing an ambitious action plan, and newsletters have been distributed.
- In Germany, a campaign oriented at the housing associations is in progress: three workshops, an Internet platform (www.soltherm.info), a conference, other presentations, a brochure and consultancy.
- In Greece, a national workshop was held discussing barriers to further market penetration and how to overcome these.
- The Italian 'Paese del Sole' initiative is supporting local promoters of solar thermal energy, with the first seven municipalities across the country starting this summer. In March 2003 a workshop was held in Verona.
- In the Netherlands, three solar water heater campaigns were started after a mailing to municipalities. A corporation is offering solar energy systems to its tenants, and has over 250 customers so far. A national workshop is in preparation.
- In the UK, a campaign was held in the East-Cambridgeshire region.

11.3.4 Co-operation

The Soltherm Europe Initiative is also integrating with existing initiatives to obtain maximal synergy. For example:

- integration with and promotion of the ESIF/ASTIG Altener Solar Keymark project: linking this EU quality label to information campaigns will speed up the acceptance of EU quality standards
- cooperation with ISES Europe to combine the knowledge collected in the WIRE facility with the experience of the project parties.
- integration of the Altener Qualisol project, aiming at an improved and broadly disseminated solar installer qualification.
- cooperation with the German BSi, for example on the follow-up of the Solarwaerme Plus campaign.
- cooperation with the French Plan Soleil, managed by Ademe (see above).
- cooperation with the government of the Wallonian parts of Belgium in the framework of their Soltherm programme, in which many elements of the Soltherm Europe Initiative are already incorporated.
- integration of the Altener SOLHAS project aiming at developing a market development approach for housing associations in Europe.
- integration of the Dutch Space for Solar Initiative, a foundation of about 60 housing associations and other organizations that collectively purchase and install collective water heating systems.
- cooperation with the Mediterra Solar Campaign, which is engaging on an action plan to stimulate the market for domestic solar thermal applications in France, Spain and Italy.

The Soltherm Initiative is geared to become a major support structure for European market actions and campaigns, and all organizations that want to participate are invited to join the initiative. More information can be found on www.soltherm.org.

11.4 A good sales discussion can be enjoyable

11.4.1 What does 'successful selling' mean?

A good sale does not mean 'completion at any price'. The focus of attention is customer satisfaction. If you advise them well your customers will recommend you, even if they themselves do not purchase anything else. Customer satisfaction is not everything, however; you also need to feel satisfied, and for example should not invest too much time unnecessarily. Successful selling therefore means striking a good balance between the cost and benefit of the sale.

Purchasing – and therefore also selling – is a matter of trust. This involves mainly honesty and respect. Good salespeople take their customers seriously, even if at first glance their interest seems strange or they give an awkward impression. Most products in one category differ only slightly. *You* as the seller make the difference. That is a demanding task – and one that can be fun, because here you can be creative!

This section offers you tools and a tailor-made conversation guide with which you can sell efficiently: from finding out the need up to conclusion. Your personal style is important, because a salesperson must be authentic; a customer can immediately see through hollow words.

11.4.1.1 PERSONAL REQUIREMENTS

Selling means handling people well. There are two observations to make on this:

- *The unconscious level will assert itself!* Remember the iceberg principle: only one seventh of communication is visible on the surface; the rest is invisible, but still very effective. This is especially shown by your gestures and facial expressions – body language does not lie. Therefore use 'tricks' only in a limited way; it is important that they are consistent. Only if you yourself are deeply convinced about yourself and your product, can you convince your customer.

Figure 11.14.
Smiling (even if only internally) is a good sales aid

- *The quickest thing you can change is yourself!* You can't change difficult customers, but you can protect yourself from them by well-managed conversations.

Good salespeople build up a close personal relationship with their customers. They show interest in the customer even beyond the immediate order. They make it easy for the customer to make contact – through friendliness and accessibility. They are of benefit to the customer, and will even tighten up the union nut on a dripping tap while they are checking the pre-requirements for a solar connection in the cellar.

During the sales discussion the central arguments are the benefits. However, for the discussion to be successful the secondary, emotional components are just as important. Arguments without the secondary components are like an apple without its flesh: the core is nourishing but doesn't taste very good.

THE FOUR PHASES OF A SALES DISCUSSION

In the following, the four phases of a typical sales discussion are discussed in more detail:

- Build a bridge.
- Establish need.
- Offer a solution.
- Achieve a result.

These phases can take place at several levels, and may partially be repeated. After the successful completion and installation of a solar system, the task of the salesperson is not complete. He or she remains the customer's partner and should offer customer care, so that the customer will help the salesperson by enthusiastic further recommendation.

11.4.2 Build a bridge

11.4.2.1 INNER ATTITUDE

Here the iceberg comes into play again. If you want to conduct a customer discussion you need to have a positive attitude to your task and to the customer. A good exercise for this is to consider: *What pleases me?* Pause for a moment before a contact. Consider what will please you in the coming meeting. You can also think about the nice evening you had yesterday, or what plans you have for the weekend. You should also fill yourself with inner sunshine.

11.4.2.2 MAKING CONTACT

Whether it be on the telephone or in a personal discussion, whenever you make contact the person you are talking to needs to know to whom he or she is speaking. Here it is particularly helpful if you can refer to mutual acquaintances or activities. You have surely already noticed how a reserved voice on the telephone relaxes when you name a mutual acquaintance. This is called a *bridge*.

CHECKLIST FOR MAKING CONTACT

Note the name of your contact and use his or her name a number of times when you speak.
Tell your contact:

- who you are, and which company you come from
- why you have made contact.

Figure 11.15.
Build a stable bridge to your customer

Find a positive approach. It is best to find things that relate to the customer, such as:

> *'I recently passed your house and was very struck by it.'*
> *'I know of you from your activities in the horticultural association.'*
> *'I got your address from Mr Green.'*

During the follow-up contact:

> *'How are you?'*
> *'Nice to see you/speak to you on the telephone.'*
> Other options: 'small talk' as a starting point (weather, travel route etc.).

Tell the customer why you are here:

> *'I am interested in finding out what you think about solar energy.'*
> *'We are contacting our customers about solar energy, and would like to hear your opinion on the subject.'*

Do not assume that your discussion partner has the time to speak to you. With one question you can show that you are not going to push too hard, and give him the choice:

> *'Do you have a couple of minutes?'*
> *'How much time have you got?'*

In this phase do not talk about the details of your services. If you make a presentation too soon, the customer will begin to ask questions – particularly questions about the price!

It is now the customer's turn: clarification of need.

11.4.3 Establish need

This is the most important phase of a sales discussion. Here you must identify the needs of the person you are talking to.

11.4.3.1 QUESTIONING TECHNIQUE

If you are able to ask good questions you already have the best prerequisites for leading a good discussion: 'He who asks, leads.' But here too the iceberg principle applies: only if you feel that the other party has a genuine interest will he or she not find your questions too 'pushy'.

Figure 11.16.
With the sharp eyes of a detective you should establish the needs of your customer

CLOSED QUESTIONS

Closed questions, to which the only answers are 'yes' or 'no', are the most frequently used form of questioning. Such questions permit you to come to clear agreements:

'Can we meet on Friday?'
'Is the 5th OK for you for the installation of the solar system?'

With semi-closed questions (to which the answer is usually one word or a fact) you can get specific information at the start of a discussion. These can also be called 'W' questions: Who, When, Where?

'How many people live in your house?'
'When will the roof be covered?'

This type of question is ideal to get to the point where your customer has to make a statement. You should ensure that the closed question is always linked to a buying argument, which makes it more difficult for the customer to say 'no':

'Would you like the anodized surrounds so that the collectors match your roof better?'
'Would you like two further modules so that you can fully utilize the subsidy?'

OPEN QUESTIONS

Open questions are designed much more to encourage your contact partner towards consideration. In this way you can tie him or her into the discussion and obtain the maximum information – occasionally also unexpected information. In the heat of the moment you may see this unexpected information as unwelcome because it needs consideration and flexibility. However, it can also be a real gold mine. It is well worthwhile following it up. Perhaps you won't sell any solar collectors, but instead a water treatment system for the swimming pool.

While you are in the 'Establish need' phase you should ask mainly open questions. In this way you can move a faltering discussion along and open up quiet people:

'How do you see the energy supplies to your home in the medium term?'
'What is of particular value to you?'
'What difficulties did you have in that?'

Throw pebbles!

Imagine you are walking along by a high wall and want to know where on the other side a pond is. How do you find that out? You keep throwing pebbles over the wall –

until you hear that you have found it by the splash. Use this detective's intuition to find out important information for yourself.

Rucksack technique
You should always keep your full range with you – in a rucksack, in a manner of speaking. If you are attending to a dripping tap at a customer's house, keep your eyes and ears open for other opportunities of working for the customer. Make it a habit always to glance at the roof and ask every customer, 'You have such a nice south-facing roof; what do you think about solar technology?'

Coffee filter
Let your discussion partner finish talking. Give him or her time to formulate complex thoughts or to become aware of something – just like a coffee filter that you don't throw away until the last drop has passed through it.

You can guide the conversation with open questions and still remain with the central theme. Open questions appear easy, but you need good preparation and some practice. Even experienced salesmen tend to ask more closed questions than open ones. Under pressure, closed questions are more spontaneous than open ones. It is therefore worth preparing and rehearsing open questions until they are almost automatic.

11.4.4 Offer a solution

Now you can prepare the completion: you propose a solution and make an offer. You should do this only when you know the customer's needs exactly, because as soon as you offer solutions the needs of the customer stop growing.

11.4.4.1 BENEFIT

At the start of the solution path lies the benefit mentioned in section 11.2.1. You have researched the customer's needs, and now you can establish the benefit of your product that matches that need.

11.4.4.2 CUSTOMER-ORIENTED FORMULATION

Speak to your discussion partner. Say 'you' instead of 'I' or 'we' at least five times once in a while. If you then combine this with the customer's name you will get his or her full attention.

Use the customer's formulations: for example, if he or she speaks of 'sun collectors' you should not then speak of 'solar collectors'.

Pay attention to *how* you speak (formulate a checklist).

11.4.4.3 PRESENTATION

Nothing is as convincing as a practical example. Take your customer to see a sample system. If that isn't possible, show him or her some (good!) photos, or take models or components with you. The more sensory channels you appeal to, the better you can anchor the product in your customer's imagination. What functions for advertising can also support you: show letters from satisfied customers or testimonials from customers whose enthusiasm has already been raised as part of your advertising. When arranging your reference folder consider the current benefit arguments. Calculate an example system, and emphasize for example the monthly payments for feeding power into the grid or the monthly savings.

The iceberg is also relevant to the presentation: feel how much you enjoy your work before you ring the customer's doorbell. Then you will be able to fire a salvo of good ideas!

11.4.5 Achieve a result

Now comes the transition from the non-binding discussion to the binding deed. Do not let the right moment pass by; frequently the concluding discussions are 'talked to death'!

11.4.5.1 THE TRANSITION

CHECKLIST OF BUYING SIGNALS

- Approving comments:
  'That sounds good.'
  'That would mean that, in future, I…'

- Detailed questions:
  'Should I then have a vacuum tube collector or a flat collector?'

- Purchase-related questions:
  'What would the complete system cost?'
  'When could you do that?'

- Non-verbal signals:
  The customer moves towards you.
  The customer goes to the product and touches it.
  Approving nods.

The checklist of buying signals helps you to recognize that you have reached this phase in the discussion. Now you can deliberately take over the lead:

FINAL CLARIFICATION OF BUYING OBSTACLES:

'Can I do anything else for you?'
'Who makes the decision?'
'What is your time schedule?'

Don't forget the iceberg. Assume the successful conclusion of the discussion within your own mind in advance. If you are completely convinced that you are making a good offer this will also transfer itself to the customer.

If you would like the customer to say 'yes', then you should set up a 'yes street' for him or her. Do this by using closed questions which can be answered with 'yes':

'Have you spoken to Mr Green about his system?'
'Did he confirm my information?'
'Have you looked at the promotional material?'
'Does the covering frame appeal to you?'

You must always remain confident of success. The purchase is a difficult decision for the customer, for he or she has to say goodbye to using the money for many other things. Remember the saying 'A trouble shared is a trouble halved': it helps the customer if he or she knows that it is also not easy for you. You should therefore understand the customer if he or she still has a query that takes time and trouble to sort out, or if he or she tries to extract another concession from you.

11.4.5.2 HANDLING OBJECTIONS

Objections are normal. You too would consider carefully the pros and cons of an approaching decision. It is natural to try to counter objections spontaneously, so as to be 'proved right' but this can degenerate into a fruitless battle. It is better to take seriously the thoughts of the person you are talking to and put forward alternative points of view:

'That's true: the larger tank is more expensive. But with it you will be able to store more solar energy on good days and hence bridge over longer periods of bad weather.'

It is important that the customer retains the choice, whether he or she is eventually convinced or not.

PRACTICAL TIPS

- *Echo technique.* This technique is ideal when you are surprised by sudden objections and counter-arguments from your discussion partner. You repeat what has been said in your own words and put a counter question:

  'You are concerned about your roof not being leak-proof? What do you mean by that?'

  In this way you gain time to 'digest' the question. The customer feels that he or she is being taken seriously and confirmed. The customer will substantiate his or her statement. You can then either dispel the worries, or you will realize that the customer really would like something else.

- *Let the garbage float past.* When you find that your customer is grumbling about really trivial things – perhaps because he or she is having a bad day – then leave it at that. Just as twigs and leaves chase about in a swollen river, you should initially not take any notice of them. If there is a genuine interest hidden among them then it will turn up again. In this way you will avoid pointless arguments.
- *Support positive statements.* If you find that the customer is more pleased about the sophisticated solar station in the cellar than the collectors on the roof, then bring it up again as a reminder during the final discussion. Perhaps you will even be able to sell him a sophisticated measured data acquisition unit.
- *The power of silence.* When everything has been said, help the customer to clear his or her thoughts by keeping quiet.
- *Stay positive.* It doesn't always lead to an immediate signature! An agreement about the next step is also a positive outcome of a discussion.

11.4.5.3 THE QUESTION OF PRICE

It is not the price that is being sold, but the offer. If you help the customer to realize his or her desires, then the price is not the main thing.

If possible, don't mention the price until the conclusion of the discussion, when it is clear what is on offer and how it can be reduced by grants or savings. If the customer asks about the price too soon, you should respond with:

> *'Would you mind waiting a moment? The price depends on various options and I first want to show them to you.'*

If the customer still keeps on asking about the price then tell him or her, otherwise he or she may feel that you are procrastinating.

A solar system is a long-term investment. Make the purchasing period really clear by, for example, giving the price per annum or month over the service life of the system. Alternatively, calculate in the grant and payback period immediately:

> *'After deducting the grant, this system will cost you €4000/$4000/£2800 and will save you about €200/$200/£140 per annum. In view of increasing energy prices the system will be amortized over 15 years, and over the remaining time of the 25-year service life you will earn money with it!'*

A solar energy system differs from other energy converters in that it does not give rise to continuing fuel costs. Make this advantage clear:

> *'With a solar system you get the added bonus of free fuel for its full service life. That is the same as buying a car and at the same time 20,000 litres of petrol.'*

If the customer fixes too firmly onto the price and says it is too expensive, you may be able to find out more about his or her reasons when you respond with:

> *'You think that this is too expensive: too expensive in relation to what?'*

Do not offer any discount. Once a customer feels that it is easy to play at price negotiations with you then he or she will get you to reduce prices time and again. If

Figure 11.17. Let the garbage float past

you *do* want to make concessions then offer more equivalent value for the same price: for example a better heat quantity meter, or an interface that enables the customer to see the yields on his or her PC.

If the customer insists on a price deduction you can ask:

> *'Which aspects of the performance would you then like to do without?'*
> *'Assuming we agree on the price what is still important to you?'*
> *'A discount is possible if you accept a larger surface area.'*

Perhaps there are still options in the fine tuning: adapting the methods of payment or the payment deadline, reducing the options, or reducing risk surcharges.

The main principle should remain: *No discount without something in return.*

You must be fully confident in your prices so that during price negotiations you can remain self-assured. The customer will immediately notice if you believe your prices to be high and are unsure. Therefore calculate everything well in advance.

So now we wish you lots of success during your next sales discussion!

11.5 Subsidies

11.5.1 General information on financial support

Subsidies are important in selling solar water heaters. It is sometimes possible to add several subsidies together and make a very attractive offer to the customer. However, the acquisition of funding from third parties, such as public institutions or electric utilities, is sometimes an art in itself. Commonly institutions giving subsidies can be classified into five groups:

- Ministries and related institutions
  - Ministry for economic affairs
  - Ministry of agriculture and/or forestry
  - Ministry for environment
  - Ministry for research and development (for innovative projects)
- Regional institutions
  - Ministries or institutions of federal states or regions
- Municipal institutions
- Independent organizations
  - e.g. environmentally focused foundations
- Energy utilities.

Usually, financial support from different state, regional or municipal support programmes can be combined in order to increase the level of support on a project. However, many support programmes limit the total rate of subsidy on a project.

Because of the frequent changes of operation of support programmes, the issuing of new programmes or the abolition of programmes, we do not give details of support programmes here but instead provide the reader with a detailed list of sources of information where overviews of support programmes and their actual conditions can be obtained for several English-speaking countries – the UK, the USA, Canada and the EU. These links will also help in acquiring information on regional programmes.

11.5.2 Sources of information in the UK

11.5.2.1 GOVERNMENT

DEPARTMENT OF TRADE AND INDUSTRY (DTI)
www.dti.gov.uk
Latest information on DTI-funded programmes for renewable energy (calls for proposals etc.).

ENERGY SAVINGS TRUST (EST)

www.est.org.uk
Information on energy efficiency and climate change; programmes for local authorities, consulting to small business.

UK GOVERNMENT NON-FOOD USE OF CROPS RESEARCH DATABASE

aims.defra.gov.uk
Lists all government-funded R&D projects on non-food crops.

OFFICE OF GAS AND ELECTRICITY MARKETS (OFGEM)

www.ofgem.gov.uk
Regulating organization for the gas and the electricity market; administers the Renewable Obligation (RO). Information on practical RO issues.

THE CARBON TRUST

www.thecarbontrust.co.uk
The Carbon Trust is developing and implementing programmes to support low carbon emitting technologies.

ENHANCED CAPITAL ALLOWANCE SCHEME (ECA)

www.eca.gov.uk
Website managed by the Carbon Trust in collaboration with DEFRA and the Inland Revenue, to provide information about the Enhanced Capital Allowance Scheme.

PRASEG

www.praseg.org.uk/
Parliamentary Renewable and Sustainable Energy Group. News on the practical implementation of support measures such as the Renewable Obligation.

11.5.2.2 FUNDING

DTI SUPPORT PROGRAMME

www.dti.gov.uk/renewable/geninfo.html
Information on UK government programmes supporting RES.

CLEAR SKIES RENEWABLE ENERGY GRANTS

www.clear-skies.org
Information on the DTI's grant scheme for renewable energy systems. The £10 million Clear Skies Initiative aims to give homeowners and communities a chance to become more familiar with renewable energy by providing grants and advice. Homeowners can obtain grants between £500 and £5000, and community organizations can receive up to £100,000 for grants and feasibility studies. Additional support on the technologies and grant applications are available through the Community Renewables Initiative or Renewable Energy Advice Centres.

Clear Skies supports projects in England, Wales and Northern Ireland. Homeowners and community groups in Scotland can apply for support under the Scottish Community Renewables Initiative.

The technologies supported are:

- solar water heating
- wind
- hydro
- ground source heat pumps
- automated wood pellet stoves
- wood fuel boilers.

Grants for photovoltaics (solar electricity) are available through the £20 million Major Demonstration Programme.

Grants will only be awarded to homeowners where an accredited installer is to be used. These installers will work to a code of practice and be vetted beforehand to ensure that the customer gets the most appropriate system for his or her needs, correctly installed at the right price.

SCOTTISH COMMUNITY AND HOUSEHOLDER RENEWABLES INITIATIVE (SCHRI)
http://www.est.org.uk/schri/
SCHRI offers grants, advice and project support to develop and manage new renewables schemes.

COMMUNITY RENEWABLES INITIATIVE (CRI)
www.countryside.gov.uk/communityrenewables/
Providing advice on how to set up RES projects, financing & funding, technology etc.

11.5.2.3 ASSOCIATIONS AND ORGANIZATIONS

SOLAR TRADE ASSOCIATION
www.solartradeassociation.org.uk
The Solar Trade Association Ltd (STA) serves as a focal point for organizations with business interests in the solar energy industry.

NATIONAL ASSEMBLY SUSTAINABLE ENERGY GROUP
www.naseg.org
Organization to promote sustainable development and renewable energy in Wales; special information on support programmes in Wales.

WESTERN REGIONAL ENERGY AGENCY & NETWORK
www.wrean.co.uk
Energy Agency in Northern Ireland promoting RES.

BRITISH ASSOCIATION FOR BIOFUELS AND OILS (BABFO)
http://www.biodiesel.co.uk/
Organization dedicated to the promotion of transport fuels and oils from renewable sources.

11.5.3 Sources of information in the USA

11.5.3.1 GOVERNMENT

DEPARTMENT OF ENERGY (DOE)
www.doe.gov
Energy policy, support programmes, links to further sources of information

11.5.3.2 FUNDING

www.science.doe.gov/grants/
Grants available through the US Department of Energy.

THE ENERGY FOUNDATION
www.energyfoundation.org
Independent foundation supported by several foundations to foster energy efficiency and clean energy; several support programmes for RES.

DATABASE ON STATE INCENTIVES FOR RES
www.dsireusa.org
Project managed by the Interstate Renewable Energy Council (IREC), funded by the DOE; information on support measures in the different states of the USA.

11.5.3.3 ASSOCIATIONS AND ORGANIZATIONS

ASES
www.ases.org
The American Solar Energy Society (ASES) is a national organization dedicated to advancing the use of solar energy for the benefit of US citizens and the global environment. ASES promotes the widespread near-term and long-term use of solar energy.

RENEWABLE ENERGY POLICY PROJECT
solstice.crest.org
Information, insightful policy analysis and news on RES; funded by DOE, EPA and several foundations.

THE CLIMATE ARK

www.climateark.org
Portal on climate change and renewable energy; extensive information and links.

SUSTAINABLE ENERGY COALITION

www.sustainableenergy.org
Lead organization of more than 30 associations active in the field of RES; provides news on RES.

11.5.3.4 R&D AND OTHER SOURCES OF INFORMATION

NATIONAL RENEWABLE ENERGY LABORATORIES

www.nrel.gov
Information on RES.

11.5.4 Sources of information in Canada

Links to governmental institutions, support programmes, associations and other sources of information.

11.5.4.1 GOVERNMENT

CANMET

www.nrcan.gc.ca
Performs and sponsors energy research, technology development and demonstration within Natural Resources Canada, a department within the Canadian federal government; information on funding.

NRCAN

www.nrcan.gc.ca
Federal government department specializing in the sustainable development and use of natural resources.

CANADIAN RENEWABLE ENERGY NETWORK (CANREN)

Created through the efforts of Natural Resources Canada (NRCan); information on all renewable energy technologies.

OFFICE OF ENERGY EFFICIENCY (OEE)

oee.nrcan.gc.ca
Centre of excellence for energy efficiency and alternative fuels information.

11.5.4.2 FUNDING

RENEWABLE ENERGY DEPLOYMENT INITIATIVE (REDI)

www.nrcan.gc.ca
Support programme for RES, among them specifically for highly efficient and low-emitting biomass combustion systems.

11.5.5 EU sources

EUROPA WEBSITE

www.europa.eu.int
European Union website with general info on policies.

CORDIS

www.cordis.lu
Information on European support programmes.

SOLTHERM EUROPE INITIATIVE

www.soltherm.org
Overview of European and local markets, grants, standards and certificates.

Appendix A: Glossary

absorber The part of the solar collector that absorbs the incoming solar radiation, converts it to thermal energy, and feeds it into the heat transfer fluid. For an optimal conversion process at higher temperatures, the absorber is selectively coated (high absorption, low emission) and provided with a pipework system (heat transportation) through which the fluid flows (normally water or a water/antifreeze mix).

absorption coefficient, α The ratio of radiation absorbed by a surface to that of the incident radiation.

albedo (or *reflectance*) The ratio of solar radiation reflected by a surface to that incident on it. Examples: snow 0.8–0.9; woods 0.05–0.18.

annuity Describes the annual cost in terms of interest and repayments, starting with investment costs, the total period of repayments and an effective interest rate so that at the end of the period the debt has been repaid.

aperture area For flat collectors, the surface area of the collector opening through which unconcentrated solar radiation can enter the inside of the collector housing. For tube collectors the aperture surface of the product is calculated from the length and width of the absorber strips and the number of strips. If the gaps between the absorbers are closed with a reflector, the aperture surface increases correspondingly.

aquifer storage An underground, natural storage system, which serves as a seasonal store and utilizes the water-filled, porous earth layers: that is, the groundwater. Apart from water, rocks or boulders can also serve as a storage medium. Thus gravel–water storage systems are also described as aquifers.

azimuth The angular deviation of the collector surface with respect to the direction due south. In solar technology the azimuth angle is defined for south as $\alpha = 0°$: deviation to the west is positive, to the east it is negative.

back-up energy The energy that the consumer uses to supplement solar energy. Examples: oil, gas or electricity.

bleeders (air vents) In solar loops, air gathers in the highest positions and can interrupt the circulation of the fluid. To remove this air, bleeder valves (air vents) are installed at the critical points (highest points) of the solar loop. There are manual and automatic bleeders. The bleeders must be suitable for the transfer fluid type and the maximum temperature in the solar loop.

bypass If the solar loop consists of long pipe runs, the installation of a short-circuit to initially circumvent the heat exchanger is recommended. During short-circuit operation the medium is heated first in the collector circuit. The pipe to the heat exchanger is only opened via a pump or motor-operated valve if the fluid has a higher temperature than the store.

check valve Prevents undesirable fluid movement. Check valves are installed in pipeline systems if an unwanted reversal in the normal fluid direction can take place under certain operating conditions. There are non-return flaps or valves and gravity brakes. In solar systems this type of device is used to prevent the store heat from thermosyphoning up to the collectors if the circulating pump is switched off. In the cold water feed pipe, a check valve is installed so that heated water cannot be forced out of the store back into the rising main as a result of thermal expansion.

collector For domestic water heating, absorbers are housed in a box or an evacuated glass tube and then covered with a highly transparent cover with thermal insulation behind. This is in order to keep the heat losses as low as possible at the higher operating temperatures. The collector is then the absorber plus these other items. If a collector is only an absorber, it may only be suitable for low-temperature applications such as swimming pools. In the case of a glazed collector a differentiation is made between the absorber surface, the aperture surface and the gross surface. The *absorber surface* is the surface area of the absorber on which radiation is converted to heat and from which the heat is taken. The *aperture surface* is the collector opening for the solar radiation, which can then reach inside the collector housing. The *gross surface* is the total surface area of the collector (important for installation).

collector efficiency, η Gives the proportion of radiation striking the absorber surface that is converted to useful heat. It is dependent on the temperature difference between the absorber and the surroundings, and on the global solar irradiance. If the collector efficiency is represented on a diagram by means of the temperature difference between absorber and surroundings, the collector characteristic curves for a specific collector are obtained in terms of the irradiance.

convection Warm liquids or gases are lighter than cold ones and ascend. They thus carry the heat upwards. Examples: convection flows in the store (heated storage water moves upwards and cold water descends); convection losses in the collector (cooler air draws out the absorber heat and climbs upwards).

corrosion The decomposition of metallic materials. Corrosion is caused mainly by the different electrochemical potentials of two metals that are connected together so that they are electrically conducting and are wetted by an electrically conductive liquid. If corrosion is expected, suitable measures, such as the addition of inhibitors to the solar fluid or coating of system parts that are subject to weathering, should be taken to protect them from corrosion damage. In enamelled containers (solar stores) sacrificial or auxiliary current anodes are used to protect against corrosion at potential weak points. This generates a protective current in the store by means of a 'potentiostat', which prevents the precipitation of copper ions on the container wall.

CPC Compound parabolic concentrator; used in evacuated tube collectors to increase the aperture surface area by means of mirror reflectors in a geometrically optimized form as parabolic channels.

degree of efficiency, η Whenever energy is converted (such as when converting solar radiation into heat in the collector), losses occur (for example heat radiation). The degree of efficiency describes the ratio of useful energy (heat) to the energy used (solar radiation). The lower the degree of efficiency, the higher the losses. If a collector has a degree of efficiency of $\eta = 0.6$ it means that, of the radiation received, 60% is converted into useful heat, and 40% is lost in the form of optical and thermal losses.

degree of transmission, τ Part of the incoming radiation does not reach the absorber owing to reflection from the glass cover and absorption as it passes through the glass material. The degree of transmission describes the transparency of the glass pane.

degree of utilization Relates to a specific period of time – normally a year – and refers to the ratio of benefit (for example useful generated heat energy) to expenditure (energy used in the form of unburnt fuel, electrical energy or solar energy). The degree of utilization is normally lower than the degree of efficiency at the normal design point (nominal load) – that is, the utilization of a boiler, where the periods in which the boiler operates under partial load or start-up are also taken into account.

drainback system In these systems the heat transfer medium (water) is only in the collector during the pump running time, as it self-drains after the pump is switched off. On the one hand this prevents evaporation; on the other hand it means that frost protection agents are not required. Advantage: in comparison with a water/antifreeze mixture, water possesses a higher heat capacity and, as a result of its lower viscosity, causes lower pressure losses.

emission coefficient, ε Gives the amount of irradiated solar energy incident on the absorber (wavelengths 0.3–3.0 μm) that is radiated back again as infrared radiation (wavelengths 3.0–30 μm). An emission coefficient of 0.12 indicates that 12% of the solar energy converted to heat is radiated back.

energy Arises in different forms: thermal energy (heat), mechanical energy or electromagnetic energy (radiation). Energy is given in different units, for example as watt-hours (Wh), kilowatt-hours (kWh) or joules (J). One joule is one watt-second (Ws). 1 kWh = 1000 Wh = 3,600,000 J (= 3.6 MJ).

expansion bellows An intermediate piece made of corrugated copper tube. The expansion bellows compensate for temperature-related length changes to avoid cracks and leaks in the solar loop.

expansion vessel Part of the safety equipment for some solar hot water systems. It is a closed container with a nitrogen cushion separated by a membrane. In the event of expansion of the volume of the heat transfer liquid (water or water/antifreeze mix) caused by heating, it takes up this expansion. For a stagnation situation it also takes up the liquid content of the collector field.

getter To maintain a vacuum, various manufacturers integrate so-called 'getters' into the vacuum tubes. These can have either a barium sulphite coating that is vapour-deposited on the glass (that is, 'Thermomax') or copper cushions that are filled with a special granulate and fixed onto the absorber. In both cases the gas molecules are absorbed and the vacuum therefore remains stable over a long period. In the case of barium sulphite getters this process is visible by a colour change in the coating (the mirror surface becomes powdery white).

global solar radiation The atmosphere of the earth reduces the radiated power of the sun through absorption and scattering (= loss) by molecules, dust and clouds. The energy therefore changes direction and reaches the earth's surface partly as diffuse radiation. Without clouds, the solar energy can reach the surface directly. The term *global solar radiation* can be described as the full radiation impacting on a horizontal surface. It therefore consists of direct and diffuse radiation. With a clear sky, global solar irradiance consists only of direct radiation and with a cloudy sky only of diffuse radiation. As an annual average in northern Europe, diffuse radiation is between 50% and 60% of the global solar radiation. Diffuse radiation also heats up solar collectors.

gross heat yield, Q_{BWE} Corresponds to the heat discharge at the collector flow or at the inlet to the store. Measured in kWh/m^2. The pipe losses are included. The gross heat yields of different solar collectors can be compared only under the same temperature conditions (average absorber temperature and ambient temperature) and with the same irradiation conditions. It is also necessary to know whether it refers to the absorber, the aperture or gross surface area of the collector.

heat loss rate, kA The product of the heat loss coefficient, k, of the store and its surface area, A. Measured in W/K. If the kA value is multiplied by the temperature difference between the inside of the tank and its surroundings, the heat loss of the hot store is obtained. The kA value already includes the heat losses through heat leakage at the connections etc. Highly thermally insulated solar stores have kA values between 1.5 and 2 W/K, depending on size.

heat production costs Describes the economic viability of solar systems and gives the cost of a kilowatt-hour generated or saved with the help of the solar system. The investment and operating costs are taken into account, together with the energy savings.

heat pipe Vacuum tube of glass in which the absorber heat is extracted from within the glass via a closed pipe containing evaporative fluid and is then transferred to the main heat transfer fluid by means of a wet or dry manifold.

hysteresis The difference between the switch-on and switch-off temperatures of a relay; often found in a temperature difference controller.

inhibitor If different metallic materials are used in the solar loop, there is a risk of electrochemical corrosion. This can be eliminated by the addition of a suitable corrosion protection agent (inhibitor) to the heat transfer fluid. In closed systems, where the fluid contains an inhibitor, all the permitted metallic materials can be used in any combination. The inhibitor should have a certificate, which has to include details of the effective life.

intrinsic safety According to the Pressure Equipment Directive solar systems must be designed to be 'intrinsically safe'. Continuous heat take-up without heat consumption must not lead to a severe fault. A severe fault exists, for example, if fluid is blown out of the safety valve or vent and the solar loop has to be refilled before it is started up again. Intrinsic safety can be achieved by suitable dimensioning of all the safety devices in the solar loop.

irradiance, G The area-related irradiated power density of the solar radiation; measured in W/m^2.

***k*-value** of the collector The heat loss factor, k (unit W/m^2K), gives the design-specific heat loss of a collector. It describes, among other things, the insulating status of the collector. The smaller the *k*-value, the smaller the heat loss.

Legionella The rod-shaped bacteria that lead to Legionnaire's disease. Legionella cause two different diseases: the potentially fatal type of pneumonia, and a feverish, influenza-like illness that is not fatal. Legionella can be found in all waters except sea water. For multiplication they require, among other things, water temperatures between 30°C and 45°C, a pH value of 6–9, and iron in dissolved or undissolved form. At temperatures above 50°C they are destroyed. Infection takes place exclusively by breathing-in finely distributed water drops (aerosol): that is, under the shower or in 'spa baths'.

lightning protection If lightning protection is provided on the building then a solar system must also be integrated into the lightning protection system. In this case the collectors should be provided with lightning protection corresponding to the BS 6651 guidelines and installed by specialist personnel qualified for the purpose. Earth cables with a cross-section of at least 10 mm^2 and suitable tube clips are necessary.

low-flow operation Significantly reduced flow rate through a collector circuit with stronger heating of the heat transfer medium (water or antifreeze fluid) compared with normal flow (volumetric flow rate 10–15 l/m^2h). Advantages: higher collector flow temperature, and hence faster availability of hot water; smaller pipeline cross-sections; lower pump power.

optical degree of efficiency (also **conversion factor**), η_0 The proportion of the radiation falling onto the collector that can be converted to heat in the absorber. It is the product of the degree of transmission of the transparent cover and the absorption coefficient of the absorber surface: $\eta_0 = \tau \times \alpha$. The optical degree of efficiency corresponds exactly with the collector degree of efficiency if the temperature of the absorber is equal to the temperature of the surrounding air and no thermal losses occur.

overheating protection If, during a long period of sunny weather, no energy is removed from the solar store, the temperature can climb to a maximum limit. In such a case the solar loop circulating pump must be switched off. As a result, the absorber temperature climbs to the stagnation temperature and part of the transfer fluid evaporates. To avoid this undesirable state as far as possible, it is recommended that an additional overheating protection device is used to ensure that the maximum temperature is not even reached in the solar storage tank under such conditions. For example, during critical operating phases excess energy can be fed via the auxiliary heating circuit. Glazed flat-plate collectors and vacuum tubes are designed to withstand the stagnation temperatures for a long time.

payback time The economical meaning is the amortization period of the solar system (capital repayment time). It also means the time after which the solar system has gained the exact amount of energy that was required for its production (ecological amortization time).

pellets Made from dry, unrefined scrap wood (sawdust and wood shavings). They ideally have a diameter of 6 mm and are 10–13 mm long. One kilogram of wooden pellets has a calorific value of about 5 kWh. They require only half the storage volume of bulk wood. Wooden pellet burners can be used as combination systems in connection with a solar thermal system for domestic water heating as well as for room heating support.

potential, electrochemical (electrochemical series of metals) The greater the electric potential of a metal, the more precious the metal. As the base metals are dissolved by the more precious metals, it is necessary to consider this electrochemical series when using different metals to avoid corrosion damage. In the flow direction the more base metals must follow the more precious metals.

Element	Lead	Tin	Copper	Zinc	Aluminium
Potential	0.13	0.14	0.34	0.76	1.66

power The energy consumed or provided over a given time. Its unit is the watt (W), kilowatt (kW), joule per second (J/s) or kilocalorie per hour (kcal/h). 1 kW = 1000 W = 1000 J/s = 860 kcal/h. Example of calculating power: if a boiler provides 1500 operating hours of heat in a year and thereby generates 30,000 kWh it has an average power of 20 kW.

primary energy The energy originally provided by nature in the form of crude oil, coal, natural gas or radiation from the sun. Primary energies are partly used directly by the end user. For the most part, however, primary energy is first converted to secondary energy and then to end energy.

priority switching Pump controllers for conventional heating systems are usually designed so that the connected domestic water heating is prioritised. If the hot water temperature in the standby store falls below a set value and the store-charging pump is switched on, the heating circuit is not supplied and rapid heating of the hot water can take place. Priority switching is also effective if a solar store is connected to the storage charging circuit of a boiler that can be conventionally charged as necessary.

secondary circulation system To increase convenience, especially in the case of long domestic water pipelines, a secondary circulation system is installed that leads domestic water by means of a circulation circuit pump past the taps to the store. In this way, hot water is directly available when necessary at the taps. The circulation system can cause considerable heat losses. The important things here are the running time of the circulating pump and the quality of the thermal insulation on the circulation lines. Pressure switches and pulse-controlled or timer-controlled pumps can be installed to reduce circulation losses.

secondary energy Arises by converting primary energy: for example, coal is processed into coke or briquettes, crude oil into petrol, diesel fuel or heating oil.

selective coating On the surface of every body the heat radiation increases significantly with increase in temperature. To reduce radiation losses due to emittance (= emission) of long-wave heat radiation, the absorbers can be coated selectively in a special process. In contrast to normal black paints, this has an alternative layer structure that optimizes the conversion from short-wave to long-wave heat radiation and keeps its radiation as low as possible.

solar degree of utilization The ratio of the solar radiation striking the collector over a given time period, which has been converted into useful heat.

system efficiency Describes the efficiency of the whole solar system (collectors, pipelines, heat exchanger and storage tank). It shows how much of the solar energy irradiated to the collector can be used as hot water. It is true that over-sized systems possess a high solar fraction but, because of the non-usable excess heat in summer, they have a low degree of system efficiency.

solar fraction Also called the solar coverage rate. Indicates the ratio of the energy used for domestic water heating that can be covered by the solar system as an annual average. It corresponds with the ratio of the solar energy yield to the total energy requirement for domestic water heating, for the coverage of the solar store losses and any other losses in the circulation system.

solar constant Gives the irradiance at the upper edge of the atmosphere. It is on average 1367 W/m^2 (variations are caused by variable distances between the earth and the sun and variations of the solar activity).

stagnation temperature If the solar loop does not transmit the energy from the collector during high radiation, the absorber heats up to very high temperatures. If then the heat losses to the surroundings are just as great as the solar gain, the absorber will reach its maximum temperature. The value of the stagnation temperature is strongly dependent on the type of collector.

stratification index Represents the retention of temperature layers of fluid during draw-off. High stratification indexes mean good retention of temperature layers. The influence of the stratification index on the solar fraction is low: an increase from 30 to 100 leads to an increase in the solar fraction of about 1%.

system performance figure The ratio between the useful solar energy gained and the non-useful, parasitic electrical energy consumption of the pump and the control system.

thermosyphon principle Because of the difference in densities between warmer and cooler water, the warm water becomes more buoyant and rises upwards. This effect is supported in good solar stores by means of internal fittings so that, even after short solar system operating time, sufficient domestic hot water is available at the top of the tank. In gravity systems this effect is used as the only means of operating the solar loop.

thermostatic mixing valve If there is a high maximum temperature of the store, a mixing valve is required to prevent scalding when a tap is turned on. This is installed between the cold water supply pipe and the hot water take-off pipe. By means of the thermostatically controlled mixing in of cold water the maximum temperature of the water at the tap can be controlled within adjustable limits.

Tichelmann principle A collector field (array) can operate at its maximum performance only if the heat transfer medium (water or antifreeze fluid) cools the whole absorber surface area uniformly. It is thus necessary to ensure that, when the collectors are connected together, no areas are created through which the heat transfer medium either does not flow or does not flow through sufficiently. This is achieved by ensuring that all flow paths through the collector field have the same flow resistance; in other words they have the same length and cross-section. If the collectors are arranged according to the Tichelmann principle these conditions are fulfilled.

thermal conductivity, λ Characteristic value for the quality of the thermal conductance in solid bodies. The unit is W/mK. Thermal insulation has λ values between 0.035 and 0.045 W/mK.

thermal conductivity, effective vertical, λer Gives information to show how a temperature stratification stage breaks down in a static storage tank. The effective vertical thermal conductivity should be as low as possible. Good stores without internal fittings (for example without an immersed heat exchanger) have values that are in the range of the thermal conductivity of the water itself (approximately 0.6 W/mK). For stores with internal heat exchangers, the effective vertical thermal conductivity is about 1–1.5 W/mK. Simulation calculations have shown that, by halving the effective vertical thermal conductivity from 2.2 to 1.1 W/mK, the annual solar fraction can be increased by 5%.

usable energy High-grade energy that, after conversion from low grade (that is, perhaps in a boiler), is then available in a more useful form of hot water under the shower or in the form of heat in the living areas.

viscosity The viscosity of fluids depends strongly on the temperature. The viscosity of the transfer fluid (water or antifreeze fluid) is also dependent on the concentration of any additives such as antifreeze.

Appendix B: Relevant UK solar regulations and technical standards

Note: For regulations in other countries, see also section 4.5.

Although starting from an initially simple premise of taking the sun's energy to heat water, there eventually builds a body of experience and knowledge that can be passed down through the generations. Often this is based on health and safety of workers and end-users but also it can be to improve cross-border trade or provide performance specifications. This information is predominantly found in regulations and standards, as these are the published end results of expert committees across the world.

As shown below, a regulation is an Act of Parliament, enforced as law in England and Wales. Sometimes there are variants for Northern Ireland and Scotland, but these are mainly found in the building regulations. A standard is mainly a British Standards Institution publication, but could be a code of practice published by a trade organization. Sometimes regulations refer to standards, and hence they also become enforceable in law. The enforcement is normally by the Health and Safety Executive, but building control officers and water utilities also have a role.

B.1 General construction regulations

Water Supply (Water Fittings) Regulations 1999
The Building Regulations 2001
The Pressure Equipment Regulations 1999
Control of Substances Hazardous to Health Regulations 1999
Gas Safety (Installation and Use) Regulations 1998

B.2 General construction standards

BS 6700: 1997 Design, installation, testing and maintenance of services supplying water for domestic use within buildings and their curtilages
BS 5449: 1990 Specification for forced circulation hot water central heating systems for domestic premises
BS 7671: 2001 16th Edition IEE Wiring Regulations
BS 6651: 1992 Code of practice for protection of structures against lightning

B.3 Solar standards

BS 6785: 1986 Solar heating systems for swimming pools
BS 5918: 1989 British Standard Code of Practice for solar water heating systems for domestic hot water
BS EN 12975 2001: Thermal solar systems and components – Solar collectors
BE EN 12976 2001: Thermal solar systems and components – Factory made systems
DD ENV 12977: Thermal solar systems and components – Custom made systems
BS EN ISO 9488: 2000 Solar energy vocabulary

B.4 Roof standards

General and pitched roofs: BS 8000, BS 6399, BS 5534, NFRC
Flat supported (felt, asphalt, liquid): BS 6229, BS 8217, BS 8218, FRCAB, NFRC, ELRA
Flat self-supporting (metal profile, fibre-cement): BS 5427, CP143, MCRMA, NFRC
Metal supported (copper, zinc, lead, aluminium): BS 6915, CP143, LSA, MCRMA

B.4.1 ACRONYMS FOR ROOF STANDARDS

BS: British Standard
CP: Code of Practice (BS)
CITB: Construction Industry Training Board
ELRA: European Liquid Roofing Association
FRCAB: Flat Roofing Contractors Advisory Board
LSA: Lead Sheet Association
MCRMA: Metal Cladding and Roofing Manufacturers Association Limited
NFRC: National Federation Roofing Contractor
NVQ: National Vocational Qualification
RIA: Roofing Industry Alliance

B.5 Working on site, height and lifting regulations

HEATH & SAFETY AT WORK ACT (HSW) 1974

Applies to all work employers, employees, self-employed.

MANAGEMENT OF HEATH AND SAFETY AT WORK REGULATIONS (MHSWR) 1999

Apply to all work employers, employees, self-employed. Assess and reduce risks.

CONSTRUCTION (HEALTH, SAFETY & WELFARE) REGULATIONS 1996

Apply to all construction work employers, employees, self-employed and all those who can contribute to the health and safety of a construction project.
Note: Regulation 6 expected to be covered under new legislation 'Work at height' – expected July 2004.

CONSTRUCTION REGULATIONS (HEAD PROTECTION) 1989

Apply to all requiring head protection.

CONSTRUCTION (DESIGN AND MANAGEMENT) REGULATIONS (CDM) 1994

Apply to all large-scale, non-domestic work.

LIFTING OPERATIONS AND LIFTING EQUIPMENT REGULATIONS (LOLER) 1998

Apply to all lifting equipment.

MANUAL HANDLING OPERATIONS REGULATIONS 1992

Apply to employers and the moving of objects by hand or bodily force.

PROVISION AND USE OF WORK EQUIPMENT REGULATIONS (PUWER) 1998

Apply to all equipment providers, including machinery that should be safe for work.

THE WORKPLACE (HEALTH, SAFETY AND WELFARE) REGULATIONS 1992

Apply to employers regarding ventilation, heating, lighting, workstations, seating and welfare facilities.

HEALTH AND SAFETY (FIRST AID) REGULATIONS 1981

Provision of suitable first aid facilities and at least one trained first aider.

REPORTING OF INJURIES, DISEASES AND DANGEROUS OCCURRENCES REGULATIONS (RIDDOR) 1995

Require employers to notify certain occupational injuries, diseases and dangerous events.

NOISE AT WORK REGULATIONS 1989

Require employers to take action to protect employees from hearing damage.

ELECTRICITY AT WORK REGULATIONS 1989

Require people in control of electrical systems to ensure they are safe to use and maintained in a safe condition.

PERSONAL PROTECTIVE EQUIPMENT AT WORK REGULATIONS 1992

Apply to employers for the provision, use and storage of appropriate protective clothing and equipment.

Appendix C: Nomenclature

C.1 Radiation

ϕ	(W)	Irradiated power
G	(W/m^2)	Irradiance, power density of radiation incident on a surface
I_0	(W/m^2)	Solar constant (1.367 ± 7 W/m^2)
G_U	(W/m^2)	Useful radiation
G_G	(Wh/m^2, kWh/m^2)	Global solar irradiance
G_{dif}	(Wh/m^2, kWh/m^2)	Diffuse radiation
G_{dir}	(Wh/m^2, kWh/m^2)	Direct radiation
G_{col}	(Wh/m^2, kWh/m^2)	Radiation on collector surface
E_G	(kWh/m^2a)	Yearly solar irradiance

C.2 Heat

$\dot{Q}$	(W/m^2)	Thermal output
$\dot{Q}_L$	(W/m^2)	Thermal loss
Q	(Wh, kWh)	Heat quantity
Q_N	(Wh, kWh)	Useful heat quantity
$\dot{Q}_N$	(W, kW)	Useful thermal output
Q_S	(Wh, kWh)	Solar heat yield, gain
Q_{aux}	(Wh, kWh)	Auxiliary heat
Q_{MW}	(Wh, kWh)	Domestic water heat requirement
Q_{BWE}	(kWh/m^2a)	Gross thermal yield

C.3 Thermal losses

k	(W/m^2K)	Thermal loss coefficient
k_1	(W/m^2K)	Linear thermal loss coefficient
k_2	(W/m^2K)	Quadratic thermal loss coefficient
k_{eff}	(W/m^2K)	Effective thermal loss coefficient
kA	(W/K)	Thermal loss rate
λ_{ev}	(W/m^2K)	Effective vertical thermal conductivity

C.4 Temperature

θ	(°C)	Temperature
θ_E	(°C)	Entry temperature into the heat exchanger
θ_A	(°C)	Exit temperature from the heat exchanger
$\theta_{absorber}$	(°C)	Average absorber temperature
θ_{air}	(°C)	Air temperature
$\Delta\theta$	(K)	Temperature difference
$\Delta\theta_{off}$	(K)	Switch-off temperature difference of controller
$\Delta\theta_{on}$	(K)	Switch-on temperature difference of controller
T	(K)	Absolute temperature

C.5 Efficiency

η	(%)	Degree of efficiency
η_O	(%)	Degree of optical efficiency

η_{sys}	(%)	Degree of system efficiency
η_{vent}	(%)	Degree of fan efficiency

C.6 Pressure

p	(bar)	Pressure
p_{in}	(bar)	Initial pressure
p_{max}	(bar)	Max. permissible pressure
p_{Bmax}	(bar)	Max. permissible operating pressure
$p_{max,SV}$	(bar)	Response pressure of safety valve
Δp	(mbar)	Pressure loss
Δp_{tot}	(bar)	Total pressure loss in solar loop
Δp_{col}	(bar)	Pressure loss in collector
$\Delta p_{heat\ exch}$	(bar)	Pressure loss in heat exchanger
Δp_{piping}	(bar)	Pressure loss in pipelines
$\Delta p_{fittings}$	(bar)	Pressure loss in fittings

C.7 Material-specific variables

α	(mm/mK)	Expansion coefficient
λ	(W/mK)	Thermal conductivity
α	(%)	Absorption coefficient
ϵ	(%)	Coefficient of emission
τ	(%)	Coefficient of transmission
ρ	(%)	Coefficient of reflection
c_W	(Wh/kgK)	Specific heat capacity of water (1.16 Wh/kgK)
$c_{G,W}$	(Wh/kgK)	Specific heat capacity of solar liquid (antifreeze fluid: 1.03 Wh/kgK)
c_L	(Wh/kgK)	Specific heat capacity of air (0.28 Wh/kgK)

C.8 Volume, volumetric flow rate and flow speed

V	(l)	Volume
V_{MEVmin}	(l)	Minimum size of expansion vessel
V_D	(l)	Expansion volume
V_{HW}	(l)	Daily hot water consumption
V_{tot}	(m^3/m^2h)	Specific volumetric flow rate of air
V_{min}	(m/s)	Minimum flow speed
V_L	(m^3/h)	Hourly airflow rate requirement

C.9 Angles

α_S	(°)	Azimuth angle of sun
γ_S	(°)	Angular height of sun
α	(°)	Azimuth angle of surface
β	(°)	Inclination angle of surface

C.10 Surfaces and lengths

A	(m^2)	Absorber surface area
A_{min}	(m^2)	Minimum absorber surface area
A_{max}	(m^2)	Maximum absorber surface area
A_Q	(m^2)	Free-flow cross-section of absorber
h_{sys}	(m)	System height (height of expansion vessel to top edge of collector)
Δl	(mm)	Temperature-related length change

l_0	(m)	Original length
h	(m)	Collector height
X	(m)	Space between collector rows
AM	(-)	Air mass factor

C.11 System variables

SF	(%)	Solar fraction
SE	(%)	System efficiency
P_F	(W, kW)	Fan power

C.12 Financing

a	(–)	Annuity factor
A_0	(€/$/£)	Procurement cost of system
T	(a)	Time period considered in years
q	(–)	Interest factor = 1 + (p/100)
p	(%)	Interest
f_k	(%,a)	Factor for acquiring the servicing (costs)
AN_o	(–)	Annuity for operation-related costs
AN_i	(–)	Annuity for investment-related costs
AN_s	(–)	Annuity for servicing-related costs
AN_c	(–)	Annuity for capital-related costs

C.13 Constants

σ	(W/m^2K^4)	Stefan–Boltzmann constant (5.67×10^{-8} W/m^2K^4)

36 *Wärmedämmwerte verschiedener Materialien*, 10th Symposium Thermische Solarenergie, 2000, S. 74
37 Ladener H and Späte F, *Solaranlagen*, 6. edition 1999, page 216
38 von Braunmühl, W, *Handbuch Contracting*, Krammer Verlag, Dusseldorf, 2000
39 Verband für Wärmelieferung e.V.: *Energielieferung*, VfW Jahrbuch 2000, Hannover
40 Bremer Energy Institute
41 FHH Umweltbehörde: Hamburg setzt auf Sonnenwarme, 2000
42 Professional experience with solar district heating in the German project Friedrichshafen/Wiggenhausen-Süd
43 *Sun in Action II: A Solar Thermal Strategy for Europe*, European Solar Thermal Industry federation, April 2003, downloadable from www.estif.org
44 Landesgewerbeamt Baden-Württemberg, Informationszentrum Energie: *Rationelle Energieverwendung in kommunalen Freibädern*, 1995
45 Schreier N, Wagner A, Orths R and RotariusT, *So baue ich eine Solaranlage*, Wagner & Co Solartechnik GmbH, Cölbe 1999
46 Hindenburg C, Henning H-M and Schmitz G, *Einsatz von Solarluftkollektoren in Sorptiongestützten Klimatisierungssystemen*, Achtes Symposium Thermische Solarenergie, OTTI, Staffelstein, May 1998
47 GBU mbH, Bensheim, Germany, www.gbunet.de
48 Bine ProjektInfo 08/2002: http: //194.175.173.199
49 ZAE Bayern: http: //www.zae-bayern.de
50 Henning H-M, Auslegung von Solaren Klimatisierungssystemen, Achtes Symposium Thermische Solarenergie, OTTI, Staffelstein, 2000
51 Hindenburg C and Dotzler W, Neuentwicklung eines Solarluftkollektors in Modulbauweise, Achtes Symposium Thermische Solarenergie, OTTI, Staffelstein, 2002
52 Binder-Kissel-Beratung und Training, Freiburg, Germany. www.binder-kissel.de
53 Prose F and Wortmann K, *Konsumentenanalyse und Marktsegmentierung der Kunden der Stadtwerke Kiel*, Universität Kiel, 1991
54 Wittbrodt E, *Kunden binden mit Kundenkarten*, Luchterhand Verlag, 1995

For further information

ISES, the International Solar Energy Society, has a large on-line library of documents called WIRE available on http://wire.ises.org.

More information is given on different countries in Section 4.5.

Index

Page numbers in *italics* refer to figures and tables